For a listing of recent titles in the
Artech House Intelligence and Information Operations Series,
turn to the back of this book.

Information Fusion and Analytics for Big Data and IoT

Éloi Bossé
Basel Solaiman

ARTECH HOUSE
BOSTON | LONDON
artechhouse.com

Library of Congress Cataloging-in-Publication Data
A catalog record for this book is available from the U.S. Library of Congress.

British Library Cataloguing in Publication Data
A catalog record for this book is available from the British Library.

ISBN-13: 978-1-63081-087-0

Cover design by John Gomes

685 Canton Street
Norwood, MA 02062

10 9 8 7 6 5 4 3 2 1

To my three children Étienne, Julianne, and Samuel; to my late mother Anne-Marie Bérubé; and to my late dear friend Pierre Valin

—Éloi Bossé

To the flowers of my life, Sausan, Razanne, Yazane, and Sarah, and to my mother Fatma, the flower I lost recently

—Basel Solaiman

Contents

CHAPTER 4

CHAPTER 5

Preface

Cyber physical social systems (CPSS) are the most general form of distributed systems that can also be called complex networks. CPSS consist of inhomogeneous, interacting adaptive agents capable of learning and large numbers of groups of people hyperlinked by information channels and interacting with computer systems, which themselves interact with a variety of physical systems to ensure conditions of good functioning. Primary examples of CPSS include command and control organizations such as 911/Emergency Response Systems and military organizations, as well as organizations that manage critical infrastructures. CPSS can be categorized into three main categories: safety-critical systems, mission-critical systems, and business-critical systems. The critical consequences of failure in these systems have an extreme impact on the society and ensuring system dependability is mandatory.

Considering the interface between the information/cyber and physical worlds, one can limit the scope of CPSS to cyber physical systems (CPS) that are integration of computation with physical processes. This is either called CPS or Internet of Things (IoT). In CPS, with embedded systems as predecessors, embedded computers and networks monitor and control the physical processes, usually with feedback loops where physical processes affect computations and vice versa. Considering the evolutions of Internet (e.g., IoT, Web 3.0, and Web 4.0), the binding and the flow of information between information sources, information processors, and information users (e.g., decision-makers and actuators) will be dynamic, cross-domain, and open. This is in contrast to the typical deployment of embedded systems where sensing assets and applications are monolithically deployed and scoped to a particular purpose. What the Internet and its evolutions are bringing to the scene is the networking of all these assets and applications.

Information overload and complexity are core problems to both military and civilian CPSS today. The problem of information overload is very well recognized has also been referred as big data in recent literature. The advances in information and communications technologies (ICT), in particular smart ICT, although providing a lot of benefits to improve dependability, efficiency, and trustworthiness in systems, have also increased tremendously the networking capabilities, thus creating the conditions of complexity by enabling richer, real-time interactions between and among individuals, objects, systems, and organizations. As a result, events that may once have had isolated consequences can now generate cascades of consequences, which can quickly spin out of control (e.g., blackouts and catastrophes) and badly affect system dependability or trustworthiness.

IoT (or CPS) is a big contributor to the big data problem. There is still a bit of confusion about the definition of big data, but IBM and other contributors developed a useful way of characterizing big data using the V dimensions: Volume, Variety, Velocity, Veracity, and Value. The inclusion of Veracity emphasizes the importance of addressing and managing the dimension of uncertainty associated to data which is in fact the overall objective of information fusion. Addressing the Big data Vs requires that fusion of information and analytics technologies (FIAT) be integrated and materialized into real-time and near-real-time decision support systems (DSS) that are capable of supporting prognosis, diagnosis, and prescriptive tasks within complex systems and organizations.

This book is not about CPSS, CPS, IoT, or big data. These are used as contextual elements to figure out the problem space. The book is about integration of FIAT. Hundreds of methods and technologies are part of FIAT. In the literature, one can find several books on those methods. For instance, there are currently more than 50 books available on either fusion of information or analytics alone, but none is addressing the integration of both fields. None of these books insisted upon the necessity to develop principles for integration of FIAT at levels from elements to systems and higher to system of systems. This is what we try to accomplish here. We say "try" because what we are proposing is incomplete. We should use "towards" instead since the *science of integration* is so immature and will require much more effort to provide a support solution to networked operations as it will be encountered in future CPSS. Integration and development of FIAT are a contribution in that direction.

In Chapters 1 and 2, we describe elements of context such as CPSS, CPS, IoT, and big data, as well as existing design methodologies for situation awareness and decision support systems. FIAT is a key enabler to support situation analysis and decisions. The challenge is to integrate appropriate technologies that will support measuring, organizing, understanding, and reasoning about situations in a given CPSS complex environment and support actuation (cybernetics functions: integration, monitoring, coordination, and control). Chapter 1 presents the general context of CPSS that emerge from the interrelation of social, cognitive, information/cyber, and physical worlds. Chapter 2 describes the various models of decision-making and situation awareness processes that become what to be supported. Understanding the process of decision-making is a prerequisite to any design of decision support solutions.

The central notions of information and its imperfection (e.g., uncertainty) are presented in Chapters 3 and 4, which review fundamental concepts and define new ones on information characterization and uncertainty representations that can be exploited in the design of intelligent DSS based upon an integration of FIAT. Analytics are more dedicated to collecting, organizing, structuring, storing, and visualizing data (variety and volume) while information fusion is more on synthesizing information (veracity and velocity). Both are complementary and need to be integrated for operational decision support. To achieve that integration, a first step is proposed as a definition of a basic element of information with which one can compose towards a computational model at system level.

Chapters 5 and 6 are proposing such a FIAT computational model. The model is framed under the three essential dimensions of archetypal dynamics: representation, interpretation, and realization (systems). These three dimensions are being considered in a holistic fashion to propose that holonic computational framework. In Chapter 7, we consider design and applications but at system and system of systems level. At a system-of-systems level, we discuss sine qua non integration notions for making systems interoperable: system openness and cooperability notions. We apply those notions to a military coalition. At the system level, we consider the integration of several potential design frameworks where abstract state machines would be the backbone. Finally, Chapter 8 is a list of prospective research activities that can contribute towards required FIAT integration for all CPSS levels for critical application domains such as energy, health, transport, and defense and security.

Chapter 9 presents a summary that this book addresses only a portion of the problem space. FIAT has been considered mainly under the Veracity and Variety dimensions of big data. The FIAT discussion focuses more on situation analysis perspectives (diagnosis and prognosis) than on decision support aspects (prescriptive) where important fields, such as multicriteria decision-making and multi-objective optimization, would have contributed. Finally, the problem is multidisciplinary by nature and several books (actual and future) would be required to explore the vast solution space that is also multidisciplinary.

We acknowledge the following scientists who have influenced this work in a way or another: Didier Guériot, Mihai Florea, Luc Pigeon, Jean Roy, Anne-Laure Jousselme, Patrick Maupin, Adel Guitouni, Roozbeh Farahbod, James Llinas, Michel Barès, Alain Appriou, Ali Khenchaf, and finally, posthumously, Pierre Valin.

CHAPTER 1

General Context of Cyber Physical and Social Systems

The purpose of this chapter is to present the general context characteristics and notions of cyber physical and social systems (CPSS) that result from the evolution of networking and the Internet (e.g., Internet of Things, semantic Web, Web 3.0). In addition, this chapter introduces *fusion of information and analytics technologies* (FIAT) as decision support solutions to crucial elements of the interface between the cyber and physical worlds. The perspective is information-centric. For instance, the social networking influence on the cyber-physical interface is considered through a social-computing layer supported by FIAT.

1.1 Sociotechnical Organizations and CPSS

Sociotechnical organizations (STOs) [1], which are increasingly more often referenced (with the evolution of the Internet) as CPSS [2], characterize our world today. Typically CPSS are composed of a large number of groups of people with multiple communications channels and a variety of computer and information systems interacting with a variety of physical systems. Four examples of CPSS are presented in Figure 1.1: defense and security organizations, energy, health, and transport networks. It is worthwhile to notice that the protection of these critical infrastructures has become a central challenge to many nations as exemplified in their emergency preparedness work [3, 4]. The use of the term protection is often related to national and international security challenges. The society can be severely affected by failures of these critical complex systems under people, economy, and national security dimensions.

1.2 CPSS Versus Cyber Physical Systems

CPSS and cyber physical systems (CPS) are tightly conjoined, coordinated, and integrated with human and social characteristics. Social and cognitive dimensions interface with the physical world through the cyber world as pictured in Figure 1.2(b): social computing with a human-computer interface. CPS [5, 6] are mainly related to the integration of computation with physical processes. In CPS, as

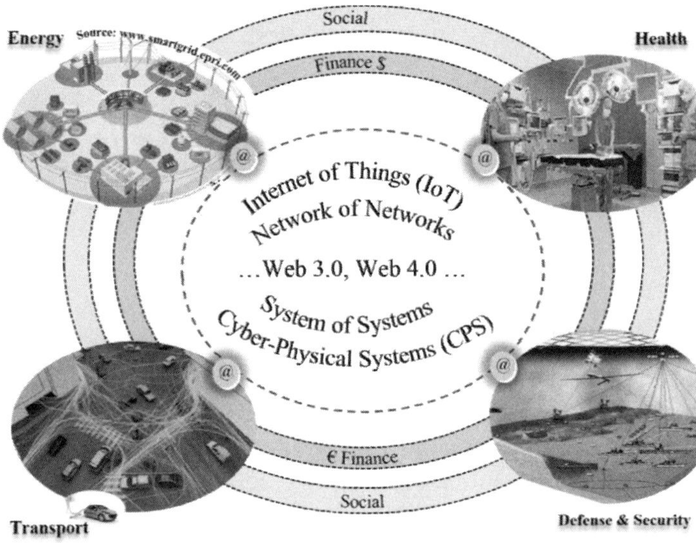

Figure 1.1 Examples of critical CPSS.

illustrated by a triad in Figure 1.2(a), embedded computers and networks monitor and control the physical processes, usually with feedback loops (where physical processes affect computations and vice versa). Human lives depend on this cyber-physical interface, and mission objectives and, hence, system performances are expected to be enhanced to meet the far-reaching goals as listed in Figure 1.3: 24/7 availability, 100% connectivity, predictability and repeatability, and real time. Poovendran [7] more explicitly expressed those high-level requirements:

> The healthy and competitive future of public and private sectors, such as transportation, power grid, medical, energy, environment, defense, law enforcement, and construction, heavily relies on CPS applications. Advances in CPS can make

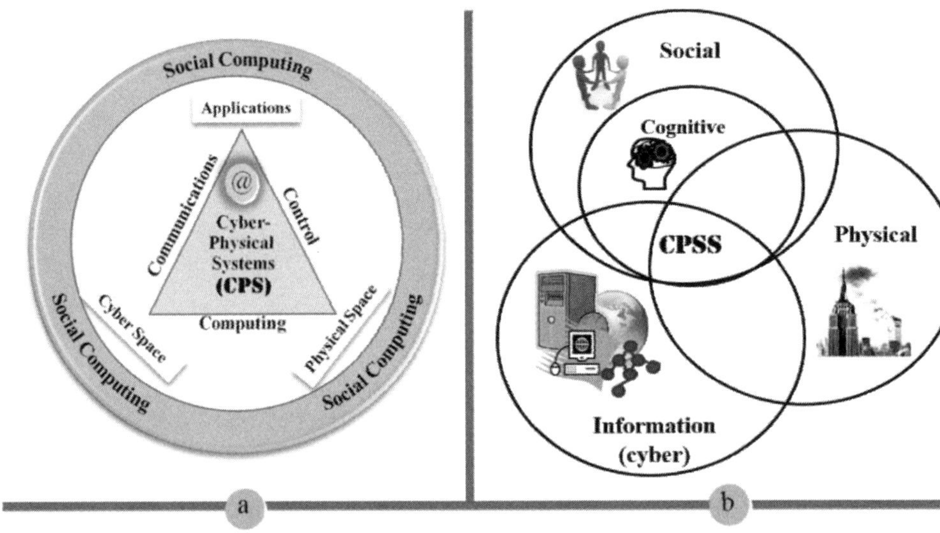

Figure 1.2 (a, b) CPSS versus CPS.

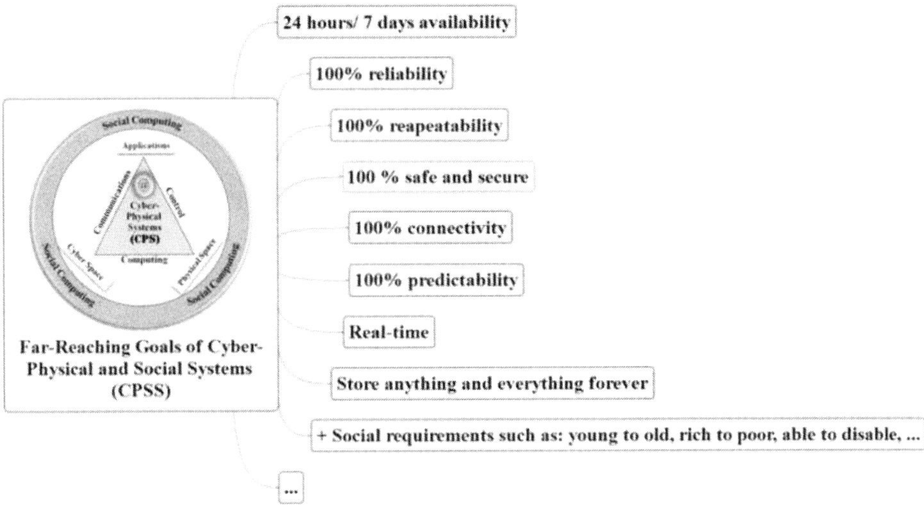

Figure 1.3 CPSS far-reaching goals.

applications faster, more spatially and temporally precise (e.g., telerobotic surgery), robust to hostile or inaccessible environments (e.g., autonomous search and rescue, disaster recovery), perform distributed coordination of large scale systems (e.g., automated road and airspace traffic control), demonstrate high efficiency (e.g., zero-net energy buildings), augment human capabilities (e.g., body sensor nets, brain–computer interfaces), and enhance quality of life (e.g., ubiquitous healthcare).

CPSS impose interdisciplinary contributions from human/social sciences, physics/engineering, and computer sciences to meet those high-level requirements (Figure 1.3). In today's context of STO or CPSS, executives or commanders claim better ways to reach the meaning of the data and consequently take action on it. In 2008, IBM first introduced the concept of a smarter planet, a world in which collaboration, systems thinking, and data analytics improve the efficiency and effectiveness of the many systems that facilitate life on Earth. The position was very well phrased in [8]:

> Every leader makes decisions. And every decision depends on information. That's been true whether someone has led a company, a government, an army or a household. Over the past 50 years, leaders have experienced a revolution in the quality and quantity of information available to them.... The industry has always sought to help leaders know with confidence all that has happened, is happening and might happen to every aspect of the enterprise. Can you spot the key patterns? Can you extract critical insights from data? Can you take latency and cost out of making and implementing a decision?

The realization of the complex interface between cyber and physical worlds is challenged by the concurrent nature and laws of physics governing our world, as opposed to the discrete and asynchronous nature of the cyber world. Adding the social dimension through the cyber world (by what is labeled social computing in Figure 1.2) evolves CPS to CPSS, sometimes alternatively referred as complex networks, system-of-systems, network-centric systems, and sociotechnical systems. Some of the following excerpts reflect challenges encountered facing this complexity.

From [7],

> CPS changes the notion of the physical systems (e.g., aircraft, vehicle) to include human, infrastructure, and platform in a system-of-systems, creating a uniquely large scope and context in which the system behavior must be predictable and provable. The resulting systems-of-systems are highly networked and dynamic in nature, with complexity, e.g., software size, growing at an exponential rate, with increasing time-critical interactions between purely physical elements and highly intangible cyber elements (e.g. social computing).

From [9],

> System integration is the elephant in the China store of large-scale Cyber-Physical Systems (CPS) design. It would be hard to find any other technology that is more undervalued scientifically and at the same time has bigger impact on the presence and future of engineered systems. The unique challenges in CPS integration emerge from the heterogeneity of components and interactions. This heterogeneity drives the need for modeling and analyzing cross-domain interactions among physical and computational/networking domains and demands deep understanding of the effects of heterogeneous abstraction layers in the design flow.

From [10], for the specific domain of Smart Grids:

> The CPS-enabled SmartGrid is in need of solutions that will support it at device, system, infrastructure and application level. ... Cyber-Physical Systems lie in the heart of the emerging SmartGrid. Their abilities on providing the "glue" between the physical world and the business side make them indispensable.

From the Canadian Chief of Force Development on Future Security Environment [11]:

> The artificial intelligence enabled networks envisioned for the future would provide a powerful toolset for commanders to rapidly make informed decisions ... while the science of the Command and Control toolset will be used to actively find and continuously collate, fuse, and synthesize all available information ... needed for developing operational and contingency plans for commanders.

1.3 CPS Versus Internet of Things (IoT)

The integration of physical and computing processes is not new as evidenced by predecessors to CPS called embedded systems [12]. However, embedded systems are rather stand-alone devices and they are usually not networked to the outside. What the Internet and its evolutions are bringing is the networking of these devices. This is CPS [13, 14]. In addition to CPS, different names have been used in the literature to refer sensibly to the same reality: "an abundance of addressable computing intelligence, with sensing and possibly actuating capabilities, will be embedded in physical (hence, cyber-physical) objects to interact with their physical surroundings, collecting information from them, disseminating to whoever needs them, and smartly (re)acting upon their desires." Those names are Internet of Things (IoT) [15, 16], ubiquitous computing [17], ambient intelligence, or pervasive networks [18]. There is no general accepted definition of CPS or IoT, but the similarities between the following two definitions suggest a sort of equivalence.

From Helen Gill, NSF, United States [14], "Cyber-physical systems are physical, biological, and engineered systems whose operations are integrated, monitored, and/or controlled by a computational core. Components are networked at every scale. Computing is deeply embedded into every physical component, possibly even into materials. The computational core is an embedded system, usually demands real-time response, and is most often distributed."

From Benghosi et al. [16], " ... define the Internet of Things as a network of networks which enables the identification of digital entities and physical objects —whether inanimate (including plants) or animate (animals and human beings)— directly and without ambiguity, via standardized electronic identification systems and wireless mobile devices, and thus make it possible to retrieve, store, transfer and process data relating to them, with no discontinuity between the physical and virtual worlds."

The overall objective of FIAT is to ease that continuity of processing from signal-data-information to actionable knowledge.

1.4 Information Overload and the Big Data Problem

There is a propensity in all organizations to produce more and more information that obviously stress the information support systems to sustain their capacities to provide decision makers and analysts with an augmented access to a deluge of data beyond human processing capabilities. This information overload is a core problem that both military and civilian organizations are facing today. The explosion came with the advent of advanced sensing and the diversity and volume of data from multiple sources and forms (unstructured and open sources, voice records, photos, video sequences). Decision-makers and analysts cannot cope with the flow of material, with potentially severe consequences on the quality of a decision and operational processes. The problem is so well recognized that *The Economist* dedicated a special report to data deluge in 2010 [19]: "Everywhere you look, the quantity of information in the world is soaring. According to one estimate, mankind created 150 exabytes of data in 2005. This year [2010], it will create 1200 exabytes." This problem has been referenced as big data in recent literature [20–24]. Big data is evidently contextual to CPSS and presents a complexity dimension mainly due to networking. Today, decision makers require better ways to communicate complex insights so they can quickly absorb the meaning of the data and take action on it.

There is still a bit of confusion about the definition of big data whether is best described by today's greater volume of data, the new types of data and analysis, social media analytics, next generation data management capabilities, or the emerging requirements for more real-time information analysis. Whatever the label is, organizations are starting to understand and explore how to process and analyze a vast array of information in new ways. To illustrate what has been just said about the definition of big data, let cite a recent editorial [25]:

> '*While there is no generally agreed understanding of what exactly is (or more importantly, what is not) big data, an increasing number of V's has been used to characterize different dimensions and challenges of big data: volume, velocity, variety, value, and veracity. Interestingly, different (scientific) disciplines highlight certain dimensions and neglect others.*' IBM [24] came first with a useful way of characterizing four dimensions of big data – Volume, Variety, Veracity and Velocity. The definitions are on Figure 1.4. A fifth V, for Value, can be added: '*the importance, worth, or usefulness of the data to those consuming it.*' The inclusion of Veracity emphasizes the importance of addressing and managing the dimension of uncertainty associated to data and information which is in fact encompassed in the overall objective of designing smart decision support systems based upon Fusion of Information and Analytics Technologies (FIAT) [18, 26, 27].

Internet of Things is a big contributor to the big data problem summarized as follows ([18]):

> *With the proliferation of devices, connectivity alternatives, applications, and services, a key challenge becomes how to orchestrate them horizontally to allow seamless interoperation of all these for the benefit of end-users. Typically referred to as the Internet of Things (IoT), an abundance of addressable computing intelligence, with sensing and possibly actuating capabilities, will be embedded in physical (hence, cyber–physical) objects to interact with their physical surroundings, collecting information from them, disseminating to whoever needs them, and smartly (re)acting upon their desires. In the true spirit of the Internet, the binding and the flow of information between information sources (e.g., sensors), information processors (e.g., fusion elements), and information users (e.g., decision makers and actuators) will be dynamic, cross-domain, and, of course, open. This is in contrast to the typical deployment of, say, current sensor-enabled applications where sensing assets and applications are "monolithically" deployed and scoped to a particular purpose.*

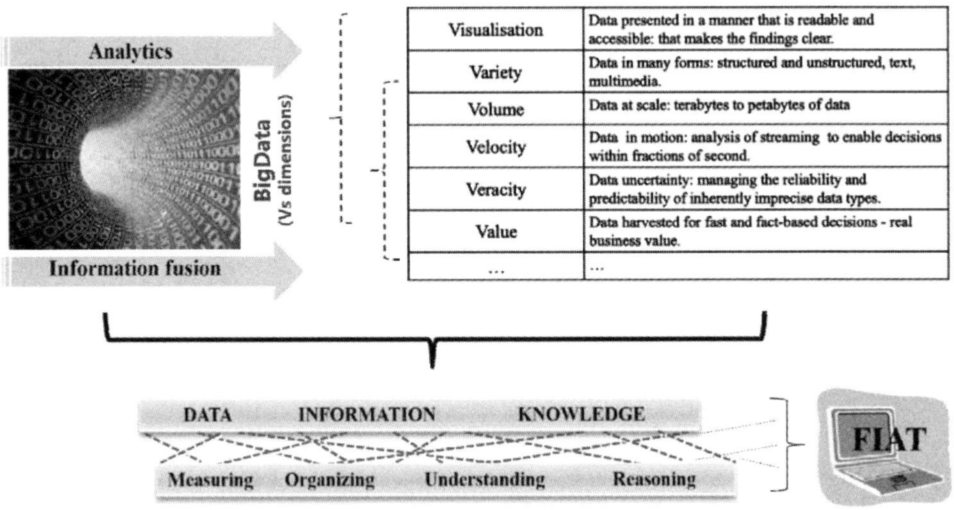

Figure 1.4 Big data Vs dimensions versus FIAT.

1.5 Complexity and System Dependability

The challenge is big to maintain CPSS under conditions of good functioning and as such labeled as dependable systems. The advances in information and communication technology (ICT), although providing a lot of benefits to improve dependability and efficiency in systems, have also increased tremendously the networking capabilities [28] so creating the conditions of complexity by enabling richer, real-time interactions between and among individuals, and between systems and organizations. As a result, events that may once have had isolated consequences can now generate cascades of consequences, consequences that can quickly spin out of control (e.g., blackouts and catastrophes) and affect badly system dependability [29].

Dependability of a system reflects the user's degree of trust in that system. In other words, it reflects the extent of the user's confidence that it will operate as users expect and that it will not 'fail' in normal use. Figure 1.5 presents definitions of five dimensions of dependability: availability, reliability, safety, security and maintainability [29–31]. Also described in Figure 1.5 are three different categories of critical CPSS: safety-critical systems, mission-critical systems, and business-critical systems. The critical consequences of failure in these systems are listed in Figure 1.5.

1.6 Fusion of Information and Analytics Technologies (FIAT)

Both military and civilian CPSS expect solutions from smart ICT for improving systems dependability and situation awareness (SA). As examples, the conceptual response to the increased world complexity for military organizations has been called Network Enabled Capability (NEC) [32, 33]. NEC exhibits a high degree of overlap with concepts derived from socio-technical systems theory [1]. Situation awareness [34] has emerged as an important concept around dynamic human decision-making in both military and civilian complex environments (e.g., defense,

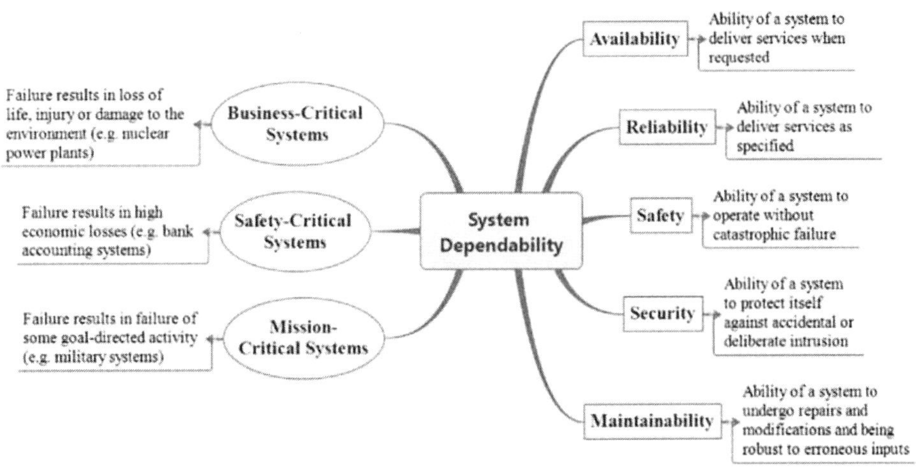

Figure 1.5 Three categories of critical systems and main dimensions of dependability.

public security, power networks, transport, business). Situation analysis is the process that provides and maintains a state of situation awareness for the user(s). In both military and civilian businesses, situation analysis provides the context and knowledge for planning. FIAT are key enablers allowing to meet the demanding requirements of situation analysis and decision support in current and future CPSS.

In the military domain, information fusion (IF) systems [35, 36] built from smart ICT have emerged as a key sense making enabler to situation analysis for complex environments (hostile and cooperative). The overall objectives of IF technologies are: (1) to support decision makers and analysts in the making sense of an ever increasing volume of data and complexity; (2) to improve the quality of information and reduce the effects of different forms of information imperfections; and (3) to ensure and improve dependability face to an ever increasing system complexity (networked aspect and heterogeneity). In the general business and industry domain, the term 'Analytics' [37, 38] is extensively used to pursue roughly the same objective. Both, information fusion and analytics refer to the application of computer sciences and technologies, operational research, cognitive engineering and mathematical approaches to allowing to understand complex situations in order to support decision-making in action [39].

1.7 Interface Between the Cyber and Physical Worlds: Four Cybernetic Functions

The complex interface between cyber and physical worlds is challenged by the concurrent nature and laws of physics governing our physical world, as opposed to the discrete and asynchronous nature of the cyber world. It is primordial to rethink interface functions, such as coordination, integration, monitoring, and control as illustrated in Figure 1.6. FIAT-based smart systems will help in that rethinking by providing, among others, two main categories of support: diagnosis and prognosis for these cybernetics functions.

FIAT is appealed to improve capabilities of measuring, organizing, understanding, and reasoning about data-information-knowledge hierarchy [40] as well as to improve system's trustworthiness/dependability and to benefit from the advances in networking (e.g., evolution of Internet). The integration of FIAT is foreseeing as a major contributor to solve the big data Vs dimensions as roughly categorized in Figure 1.7. The FIAT integration framework is then required to provide the analysis (analytics) and synthesis (information fusion) intelligent support to decision-makers.

Figure 1.6 Crucial elements of the interface between cyber and physical worlds.

Analytics that is more dedicated to collecting, organizing, structuring, storing and visualizing data (variety and volume) whereas information fusion is more concerned by synthesizing information (veracity and velocity) for operational decision support. Both are complementary as illustrated in Figure 1.7 where one can see, as well, the respective dominant technologies that have accompanied the evolution of these two emerging domains that now really needs to converge.

1.8 General Discussion on Decision Support Systems in Complex Dynamic CPSS

The traditional systems engineering approach is ineffective when confronted with CPSS diverse functional requirements to engineer, for instance, FIAT-based decision support solutions. Diversity in requirements comes from real-time embedded processing on-board platforms (embedded systems [12]) to the cyber-physical Internet or systems [14, 41, 42] connected through the Internet with large-scale enterprise systems comprising application servers, databases, and service-oriented architectures and connected to human organizations and social networks (future CPSS [43, 44]). From an information standpoint, CPSS, also referred as network-centric [33] or network-enabled systems [2, 45], present the following three main facets: computing engines, networking infrastructure, and the underlying cybernetics functions of Figure 1.6 (i.e., coordination, integration, monitoring, and control).

FIAT-based decision support systems are built mainly from computational intelligence techniques to ensure and improve system trustworthiness and dependability. As an example, we can think about a defense and security program where the objective is to preempt threats and attacks on the nation through processing a tremendous volume of data from a large number of miscellaneous sources that are dispersed geographically. In this example, the objectives are to avoid becoming

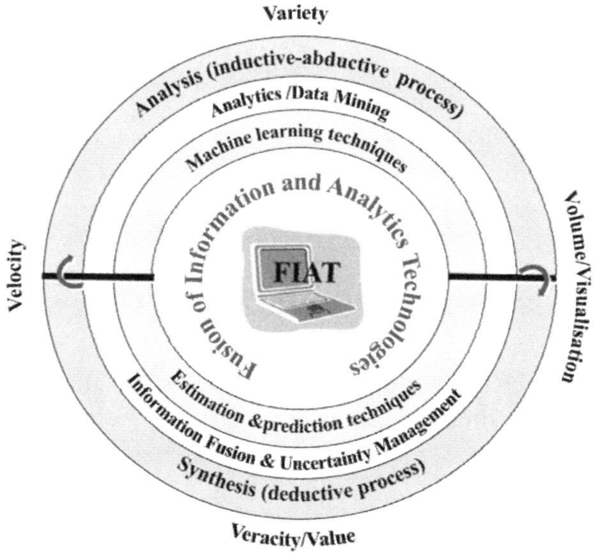

Figure 1.7 Analytics and information fusion emerging domains.

overwhelmed from information overload and to be capable of coordinating the activities of geographically dispersed public safety agencies in a precise, accurate, and timely manner. To meet these objectives, first, FIAT-based decision support systems [46] will identify and acquire data from relevant sources of information. Second, raw data and information are transformed into actionable information. Third, information is disseminated to geographically dispersed regions. Fourth, actionable information is used to support distributed decisions and actions [47].

1.8.1 The Ideal Support System

A computer-based support system (CBSS) is a computerized system that is intended to interact with and complement a human user [48]. Such support systems go from formulae embedded within a spreadsheet to sophisticated autonomous reasoning agents. Whatever the nature of the CBSS, the objective is to develop CBSS features that intuitively fit the perceptual and cognitive processes of the human user. The ideal CBSS is one that [35]:

1. Provides the information needed by the human decision-maker, as opposed to raw data that must be transformed by the user(s) into the information needed. If the data can be perfectly transformed into information by the CBSS, then no human cognitive effort is required for the transformation and no cognitive work is expended on the data, allowing total focus on the domain's problem-solving.
2. Can be controlled effortlessly by the human user. It presents the information to the user as effortlessly as a window allows a view of the physical world outside. In this sense, the CBSS is considered as transparent to the user. If the interaction with the CBSS is completely effortless, no human cognitive effort is required to manage and interact with the CBSS, allowing total focus on the domain's problem-solving.
3. Complements the cognitive power of the human mind. In this way, the CBSS does not only avoid creating a world (that is ripe for human decision-making errors), but it can include features that complement the trends and power of human cognitive processes. Decision-making errors categorized by such terms as fixation and garden path are avoided by the form of the decision-making world embodied by the CBSS.
4. Supports a wide variety of problem-solving strategies, from nearly instinctive reactions to events to knowledge-based reasoning on fundamental principles in a situationally independent manner.
5. Effective CBSS are the ones that make the problem transparent to the user.

In the case of a CBSS for situation analysis and decision-making, two main domains of characterization are required: information and decision domains. Characterization of information and its imperfections is absolutely essential in any design of decision support systems (DSS) and ultimately true for FIAT-based systems [35, 49, 50].

Moreover, quality of information (QoI) [51] criteria must be identified. These criteria are means that can be used to manage CBSS in a networked environment [47] to meet overall system dependability and to improve system's trustworthiness.

Among the set of critical QoI criteria is the notion of information relevance that, if appropriately formalized, can frame the intelligent processing functions of CBSS. For instance, relevance is a key concept to sense making activities [52], information foraging [53], and information/knowledge filtering [54].

Characterization of the decision environment is a sine qua non condition for proper DSS design. FIAT-based systems are designed to support decisions and consequent actions. A CPSS presents a highly dynamic, heterogeneous, and distributed decision environment. CPSS consist of interlocking collective of constituent subsystems, inhomogeneous, interacting adaptive agents capable of learning. Actions are carried out by agents in order to execute their actions either concurrently, dependently or independently. The decision-making process takes place in an environment that changes over time (dynamic) either due to the previous actions or due to events that occur in real time [47]. The decision environment is then characterized as being asynchronous, temporally dynamic, and decentralized.

1.8.2 Complex Situations

One way to define complex situations could be conducted by considering the level of complexity of such situations. That is, the approach reported in Guitouni [55] as part of an analysis conducted to support the Canadian Crisis and Emergency Management community with respect to complexity: "compositions and interactions of variables in a situation that affect the cause and the effects, the sources and accuracy of information, the communication and decision making processes, and the activities/actions that are required to achieve a desired end-state." Guitouni [55] presented the key variables (Figure 1.8) that have been identified as being influential for situations complexity. For instance, crisis and emergency response plans revealed a number of common variables in emergency situations resulting from terrorist activities and asymmetric threats. They can be grouped into four categories (Figure 1.8) of response team, adversary, environment, and incident. Further details on these four categories can be found in [55].

1.8.3 General Overarching System Engineering Concepts in CPSS

System of systems engineering methodologies and enterprise architecture (EA) [56–59] are emerging approaches to deal with the development and maintenance of CPSS. Architectures help in managing complexity through techniques such as modularization and abstraction at various levels, and facilitate changes by providing a documentation of the components, relationships, and constraints of the systems or processes. In the military domain, current EA approaches [55, 56] lack two important issues required in adopting a practice of EA within an enterprise: (1) a set of fundamental guiding principles that state how the practice of EA is expected to support the enterprise's business direction and processes, and (2) a proper methodology (or at least a process) guiding the development and maintenance of architectural artifacts (e.g., exploitation of FIAT).

This book is not about complex systems engineering but intends to intervene with decision support within a CPSS environment. It is worth considering or, perhaps, anticipating the general concepts or principles from which such complex

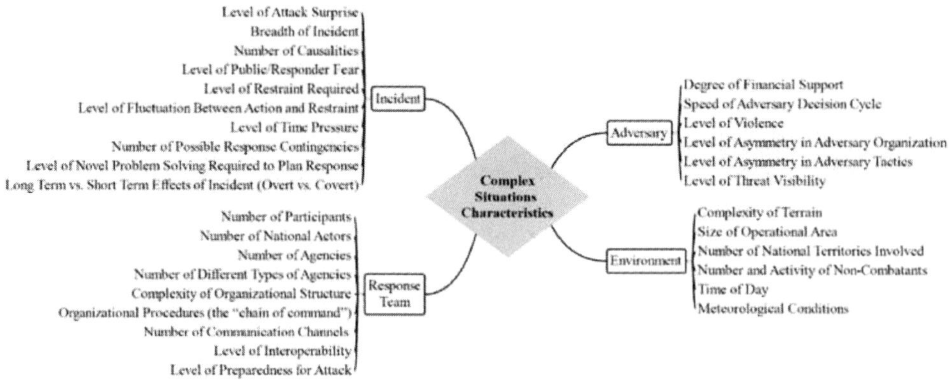

Figure 1.8 Framework listing the common variables characterizing complex situations.

systems might be constructed in the future. The following two general concepts and principles are emerging from literature:

1. The concept of composition, defined in Figure 1.9 as composability and compositionality, that is very important to deal with complexity through system scalability, modularity, and abstraction and to exploit homomorphism in constructing layered systems and services.
2. The principles of a transcendence/emergence pair, also defined in Figure 1.9, where self-organization concepts such as the autonomic principles brought by IBM [60, 61] are guided by a transcendent concept task to be performed: the goal here is to realize software systems and applications that can manage themselves under high-level supervision from humans.

In CPSS, the combined scale, complexity, heterogeneity, and dynamics of networks, systems, and applications impose ever increasing constraints on information infrastructure that alternate paradigms and strategies need to be used to ensure system dependability and trustworthiness. Autonomic systems [62] are characterized by their self-properties: self-configuration, self-healing, self-optimization, and self-protection.

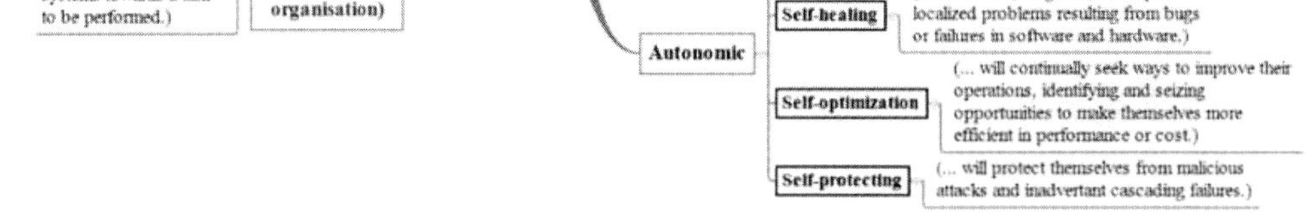

Figure 1.9 General concepts of complex systems engineering.

A short discussion is required at this level about the principles of systems transcendence and emergence that impact greatly on the principle of composition for complex systems. Bernard-Weir [63] discussed the principle of transcendence and proposed points about the system's evolution that could be directed by the concept of a task to be performed (a sort of blueprint) and stated that random causality and unpredictable unfolding may not be sufficient to explain such evolution. He presents remarks concerning the concepts of self-organization (SO) and emergence, which remains a valid attainment, provided they are associated with the concepts of hetero organization and immergence: "SO does not (self-) organize at random.... Self Organisation (SO) is a very important concept, but it is not sufficient in itself. Too much stress has been placed on the unpredictability and the novelty of the new states or levels that would emerge by virtue of SO." Bernard-Weir [63] made reference to Agonistic Antagonistic System Sciences (AASS) for the transcendence versus immanence pair that proved its efficiency in various domains from biomedical to social sciences [64, 65]. He presented a series of examples from the biological field where the recurrence of the same general mechanisms may be observed at every level of evolution (a kind of life model): "actions of pairs such as stimulation/inhibition, gene expression/repression, trophotrope/ergotrope regulation, information/catalysis, phenotype/genotype ..."

Highly relevant to this discussion on CPSS, Bernard-Weil [63] continued:

> ...when socially complex organizations and firms have "emerged," the pairs of the life model, or their corresponding properties, are found again.... According to this concept, a social system, for instance, would reach an optimal equilibration between ruling powers and developmental trends, if self-organizational processes were introduced in its functioning. This is, however, far from being certain. The specialists of management are quite right in claiming that "strategies of emergence," though they might well contribute to generate new types of organization, will not necessarily bring forth solutions to the dysfunctions within the firms, if they are not combined with deliberate strategies.

1.8.4 General Comments About Decision Support for a Smart Grid Environment

Let us further illustrate the above discussion by using an example from the energy sector: a Smart Grid infrastructure. In [66], Smart Grid was defined as: "an electric system that uses information, two-way, cyber-secure communication technologies, and computational intelligence in an integrated fashion across the entire spectrum of the energy system from the generation to the end points of consumption of the electricity." Developing a FIAT-based decision support system specifically designed to cope with such complexity is quite a challenge.

FIAT-based decision aids can be distributed all over the grid to help reaching the objectives of a fully realized Smart Grid having the following desired characteristics [66]:

1. *Self-healing:* Automatic repair or removal of potential faulty equipment from service before it fails, and reconfiguration of the system to reroute supplies of energy to sustain power to all customers [67].

2. *Flexible:* The rapid and safe interconnection (self-protection) of distributed generation and energy storage at any point on the system at any time.
3. *Predictive:* Use of machine learning, weather impact projections, and stochastic analysis to provide predictions of the next most likely events so that appropriate actions are taken to reconfigure the system before next worst events can happen (self-configure).
4. *Interactive:* Appropriate information regarding the status of the system is provided not only to the operators, but also to the customers to allow all key participants in the energy system to play an active role in optimal management of contingencies.
5. *Self-optimization:* Knowing the status of every major component in real or near real time and having a control equipment to provide optional routing paths provides the capability for autonomous optimization of the flow of electricity throughout the system.
6. *Secure (self-protection):* Considering the two-way communication capability of the Smart Grid covering the end-to-end system, the need for physical and cyber security of all critical assets is essential [68].

The question is: how do we achieve and distribute the required FIAT-based decision support to the four cybernetic functions (integration, monitoring, coordination, and control) in a CPSS complex environment like a smart grid? As correctly pointed out in [69, 70], there is a lack of multidisciplinary approaches and no well-established methodology (e.g., modeling, analysis, and decision-making paradigms) to support the deployment of information and communications technologies (ICTs) such as FIAT that are essential to evolve current power grids to smart grids. What is available are fragmented coarse models of socio-ecological systems (SESs) [71], climate change energy models, man-made electric power grids, and fragmented approaches to ICT developed for other applications and believed to be directly applicable to Smart Grid design and operations. The same situation can be observed in today's Command and Control Organizations (CCOs) such as military organizations with their conceptual response to a CPSS problem called network-enabled capability (NEC) [27, 31, 32].

1.9 Summary

The high-level notions presented in this chapter are necessary to understand the overall context and get a flavor of what level of complexity the problem of intervening in CPSS can be. The tangent that is advocated in the rest of this book is that the integration of FIAT is a sine qua non enabler to cope with CPSS facing information overload and complexity. FIAT will support rethinking of cyber interface functions, such as coordination, integration, monitoring, and control and progress decision support concepts to be operational through these interface functions.

The high-level requirements for all CPS or IoT, composing a CPSS, are expected to be dependable, secure, safe, and efficient and operate in real time in addition to be scalable, cost-effective, and adaptive. These high-level requirements are important here since they raise the question to what extent FIAT-based decision support systems may take them into consideration as built-in information processing capability. At the end, FIAT exist to support decisions mainly in: (1) making sense of an ever increasing complexity of the 4Vs of big data; (2) improving the quality of information (QoI) to reduce and cope with uncertainty; (3) supporting distributed decision-making and execution of actions; and (4) ensuring a proper overall system dependability and trustworthiness.

References

[1] Walker, G. H., et al., "A Review of Socio-Technical Systems Theory: A Classic Concept for New Command and Control Paradigms," *Theoretical Issues in Ergonomics Science*, Vol. 9, November 2008, pp. 479–499.

[2] Liu, Z., et al., "Cyber-Physical-Social Systems for Command and Control," *IEEE Intelligent Systems*, Vol. 26, 2011, pp. 92–96.

[3] Rinaldi, S. M., J. P. Peerenboom, and T. K. Kelly, "Identifying, Understanding, and Analyzing Critical Infrastructure Interdependencies," *IEEE Control Systems*, Vol. 21, 2001, pp. 11–25.

[4] Doorman, G. L., et al., "Vulnerability Analysis of the Nordic Power System," *IEEE Trans. on Power Systems*, Vol. 21, 2006, pp. 402–410.

[5] Lee, E. A., "Cyber Physical Systems: Design Challenges," *11th IEEE Intl. Symp. on Object Oriented Real-Time Distributed Computing (ISORC)*, 2008, pp. 363–369.

[6] Sridhar, S., A. Hahn, and M. Govindarasu, "Cyber-Physical System Security for the Electric Power Grid," *Proc. of the IEEE*, Vol. 100, 2012, pp. 210–224.

[7] Poovendran, R., "Cyber-Physical Systems: Close Encounters Between Two Parallel Worlds [Point of View]," *Proc. of the IEEE*, Vol. 98, 2010, pp. 1363–1366.

[8] IBM, "A New Intelligence for a Smarter Planet," 2012. http://www.ibm.com/smarterplanet/us/en/?ca=v_smarterplanet.

[9] Sztipanovits, J., et al., "Toward a Science of Cyber-Physical System Integration," *Proc. of the IEEE*, Vol. 100, 2012, pp. 29–44.

[10] Karnouskos, S., "Cyber-Physical Systems in the Smartgrid," *9th IEEE Intl. Conf. on Industrial Informatics (INDIN)*, 2011, pp. 20–23.

[11] CFD, "The Future Security Environment 2008-2030," January 27, 2009. http://publications.gc.ca/collections/colecction_zoll/dn-nd/D4-8-1-2-10-eng.pdf.

[12] Malinowski, A., and H. Yu, "Comparison of Embedded System Design for Industrial Applications," *IEEE Trans. on Industrial Informatics*, Vol. 7, 2011, pp. 244–254.

[13] Shi, J., et al., "A Survey of Cyber-Physical Systems," *2011 Intl. Conf. on Wireless Communications and Signal Processing (WCSP)*, 2011, pp. 1–6.

[14] Gill, H., "From Vision to Reality: Cyber-Physical Systems," *HCSS National Workshop on New Research Directions for High Confidence Transportation CPS: Automotive, Aviation and Rail*, 2008.

[15] Miorandi, D., et al., "Internet of Things: Vision, Applications and Research Challenges," *Ad Hoc Networks*, Vol. 10, 2012, pp. 1497–1516.

[16] Benghozi, P. -J., S. Bureau, and F. Massit-Folléa, *L'internet des objets: quels enjeux pour l'Europe* : Maison des Sciences de l'homme, 2009.

[17] Bohn, J., et al., "Social, Economic, and Ethical Implications of Ambient Intelligence and Ubiquitous Computing," in *Ambient Intelligence*, New York: Springer, 2005, pp. 5–29.

[18] Conti, M., et al., "Looking Ahead in Pervasive Computing: Challenges and Opportunities in the Era of Cyber-Physical Convergence," *Pervasive and Mobile Computing*, Vol. 8, 2012, pp. 2–21.

[19] Cukier, K., "Data, Data Everywhere: A Special Report on Managing Information," *The Economist*, 2010.

[20] McFedries, P., "The Coming Data Deluge [Technically Speaking]," *IEEE Spectrum*, Vol. 48, 2011, p. 19.

[21] Bryant, R., R. H. Katz, and E. D. Lazowska, "Big-Data Computing: Creating Revolutionary Breakthroughs in Commerce, Science and Society," December, 2008. http://cra.org/ccc/wp-content/uploads/sites/2/2015/05/Big_Data.pdf.

[22] Marx, V., "Biology: The Big Challenges of Big Data," *Nature*, Vol. 498, 2013, pp. 255–260.

[23] Zikopoulos, P., and C. Eaton, *Understanding Big Data: Analytics for Enterprise Class Hadoop and Streaming Data*, New York: McGraw-Hill Osborne Media, 2011.

[24] Schroeck, M., et al., "Analytics: The Real-World Use of Big Data," 2012.http://www-03.ibm.com/systems/hu/resources/the_real_word_use_of_big_data.pdf.

[25] Hitzler, P., and K. Janowicz, "Linked Data, Big Data, and the 4th Paradigm," *Semantic Web*, Vol. 4, 2013, pp. 233–235.

[26] Pedrycz, W., and S. -M. Chen, *Information Granularity, Big Data, and Computational Intelligence*, New York: Springer, 2014.

[27] Blasch, E., É. Bossé, and D. A. Lambert, (eds.), *High-Level Information Fusion Management and Systems Design*, Norwood, MA: Artech House, 2012.

[28] Alberts, D. S., "The Agility Imperative: Precis," Unpublished white paper, http://www.dodccrp.org, 2010.

[29] Petre, L., K. Sere, and E. Troubitsyna, *Dependability and Computer Engineering: Concepts for Software-intensive Systems*, Engineering Science Reference, Hershey, PA, 2011.

[30] Avizienis, A., J.-C. Laprie, and B. Randell, Fundamental Concepts of Dependability, University of Newcastle upon Tyne, Computing Science, Newcastle University Report no. CS-TR-739, 2001.

[31] Nicol, D. M., W. H. Sanders, and K. S. Trivedi, "Model-Based Evaluation: From Dependability to Security," *IEEE Trans. on Dependable and Secure Computing*, Vol. 1, 2004, pp. 48–65.

[32] Alberts, D. S., *The Agility Advantage: A Survival Guide for Complex Enterprises and Endeavors*, CCRP Publication Series, Washington DC, 2010.

[33] Alberts, D. S., R. K. Huber, and J. Moffat, "NATO NEC C2 Maturity Model," NATO SAS-065, www.dodccrp.org, 2010.

[34] Endsley, M. R., and D. J. Garland, *Situation Awareness Analysis and Measurement*, Boca Raton, FL: CRC Press, 2000.

[35] Bossé, É., J. Roy, and S. Wark, *Concepts, Models, and Tools for Information Fusion*, Norwood, MA: Artech House, 2007.

[36] Liggins, M., D. Hall, and J. Llinas, *Handbook of Multisensor Data Fusion: Theory and Practice*, 2nd ed., London, U.K.: Taylor & Francis, 2008.

[37] Apte, C. V., et al., "Data-Intensive Analytics for Predictive Modeling," *IBM Journal of Research and Development*, Vol. 47, 2003, pp. 17–23.

[38] Laursen, G. H. N., and J. Thorlund, *Business Analytics for Managers: Taking Business Intelligence Beyond Reporting*, New York: Wiley, 2010.

[39] Klein, G. A., et al., *Decision Making in Action: Models and Methods*, Stamford, CT: Ablex Publishing, 1993.

[40] Waltz, E., *Knowledge Management in the Intelligence Enterprise*, Norwood, MA: Artech House, 2003.

[41] Jeschke, S., "Cyber-Physical Systems - History, Presence and Future," Aachen, Germany, 2013.

[42] Koubâa, A., and B. Andersson, "A Vision of Cyber-Physical Internet," *Proc. of the Workshop of Real-Time Networks (RTN 2009), Satellite Workshop to ECRTS 2009*, July 2009.

[43] Liu, Z., et al., "Cyber-Physical-Social Systems for Command and Control," *IEEE Intelligent Systems*, July–August 2011, pp. 92–96.

[44] Wang, F.-Y., "The Emergence of Intelligent Enterprises: From CPS to CPSS," *IEEE Intelligent Systems*, Vol. 25, 2010, pp. 85–88.

[45] Conley, S. F., "Test and Evaluation Strategies for Network-Enabled Systems," DTIC Document, ITEA Journal, 2009, Vol. 30, pp. 111–116.

[46] Solano, M. A., S. Ekwaro-Osire, and M. M. Tanik, "High-Level Fusion for Intelligence Applications Using Recombinant Cognition Synthesis," *Information Fusion*, Vol. 13, 2012, pp. 79–98.

[47] Ghosh, S., *Algorithm Design for Networked Information Technology Systems*, New York: Springer, 2004.

[48] Elm, W. C., et al., "Applied Cognitive Work Analysis: A Pragmatic Methodology for Designing Revolutionary Cognitive Affordances," *Handbook of Cognitive Task Design*, 2003, pp. 357–382.

[49] Solaiman, B., et al., "A Conceptual Definition of a Holonic Processing Framework to Support the Design of Information Fusion Systems," *Information Fusion*, 2013.

[50] Blasch, E., et al., "High-Level Information Fusion Developments, Issues, and Grand Challenges," in *Proceedings of the 13th International Conference of Information Fusion*, Edinburgh, Scotland, 2010.

[51] Rogova, G. L., and E. Bosse, "Information Quality in Information Fusion," *13th Conf. on Information Fusion (FUSION)*, 2010, pp. 1–8.

[52] Llinas, J., "A Survey of Automated Methods for Sensemaking Support," *SPIE Sensing Technology+ Applications*, 2014, pp. 912206-1–912206-13.

[53] Pirolli, P. L., *Information Foraging Theory: Adaptive Interaction with Information*, Oxford, U.K.: Oxford University Press, 2007.

[54] Barès, M., *Maîtrise du savoir et efficience de l'action*, Pais, France: Editions L'Harmattan, 2007.

[55] Guitouni, A., "A Time Sensitive Decision Support System for Crisis and Emergency Management," NATO RTO-IST-086, C3I for Crisis, Emergency and Consequence Management, Bucharest, Romania, May 2009.

[56] U.S. Department of Defense, "The DoDAF Architecture Framework Version 2.02," http://dodcio.defense.gov/Portals/0/Documents/DODAF/DoDAF_v2-02_web.pdf2010, 2010.

[57] Alghamdi, A. S., "Evaluating Defense Architecture Frameworks for C4I System Using Analytic Hierarchy Process," *Journal of Computer Science*, Vol. 5, 2009, p. 1075.

[58] Atzori, L., et al., "The Social Internet of Things (SIOT)–When Social Networks Meet the Internet of Things: Concept, Architecture and Network Characterization," *Computer Networks*, Vol. 56, 2012, pp. 3594–3608.

[59] Solano, M. A., and G. Jernigan, "Enterprise Data Architecture Principles for High-Level Multi-Int Fusion: A Pragmatic Guide for Implementing a Heterogeneous Data Exploitation Framework," *15th Intl. Conf. on Information Fusion (FUSION)*, 2012, pp. 867–874.

[60] Agoulmine, N., *Autonomic Network Management Principles: From Concepts to Applications*, Elsevier Science, 2010.

[61] IBM, "Autonomic Computing," white paper, June 2005. http://www-03.ibm.com/autonomic/pdfs/AC%20Blueprint%20whote%20paper%20V7.pdf.

[62] Parashar, M., and S. Hariri, *Autonomic Computing: Concepts, Infrastructure, and Applications*, Boca Raton, FL: CRC Press, 2006.

[63] Bernard Weil, E., "Transcendance, an Essential Concept for System and Complexity Sciences to Spread Out," *Complexity*, Vol. 6, pp. 23–35, 2000.

[64] Bernard-Weil, E., "Reconsidérer la nature de la violence et les objectifs de son contrôle grâce à la science des systèmes ago-antagonistes," *Colloque La violence: du biologique au social*, 2000.

[65] Bernard-Weil, É., "Ago-Antagonistic Systems," in *Quantum Mechanics, Mathematics, Cognition and Action*, New York: Springer, 2002, pp. 325–348.

[66] Gharavi, H., and R. Ghafurian, Smart Grid: The Electric Energy System of the Future, *Proceedings of the IEEE 99(6)*: pp.917–921, May 2011, New York: IEEE Press.

[67] Anderson, R. N., et al., "Adaptive Stochastic Control for the Smart Grid," *Proc. of the IEEE*, Vol. 99, 2011, pp. 1098–1115.

[68] Overman, T. M., et al., "High-Assurance Smart Grid: A Three-Part Model for Smart Grid Control Systems," *Proc. of the IEEE*, Vol. 99, 2011, pp. 1046–1062.

[69] Ilic, M. D., "Dynamic Monitoring and Decision Systems for Enabling Sustainable Energy Services," *Proc. of the IEEE*, Vol. 99, 2011, pp. 58–79.

[70] Ilic, M. D., et al., "Modeling of Future Cyber–Physical Energy Systems for Distributed Sensing and Control," *IEEE Trans. on Systems, Man and Cybernetics, Part A: Systems and Humans*, Vol. 40, 2010, pp. 825–838.

[71] Ostrom, E., "A General Framework for Analyzing Sustainability of Social-Ecological Systems," *Proc. R. Soc. London Ser. B*, 2007, p. 1931.

CHAPTER 2

Situation Awareness and Decision Support

The purpose of this chapter is to describe the various models of decision-making and situation awareness processes that become what needs to be supported. Understanding the process of decision-making is a prerequisite to any design of decision support solutions. We review the models and frameworks for fusion of information and analytics (FIAT) whose associated technologies can be integrated as decision support solutions to problems (e.g., big data and IoT) anticipated with complex environments such as cyber physical and social systems (CPSS) and the evolution of the Internet.

2.1 Introduction

Decision-making is involved in all aspects of our lives, and it is of particular importance for the critical CPSS illustrated in Figure 2.1: health, transport, energy, and defense and security. With the advancement of information and communications technologies (ICTs), these four environments become more and more complex and constitute real challenges in terms of decision-making. The oversimplified Boyd's Observe-Orient-Decide-Act (OODA) loop [1], illustrated in Figure 2.1, is used to describe the decision process. Although the OODA loop might give the impression that activities are executed in a sequential way, in reality, the activities are concurrent and hierarchically structured. The processes of the loop are typically performed in a very dynamic and complex environment and are heavily influenced by factors such as uncertainty, information and knowledge imperfections, and time stress.

The four CPSS environments in Figure 2.1 present cases where interdependent decision-making takes place in an environment that changes over time either due to the previous actions of the decision-maker or due to events that are outside of the control of the decision-maker [2]. In this sense, CPSS present, in addition to conventional one-time decisions, dynamic decisions. Dynamic decisions are typically more complex than one-time decisions. They occur in real time and involve observing the extent to which people are able to use their experience to control a particularly complex system, including the types of experience that lead to better decisions over time. Among the four CPSS, the most constraining environment could be argued to be command and control activities in military and public security

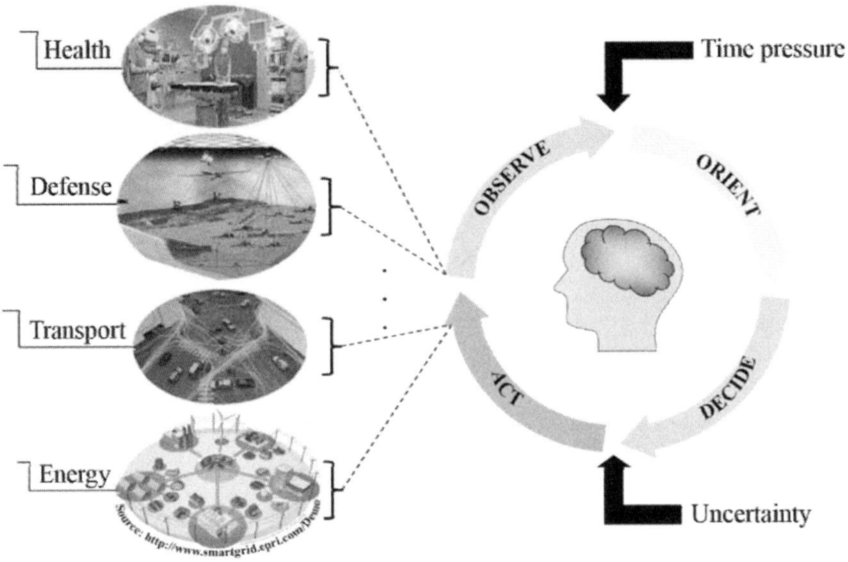

Figure 2.1 Boyd's OODA loop and four examples of CPSS.

settings. This will be considered as our major reference environment for different discussions in this book. If we can provide solutions to this particular environment, then transposition within the other three might be facilitated.

In a problem-solving approach for complex environments such as CPSS, the most important step is to understand and frame the problem. Over the years, multiple efforts have thus been deployed to better understand and explain the decision-making process. Fields like decision sciences, management sciences, administrative sciences, social choice, psychology, or naturalistic decision-making are examples of the growing effort in modeling and understanding the individual as well as the organizational decision-making. Knowledge of the decision-making process is essential to guide the development of appropriate decision support through the provision of ICT-based support such as FIAT.

An organization should work towards identifying the demands and constraints (lack of information, lack of time) placed on decision-making in different aspects of its operations prior to the determination of specific decision-making strategies best suited to these conditions. Tools and procedures that clarify the link between various decision strategies (e.g., analytic and intuitive) have the potential to enhance both plans and decisions made through all levels (e.g., strategic, operational, and tactical) of the organization. A decision-making process is defined as a set of activities and tasks that directly or indirectly support the decision-maker. The process and its activities could be well or poorly defined, rational or irrational, or based on explicit assumptions or tacit assumptions.

2.2 Decision-Making in Defense and Security CPSS

Modern military operations take place within an enormously complex environment to accomplish missions across the spectrum of conflicts from humanitarian assistance to high-intensity combat. In the past several decades, the battlespace has expanded enormously to cope with increasingly potent and accurate weapons capable of being launched at progressively further ranges from their targets. Concomitantly, the pace and scope of operations have continued to increase with each successive advance in warfare. In response to these challenges, powerful new sensors have been deployed at sea, ashore, and in space, while the capacity of communication systems has multiplied to make available huge volumes of data and information to commanders and their staffs. In short, technological improvements in mobility, range, lethality, and information acquisition continue to compress time and space, forcing higher operating tempos and creating greater demands on decision-making. That is likely the most constrained environment for decision-making.

Usually, a defense and security CPSS is being called C4ISR (Command Control, Communications, Computer, Intelligence, Surveillance, and Reconnaissance). The term cyber physical and social system is not used in this community but this is a good representative example of what a CPSS is. For the military community, C4ISR is: "A system to collect, analyze and communicate information, plan and coordinate operations, and provide the capabilities necessary to direct forces to achieve assigned missions." The C4ISR world can be summarized using Figure 2.2 with the link and similarities between the OODA loop [used to describe the Command and Control (C2) process during military or public security operations] and the intelligence cycle. Commanders exercise C2 by synchronizing military actions in time, space, and purpose to achieve unity of effort within a military force under two main constraints (uncertainty and time) that dominate the environment in which military decisions are made.

The OODA loop (OPS: operations) and the intelligence cycle are the two central models for the information and decision-making processes. The two loops

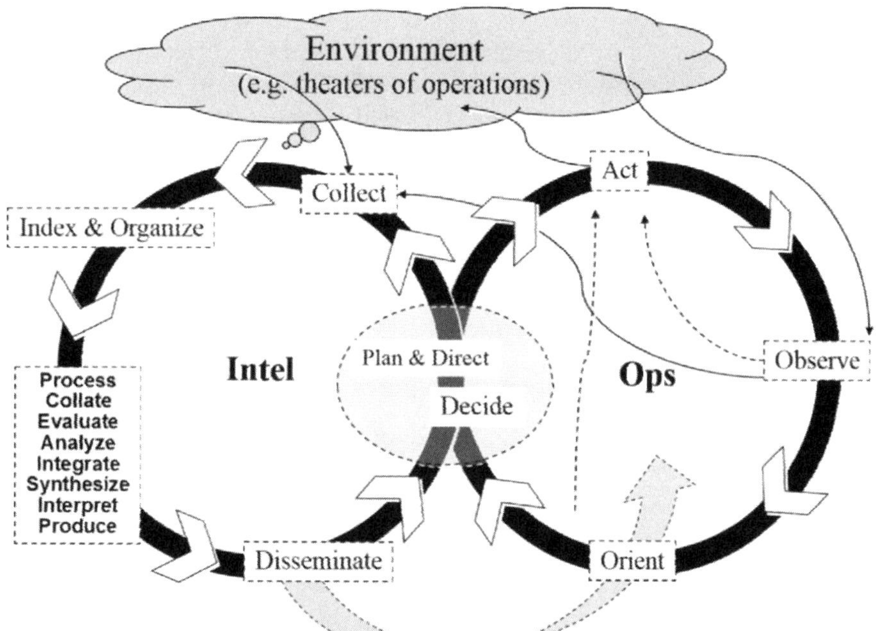

Figure 2.2 C4ISR decision and intelligence loops.

hinge on the fulfillment of two broad functions: first, all commanders within a force arrive at a shared and consistent understanding of the battlespace arising through battlespace (situation) awareness, and second, unity of effort is achieved (decision-making) throughout a joint and combined force through commonly held intent.

Several models of the C2 process have been proposed over the years and many documents have been published on the subject under the Command and Control Research Program (CCRP) (http://www.dodccrp.org/index.htm). In those documents, C2 is considered as "the exercise of authority and direction by a designated commander over assigned forces in the accomplishment of the force's mission." Another definition states that C2 is "the process by which commanders can plan, direct, control and monitor any operation for which they are responsible" (http://www.dodccrp.org/index.htm). Whatever the definition, it is generally accepted that C2 is composed of a number of dynamic and cyclic perceptual, procedural, and cognitive activities, achieved by humans, computer systems, or both of them. At a very high level, these activities can be summarized as the perception of the environment, the appraisal of the situation, the decision-making about a course of action, and the implementation of the chosen plan.

The process that delivers strategic and operational intelligence products is generally depicted in a cyclic form, with distinct phases [3]. This is illustrated in Figure 2.2 adjacent to the OODA loop. The process begins as policy and decision-makers define, at a high level of abstraction, the knowledge that is required to make decisions. The requests are parsed into information required and then to data that must be collected to estimate or infer the required answers. Data requirements are used to establish a plan of collection, which details the elements of data needed and the targets (people, places, and things) from which the data may be obtained. Following the plan, human and technical sources of data are tasked to collect the required raw data. The collected data is processed (e.g., machine translation, foreign language translation, or decryption), indexed, and organized in an information base. Progress on meeting the requirements of the collection plan is monitored and the tasking may be refined on the basis of the received data.

The organized information base is processed using techniques that combine all-source data in an attempt to answer the requestor's questions. The data is analyzed and solutions are synthesized. The topics or subjects (intelligence targets) of study are modeled, and requests for additional collection and processing may be made to acquire sufficient data and achieve a sufficient level of understanding to answer the consumer's questions. Finished intelligence is disseminated to consumers in a variety of format, providing answers to queries and estimates of accuracy of the product delivered. Although introduced here in the classic form of a cycle, in reality the process operates as a continuum of actions with many more feedback (and feedforward) paths that require collaboration between consumers, collectors, and analysts.

The Boyd's OODA loop [1], shown in Figure 2.3, presents what Boyd really proposed [4]. In the Orient analysis, there is additional information that a user engages in analyzing and synthesizing a situation. The Orient phase is the repository of our genetic heritage, cultural tradition, and previous experiences. Boyd's Orient is the most important part of the OODA loop since it shapes the ways we observe, decide, and act. The OODA loop model has been widely used to represent

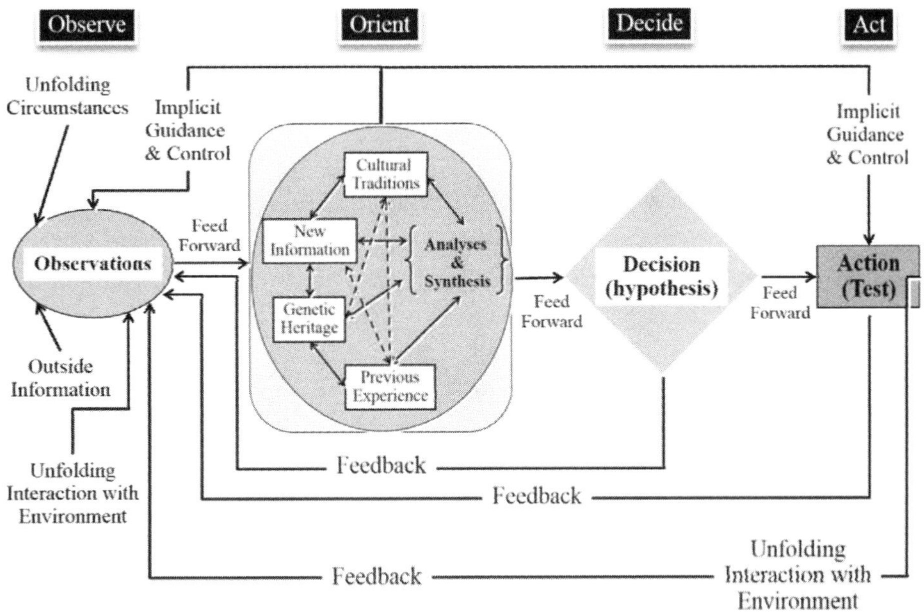

Figure 2.3 The extended OODA loop. (Adapted from [1].)

decision-making in military environments as well as in civilian domains. The OODA applications include information fusion [5–7], analytics [8], autonomic systems [9–12], cultural modeling [13], business intelligence [14, 15], cyber security [16], and semiautomated decision-making [7, 17].

The classical military decision-making model has been referred to as the operational planning process (OPP) as illustrated in Figure 2.4. The OPP can be characterized as contemplative, analytic, or involving nomination and evaluation of options that are weighed in some context, eventually leading to a choice of a course of action (COA). Traditionally, providing decision support means to support the OPP so it is mainly supporting the bottom half of the OODA loop. Operational research and optimization communities made very significant contributions to help solve important problems in different application domains such as logistics, allocation of resources, and scheduling.

Developing COAs implies the exploitation of the intelligence cycle (Intel) products of Figure 2.2. For many, the military decision-making process might be seen as the full integration of both the OPP and the intelligence preparation of the battlespace (e.g., a result of the Intel loop). However, to face challenges of today complex situations, a full integration implies a lot. Decision-making in CPSS will be highly and heterogeneously dynamic (e.g., different tempos). Dynamic decisions, unlike simple and conventional one-time decisions, are typically more complex and occur in real time and involve observing the extent to which people are able to use their experience to control a particular complex system, including the types of experience that lead to better decisions over time.

Decision-making is a continuous, streaming process where goals, resources, uncertainty, and risks are being constantly appraised and decisions taken. Full integration addresses the two halves of the OODA loops with all their interdependencies

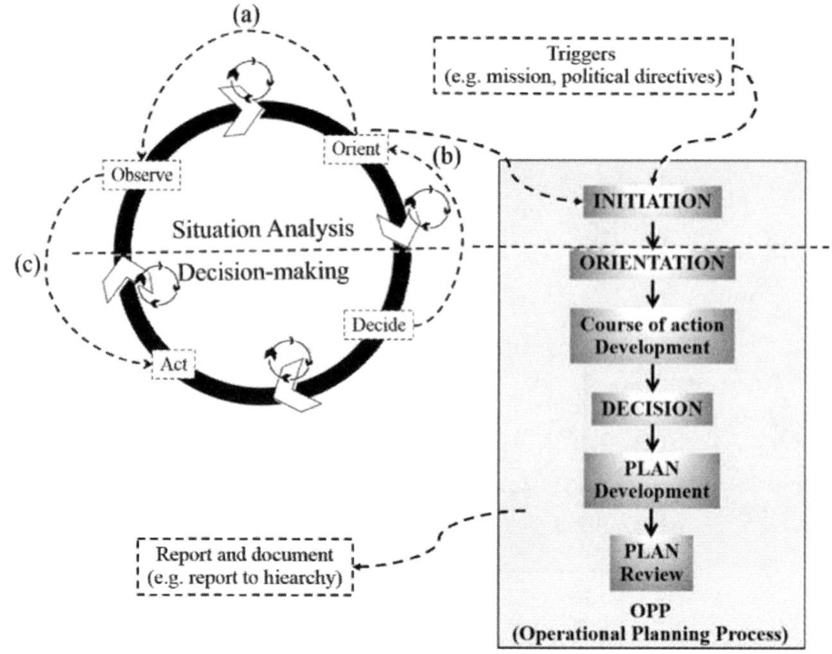

Figure 2.4 Military OPP versus OODA loop.

[18]. It assumes that many feedback loops, different tempos, and distributed control and tasking protocols have to be managed as pictured by the multiple loops in Figures 2.4 and 2.5. The feedback loops are particularly significant as execution may alter situational understanding and require plans to be adjusted, new decisions to be taken, and direction to be amended. This is called a dynamic decision-making environment [2], which is characteristic of most current military theaters of operations of today and tomorrow [19–22]. In fact, the advancement of networking technologies and ICTs in general creates the conditions of complexity (e.g., the big data issue). In the military CPSS and similar in civilian CPSS, ICTs, particularly networking technologies, have an impact on the organizational structures. For instance, the tactical, operational, and strategic boundaries tend to change as illustrated in Figure 2.5, leading to more flat organizations that operate differently. Flat organizations can present more agility in the Alberts' sense [20] to face the unknown but consequently require more autonomy. The operations cycle (Ops) and intelligence (Intel) tend to work more jointly, being on the same temporal page, in a networked warfare environment to finally be represented as a network of coordinated OODA loops.

2.3 Decision-Making Models

Over the years, multiple efforts have been deployed to better understand and explain decision-making. The research community has proposed different decision-making models based on decision sciences, human factors, cognitive sciences, organizational behavior, and social sciences. Normative, descriptive, prescriptive or

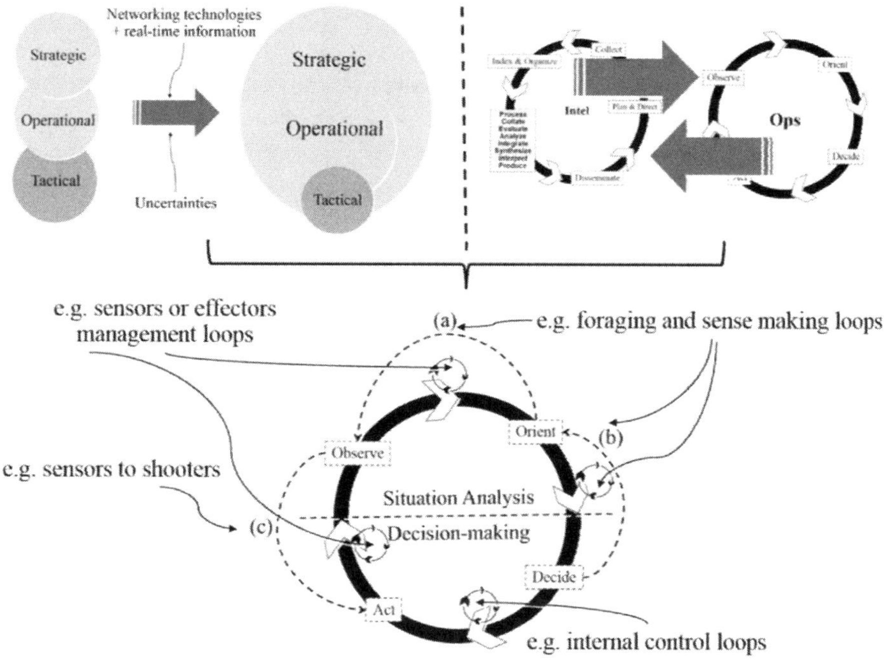

Figure 2.5 A network of OODA loops in future military CPSS.

analytical, naturalistic or intuitive, behavioral, social, team, and heuristics models are a few examples of such diversity of perspective used to represent and understand decision-making. Among these models, two main influential streams [23–26] are generally recognized to understand decision-making. The first stream is based on the premise that human decision-making can be modeled in terms of formal analytic processes predicted by normative theories of probability and logic. The theories under the first group claim that good decision-making follows a rational approach in which decisions are based on expected outcomes and there is an attempt to select a course of action that will bring the optimal outcome. The second stream, called naturalistic or intuitive theories, is based on the premise that people use informal procedures or heuristics to make decisions within the restrictions of available time, limited information, and limited cognitive processing.

Guitouni et al. [26] classified the two streams as descriptive and prescriptive. They reviewed a significant number of decision-making models and present a comprehensive view of the seemingly disparate approaches reported in the literature. They realize that each of the models has a valid contribution to the understanding of the commander's decision-making process but lack a common context. They propose a unified decision-making framework that provides a setting to appreciate many contributions to the decision-making process modeling. It is developed around four domains of the decision-making activities: cognition, knowledge (information), organization, and observable effects. The framework could be mapped around four fundamental stages of the decision-making process: perception (observation), understanding (awareness), decision, and action. They assume that many feedback loops and control and tasking protocols govern the framework.

The analytic or rational decision-making process corresponds closely to the military estimate of a situation. According to this analytic approach, the commander generates several options, then identifies criteria for evaluating these options, rates the options against these criteria and then selects the best option as the basis for future plans and action. It aims at finding the optimal solution, but it is time-consuming and information-intensive. Rational models range from purely normative models, like the expected utility models, to more descriptive models. Chapter 2 of [23] presented a description of the most known approaches: the expected utility theory [27], the prospect theory [28], the regret theory [29], bounded rationality [24, 30, 31] and multi-attribute heuristic [32, 33] approaches.

Naturalistic decision-making (NDM) [34] emphasizes the acquisition of knowledge, the development of expertise, and the ability of humans to generalize from past experience. It stresses pattern recognition, creativity, experience, and initiative. Intuitive or naturalistic decision-making relies on the commander's ability to recognize the key elements of a problem based on experience and judgment. This approach focuses on situation assessment and strives to find the first solution that solves the considered problem. The intuitive model works on the assumption that, by drawing on upon personal experience and judgment, the commander will generate the first workable solution. If time permits, the commanders may evaluate their decision in a more analytic way; if they find it defective, they move on to the next reasonable solution. Chapter 2 of [23] presents a description of the most known models: the recognition-primed decision model [35], the image theory [36], and argument-driven models [37, 38].

The analytic approach has a wide military applicability when time is not a constraint and extensive information may be gathered. It is appropriate for contingency planning and in operations preparing. This is the traditional operational research approach. It may also have merit in situations falling outside the scope of the commander's previous experience. Nevertheless, the intuitive approach appears to be more applicable for a majority of typical tactical or operational decisions: decisions made in highly fluid and dynamic conditions of combat, when time and uncertainty are the dominant factors affecting decisions. However, we must say that naturalistic decision-making using the recognition-primed decision (RPD) model fails in theory if there is a lack of experience or when encountering a completely novel scenario [39]. Any presented situation that is not adequately similar to this body of experience will require adaptation and learning.

Bryant et al. [25] insisted upon a continuum in decision strategy to adopt the approach that is best tailored to the situation and may use elements of the two approaches at the same time as illustrated in Figure 2.6. As explained in [25], "the continuum is based on the formalism of the decision making method, which is defined by how exhaustive, quantitative, and compensatory the procedure is, and by how many options are evaluated, and whether the options are considered sequentially against a criterion or concurrently through feature-by-feature evaluation. Analytic methods are those in which multiple options are evaluated through an exhaustive, quantitative, and compensatory feature-based analysis. Intuitive methods are those in which alternatives are evaluated holistically through a sequential comparison process that is generally qualitative, noncompensatory and not exhaustive." Finally, Bryant et al. [25] made the following remarks on decision-making that are

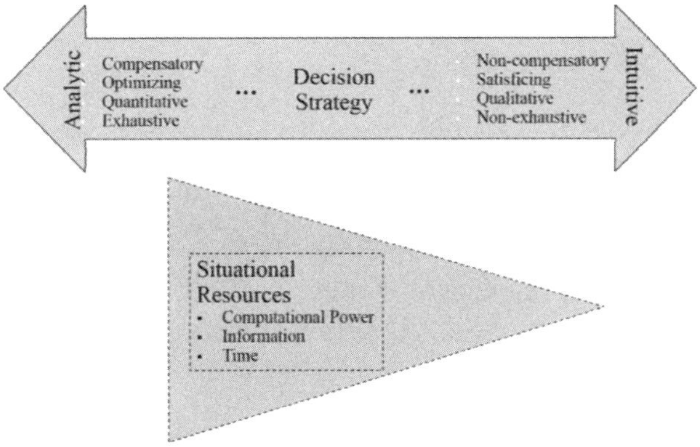

Figure 2.6 A continuum of approaches as decision strategy. (Adapted from [25].)

in agreement with the discourse in Guitouni et al. [26] and contribute to provide useful overall guidelines for decision support:

> It seems unlikely that an entirely analytic approach or an entirely intuitive approach will serve as a useful model for every aspect of military decision making. Consequently, effort should be directed at employing the advantageous aspects of each approach....recognize the synergy between analytic and intuitive strategies to enhance overall decision making success. Specifically, analytic strategies could be used in planning to generate high quality courses of action to anticipated problems so that these courses of action can be rapidly selected and implemented by intuitive, recognition-based decision making during critical events. In this way, the benefits of analytic decision making can be indirectly imported to intuitive decision making under stress and time pressure. Thus, a major area of opportunity for the design of decision support is improving the connection between planning and action.... analytic models might apply in situations in which there are well-defined goals and in which factors and options can be explicitly defined....Intuitive theories have the advantages of being closely linked to what expert decision makers actually do in real-world situations and of being applicable to dynamic, uncertain, and high risk environments...Intuitive approaches certainly are better suited to situations where time and data are limited.... intuitive theories are, by their nature, more descriptive than prescriptive, making them effective for understanding human decision making but less so for developing systems to aid/support actual decision making performance.

2.3.1 OODA Loop Interdependencies: Situation Analysis and Decision-Making

Llinas [18] raised issues involving interdependencies among the situation analysis processes (e.g., fusion, sense-making) and decision-making processes for complex decision-making environments. He advocated that a critical examination of interprocess (situation analysis and decision making) interdependencies is needed to design decision support systems (DSS) for optimum performance. He recommended an integrated, multidisciplinary approach (cognitive sciences, human factors, decision sciences and computer sciences); otherwise, DSS designs will remain disconnected and suboptimal. In Figure 2.7, there is an attempt to illustrate the interconnected processes identified by Llinas. The following list of excerpts, slightly adapted from [18], describes these processes:

1. *Automated FIAT-based situation analysis process:* Llinas used information and data fusion (IDF) to name that process and he included analytics in his information fusion process definition, a largely automated inferencing/estimation providing process that offers algorithmically developed situational estimates, organized raw observational data (structured and unstructured), and controllable collection management of observational data.
2. *Sense-making and information foraging:* For further reading, see [40–42]). This is a semiautomated human-on-the-loop process that considers the automated situation analysis-provided estimates, forages over these hypotheses as well as the data, assesses the nature of the problem at hand, considers possible policy, authority, and mission factors, and culminates in a final projected situation that is also judged as to acceptability; if not, this hypothesis is the starting point for decision-making and action-taking to manage the situation.
3. *Decision-making:* This is a semiautomated, human-on-the-loop process that operates in a System 1 (analytic), System 2 (intuitive), or hybrid/mixed decision-making mode, yields a selected course of action (COA), and triggers a resource optimization process to define specific resources that physically enable the selected COA onto the real-world situation.

Situational assessment is an analytical, estimation, and inferencing process that can be supported by automated methods (e.g., FIAT-based), and ideally yields situational estimates as hypotheses to be considered by a human analyst [42]. In complex environments like CPSS, even the best current automated FIAT [Figure 2.7(a)] are of limited capability and can consequently produce only what Llinas [42] called situational fragments: partial hypotheses representing situational substructures as patterns. Llinas also said that the current evolution of tool suites is largely directed to structuring these situational fragments for almost entirely human-based hypothesis synthesis into a complete situation. Even the automated methods that at least would support synthesis of these fragments are not yet adequately mature.

The sense-making process [Figure 2.7(b)] is a stage wherein final situation assessments and understandings (in the human mind) are developed. Sense-making process (see Pirolli's framework [41]) is organized into two major loops of activities: a foraging loop that involves processes aimed at seeking information, searching and filtering it, and reading and extracting information, possibly into some schema; and a sense-making loop that involves iterative development of a mental model from the schema that best fits the evidence. The foraging function within sense-making implies that the situation analysis FIAT-based process will have to be open to and enable a range of queries that will be in regard to raw or processed observational data, information fusion and analytics functional operations (e.g., data association, data prediction), and nominated situational hypotheses, among possibly other runtime interactive operations. For further reading, a significant amount of sense-making work in line with the discussion above was conducted by Roy et al. [43–45].

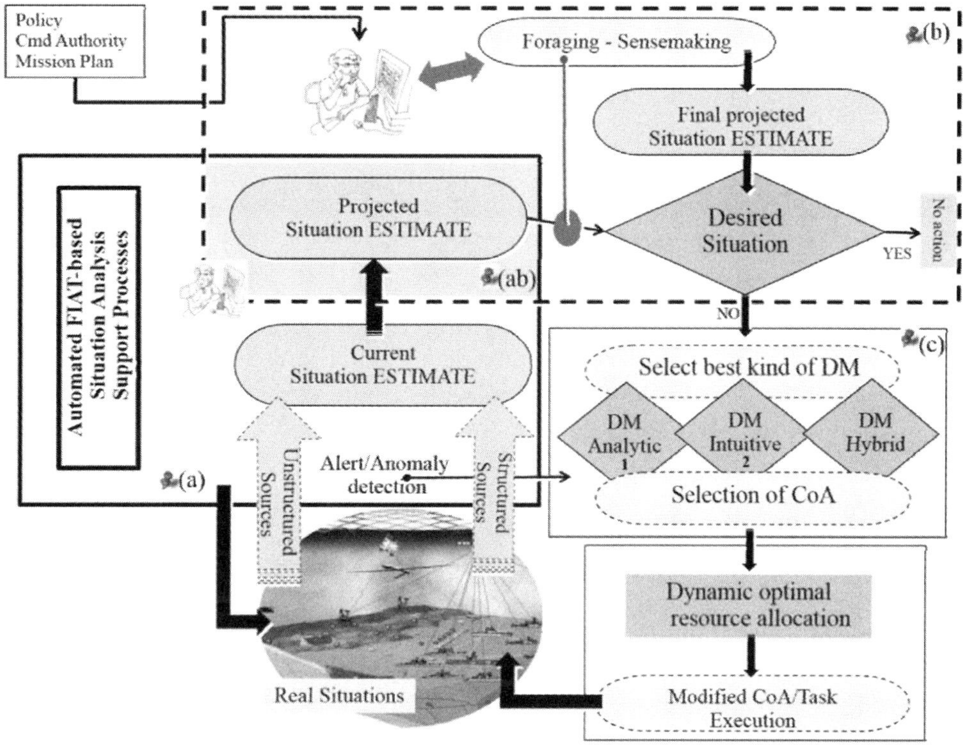

Figure 2.7 Interconnected/dependent processes in situation analysis and decision-making. (Adapted from [18]).

2.3.2 Selection of the Best Decision-Making Models

Multiple models exist to represent the decision process as suggested in Figure 2.7(c), each approaching the unknown from a slightly different perspective. The decision model chosen can influence the technical solution. For instance, the support required from an automated FIAT-based situation analysis is greatly shaped by the tasks to accomplish. It is important to look at the challenges facing the decision-maker from multiple perspectives (physical, information, social and cognitive). Understanding the decision process from only one perspective will lead to get support only from that perspective. Relating multiple models helps to get a true appreciation of the complex environment and then can result in better support solutions for the decision-maker.

Among multiple model approaches, Guitouni et al. [26] mentioned social theory models and game theory where there is a collective welfare function. They discussed very briefly the multicriteria decision aid methodology invented by Bernard Roy [46] with different decision models called constructive models. They concluded their literature review with a series of open and still valid questions related to pin (c) of Figure 2.7: Is a single model needed, or multiple models? Is there a better way to represent the decision-making process? Do we need models to understand and other models for the application? What relationship exists between understanding and application?

2.4 Situations and Situation Analysis

The definitions of situation and situation analysis are based on the numerous works of J. Roy mainly from [23, 47]. He defined situation analysis as: "a process, the examination of a situation, its elements, and their relations, to provide and maintain a product, i.e., a state of situation awareness, for the decision maker" and a situation as "a specific combination of circumstances, i.e., conditions, facts, or states of affairs, at a certain moment."

The situation analysis process is concerned with understanding the world. There is a real situation in the environment, and the situation analysis process will create and maintain a mental representation in the mind of the decision-maker(s). The idea of awareness has to do with having knowledge of something. In addition to the cognition facet, awareness is also linked with the notions of perception and understanding/comprehension. The two basic elements involved in situation awareness are the situation and the person as pictured in Figure 2.8. The situation can be defined in terms of events, entities, systems, other persons, and their mutual interactions. The person can be defined according to the cognitive processes involved in situation awareness (SAW) or simply by a mental or internal state representing the situation.

Roy [47] stated that the main two basic situation elements are entity and event. He pursues the description of elements of a situation, illustrated in Figure 2.9, the following way: "...an entity is an existing thing (as contrasted with its attributes), i.e., something that has independent, separate, self-contained, and/or distinct existence and objective or conceptual reality. An event is something that happens (especially a noteworthy happening). Hence, entities exist, while events occur. A scenario is defined as a sequence of events. A group represents a number of individuals (entities and/or events) assembled together or having some unifying relationship, i.e., an assemblage of objects/events regarded as a unit. The term activity refers to the notions of action, movement, and motion. It is appropriate when something

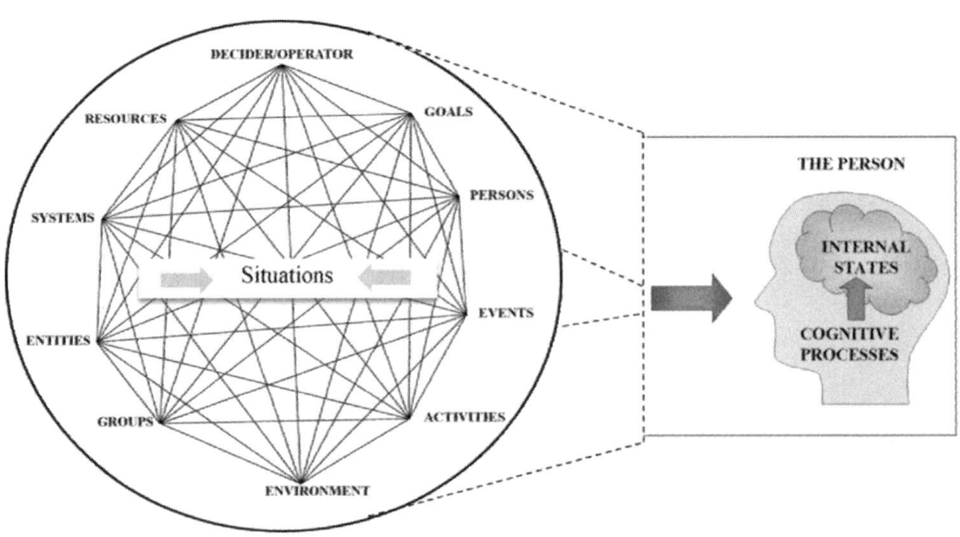

Figure 2.8 Elements of situations. (Adapted from [23].)

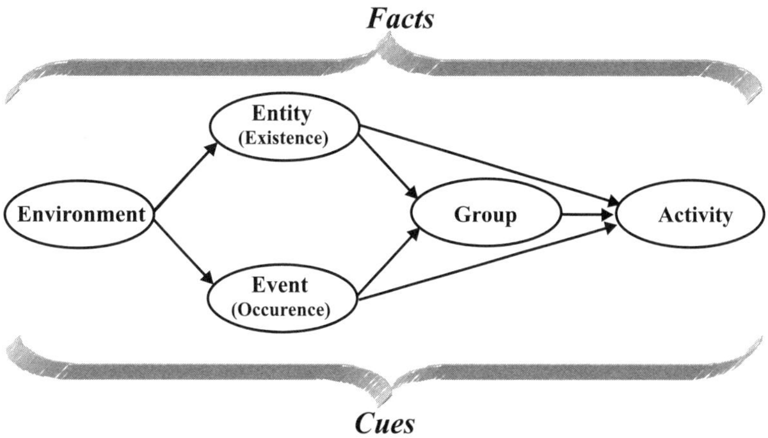

Figure 2.9 Basic elements of a local situation. (From [23].)

has the quality or state of being active, i.e., when something is characterized by action or expressing action as distinct from mere existence or state. Finally, one may think of a global situation as being composed of a set of local situations. The term scene, i.e., a single situation in this set, could be used to refer to a local situation."

Roy [47] terminated the description of basic elements of a situation (Figure 2.9) by: "A fact is defined as something that has actual existence, or an actual occurrence. It is a piece of information presented as having objective reality. Clearly, a fact has to do with the notions of truth, verity, and reality. Data is indeed factual information used as a basis for reasoning, discussion or calculation. Cues, i.e., features indicating the nature of something perceived, are generated and received from the various data/information sources. Hence, the SA process manipulates data and information that include cues, about facts, obtained from external sources, and inferences drawn by the system itself." If we are talking about situation analysis support systems, we refer to systems (e.g., FIAT-based) that can provides answers, in whole or in part, to the six basic epistemic questions excerpted from Sulis [48] that have been listed in Figure 2.10. Phrased in a different way, similar questions appeared in Nicholson [49] with respect to situation awareness: "What are the objects of interest? Where are they? How are they moving? Where have they been? Where might they be going?" Objects could refer to either physical objects such as vehicles or symbolic objects such as terrorist plans.

2.4.1 Complex Events and Information Granules

A complex event is a composition of events that suggest more complicated circumstances. These events may be happening across the various layers of an organization as "sales leads, orders or customer service calls. Or, they may be news items, text messages, social media posts, stock market feeds, traffic reports, weather reports, or other kinds of data." Complex event processing (CEP) methods [50, 51] track and process streams of information (data) from multiple sources about events, establish temporal and causal dependencies between them, infer patterns, and derive conclusions (such as opportunities or threats) from all that. This is quite an important set of analytics methods and technologies part of FIAT that have applications for system health monitoring, anomalies and fraud detection, Internet of Things, and

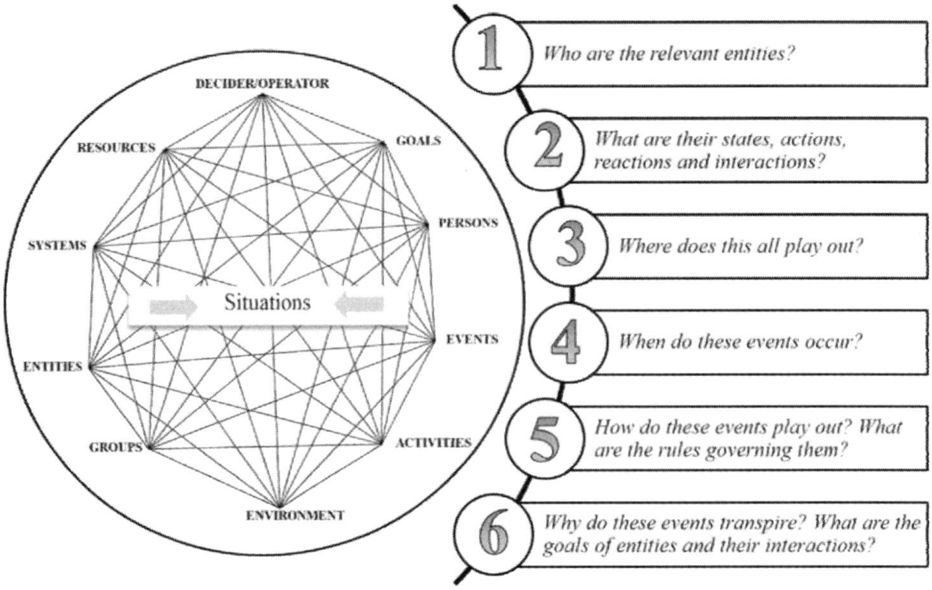

Figure 2.10 Six epistemic questions for situation assessment.

situation analysis in general. CEP provides ways to analyze patterns in real time and can be of prime importance in the composition of services for the enterprise. Note also that an event may also be defined as a change of state and so CEP methods share commonalities with change detection methods [52–54], a terminology more often used in the C4ISR domain.

Granular computing [55–57] is another set of analytics methods and technologies that has emerged in recent years. It concerns the representation, construction and processing of complex information entities called information granules. Information granules are the product of a process of data abstraction and derivation of knowledge from information or data. They can be treated as linked collections of objects or entities that usually originate at the numeric level, are drawn together by the criteria of indistinguishability, similarity, proximity or functionality, coherency, or the like. "As a theoretical perspective, it encourages an approach to data that recognizes and exploits the knowledge present in data at various levels of resolution or scales. In this sense, it encompasses all methods which provide flexibility and adaptability in the resolution at which knowledge or information is extracted and represented." Zadeh [58] pointed out that granulation is one of the three basic concepts that underlie human cognition: granulation, organization, and causation. Informally, granulation involves decomposition of whole into parts; organization involves integration of parts into whole; and causation involves association of causes with effects.

2.4.2 Sense-Making

Pirolli's sense-making framework [41, 59] is important to be considered in the development of FIAT-based situation analysis support system. Roy et al. [45, 60] recognized Pirolli's work that inspired their multiple hypothesis sense-making approach for situation analysis. Sense-making is the process of creating situational awareness and understanding in situations of high complexity. Sense-making is an iterative operation involving a hopefully converging dynamic between a supporting information/evidence-space and an evolving situation hypothesis space. Here, the former is considered to be automatically constructed by an automated FIAT process and the latter is seen as occurring in the human mind, possibly aided by automated utilities (e.g., FIAT-based). In Pirolli [59], the overall sense-making process is "organized into two major loops of activities": (1) a foraging loop that involves processes aimed at seeking information and/or hypotheses, searching and filtering them, possibly into some situational schema, and (2) a sense-making loop that "involves iterative development of a mental model (a conceptualization) from the schema that best fit the evidence." The data-frame model of Klein et al. [61, 62] has many similarities with Pirolli's framework.

In the model of Figure 2.11, the rectangular boxes represent an approximate data flow. The circles represent the process flow. This is a process with lots of back loops and it seems to have one set of activities that cycle around finding information and another that cycles around making sense of the information, with plenty of interaction between these. The data flow shows the transformation of information as it flows from raw information to reportable results by an analyst [59]. The "shoebox" is the much smaller subset of that external data that is relevant for processing. The "evidence" file refers to snippets extracted from items in the shoebox. Schemas are the representation, or organized marshalling, of the information so that it can be used more easily to draw conclusions. Hypotheses are the tentative representation of those conclusions with supporting arguments. Ultimately, there is a presentation or other work product. Basically the data flow represents the

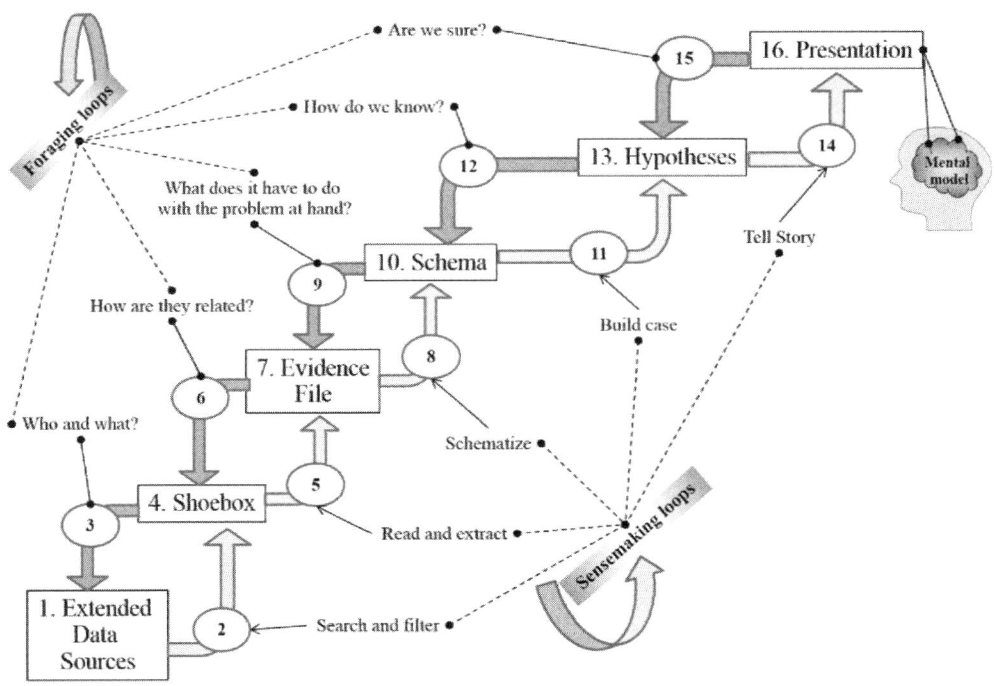

Figure 2.11 Sense-making framework according to Pirolli [41, 59].

transducing of information from its raw state into a form on which expertise can be applied and then out to another form better suited for communication.

In today complex situations (e.g., asymmetric threat, terrorism), where it is difficult to get reliable a priori models or procedural knowledge of the expected adversarial dynamics then deductively based analyses (fusion approaches) can be pretty restraint. As a result, analysis approaches to such problems require mixed paradigms where any available (and reliable) a priori knowledge can be employed in a deductive framework but at the same time inductive and adductive methods involving learning and discovery operations and tools are also required parts of any modern analytics and information fusion tools suite (FIAT). The FIAT-based system and its underlying automated technologies should support the functions of hypothesis synthesis, situation iteration and nomination, and story development. These three main categories of functions identified in Llinas [42] are dynamic and iterative and are dependent on effective interfaces to the automated FIAT to support foraging and sense-making iterations of Figures 2.7 and 2.11. Human intelligence is seriously involved in these operations and so effective and efficient interfaces and visualization designs are required to interface FIAT-based system to the human analysts or decision-makers.

2.5 Situation Awareness

Situation awareness (SAW) [63, 64] has emerged as an important concept around dynamic human decision-making in both military and civilian complex (e.g., public security, power networks, health networks, transport) environments. Situation analysis is defined as the process that provides and maintains a state of situation awareness for the decision-maker(s), and intelligent systems (e.g., data mining, machine learning, information fusion, analytics) is a key enabler to meeting the demanding requirements of situation analysis in current and future complex environments. SAW is a state in the mind of a human (Figure 2.8). At the heart of this process is the provision of decision quality information to the decision-maker, thereby enabling timely situation awareness. This is on timeless requirement that is being immeasurably complicated by the big data Vs (Variety, Volume, Velocity, Veracity, and Value) discussed in Chapter 1. Blasch in Chapter 2 of [65] entitled "Situation Assessment and Situation Awareness" provided a recent state of the art on this topic.

Figure 2.12 illustrates a theoretical model, derived by Endsley [64], of SAW based on its role in dynamic human decision making. SAW is defined as the perception of the elements in the environment, within a volume of time and space, the comprehension of their meaning, and the projection of their status in the near future. The first level of SAW yields in the perception of the status, attributes and dynamics of relevant elements in the environment. Endsley described the comprehension process as follows: "Comprehension of the situation is based on a synthesis of disjoint level 1 elements." Level 2 of SAW goes beyond simply being aware of the elements that are present, to include an understanding of the significance of those elements in the light of pertinent operator goals. Based on knowledge of Level 1 elements, particularly when some elements are put together to form patterns with other elements, the decision-maker forms a holistic picture of the environment, comprehending the significance of objects and events. The third and last step in

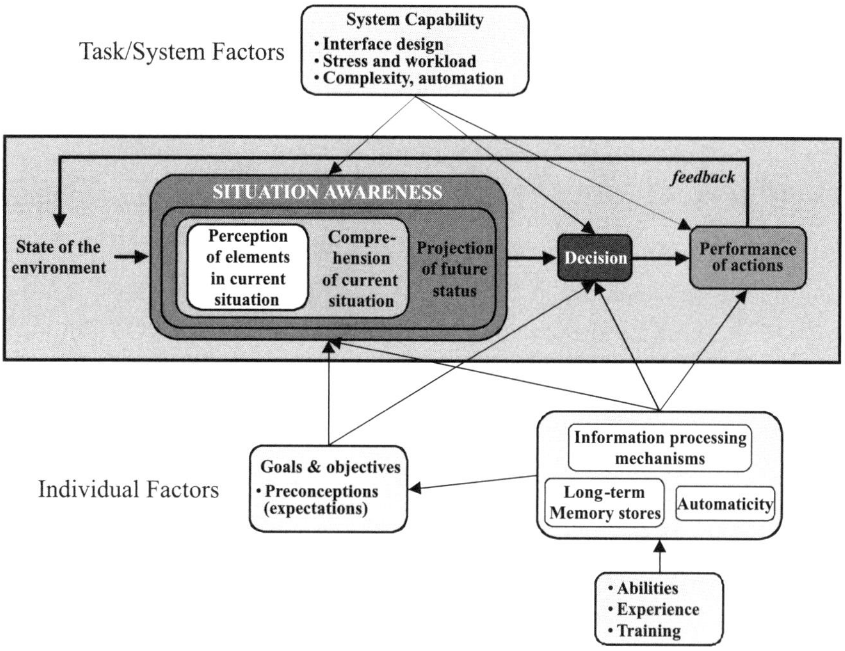

Figure 2.12 Endsley's situation awareness model. (From: [23].)

achieving situation awareness is the projection of the future actions of the elements in the environment. This is achieved through knowledge of the status and dynamics of the perceived and comprehended situation elements.

The situation awareness processes described by Endsley are initiated by the presence of an object in the perceiver's environment. However, processes related to situation awareness can also be triggered by a priori knowledge, feelings, or intuitions. In these situations, the picture is understandable, and projections in the future are possible, if any event, which has not been perceived at this time, can be found in the environment. Hence, hypotheses related to the possible presence of an object are formulated. The perceiver then initiates search processes in the environment that confirm or invalidate these hypotheses. Note that this type of SAW is possible only if mental models related to the possible objects are available.

If one compares the OODA loop with the SAW model of Endsley, it is straightforward to witness a close resemblance. In both models one finds a decision-making part and an action part. In Endsley's model, SAW is one of the main inputs for decision-making. In the OODA loop, the processes Observe and Orient provide inputs for the decision-making process. However, one should recall that SAW in Endsley's model is a state of knowledge and not a process. In her theory of SAW, Endsley clearly presumed patterns and higher-level elements to be present according to which the situation can be structured and expressed. Thus, SAW can be interpreted as the operator's mental model of all pertinent aspects of the environment (processes, states, and relationships).

There is a tight link between this mental model used to structure and express situation elements and the cognitive processes involved in achieving the levels of awareness. This link is known as the cognitive fit and requires an understanding of how the human perceives a task, what processes are involved, what are the human needs and what part of the task can be automated or supported. This understanding is crucial and only achieved via a number of specialized human factor investigations known as cognitive engineering analyses.

2.6 Cognitive System Engineering (CSE)

Among the procedures developed to identify cognitive processes, there are the cognitive task analysis (CTA) [59, 66–68] and the cognitive work analysis (CWA) [69–73]. There are only subtle and ambiguous differences between these two procedures. Moreover, their labels are frequently used in an interchangeable manner in the literature. However, the CWA can be seen as a broader analysis than the CTA. According to Vicente [74], traditional task analysis methods typically result in a single temporal sequence of overt behavior. This description represents the normative way to perform the task. Nevertheless, traditional methods cannot account for factors like changes in initial conditions, unpredictable disturbances, and the use of multiple strategies. The use of the traditional task analysis brings an artifact that will only support one way to perform the task.

Cognitive system engineering (CSE) [75, 76], analyses are defined as approaches that aim to develop knowledge about the interaction between human information processing capacities and limitations, and technological information processing systems. The usefulness of a system is closely related to its compatibility with the human information processing. Therefore, CSE analyses focus on the cognitive demands imposed by the world to specify how technology (e.g., FIAT) should be exploited to reveal the problems intuitively to the decision-maker's brain.

2.6.1 Cognitive Work Analysis (CWA) and Applied CWA

Vicente [74] proposed an ecological approach, which can be seen as a CWA, takes its origin in psychological theories. This research raised the importance to study the interaction between the human organism and its environment. The perception of an object in the environment is a direct process, in which information is simply detected rather than being constructed. The human and the environment are coupled and cannot be studied in isolation. A central concept of this approach is the notion of affordance. The affordance is an aspect of an object that makes it obvious how the object is to be used. Examples are a panel on a door to indicate push and a vertical handle to indicate pull. When the affordance of an object is obvious, it is easy to know how to interact with it. The environment in which a task is performed has a direct influence in the overt behavior. Hence, the ecological approach begins by studying the constraints in the environment that are relevant to the operator. These constraints influence the observed behavior.

The ecological approach is comparable to and compatible with Rasmussen's abstraction hierarchy framework [75, 77]. This abstraction hierarchy is represented by means-ends relations and is structured in several levels of abstraction that represent functional relationships between the work domain elements and their purposes. Rasmussen has developed a comprehensive methodology that overcomes the limitations of traditional cognitive task analysis (CTA) by taking into account the variability of performance in real-life, complex work domains. CWA seems to be the best choice to answer questions related to understanding the C2 (decisions) task. A feasibility study to investigate the applicability of CWA for C2 was performed to confirm this issue [78, 79]. The study revealed that the methodology is well suited to deal with decision support design issues but is in practice, if done in a full scale (all sequential CWA phases) for a small problem, time-consuming and very expensive to conduct, and the quality of the findings is dependent on subject matter experts' availability and on the skills of the people conducting it. In addition, the study did not show the gap between cognitive analyses and design, making the DSS engineering process inefficient.

The applied cognitive work analysis (ACWA) methodology [70, 80] emphasizes a stepwise process to reduce the gap to a sequence of small, logical engineering steps, each readily achievable. At each intermediate point, the resulting decision-centered artifacts create the spans of a design bridge that link the demands of the domain as revealed by the cognitive analysis to the elements of the decision aid. The ACWA approach is a structured, principled methodology to systematically transform the problem from an analysis of the demands of a domain to identifying visualizations and decision-aiding concepts that will provide effective support. A more detailed description of the methodology steps, illustrated in Figure 2.13, can be found in [81]. The steps in this process include:

- Using a functional abstraction network (FAN) model to capture the essential domain concepts and relationships that define the problem-space confronting the domain practitioners;
- Overlaying cognitive work requirements (CWR) on the functional model as a way of identifying the cognitive demands, tasks, and decisions that arise in the domain and require support;

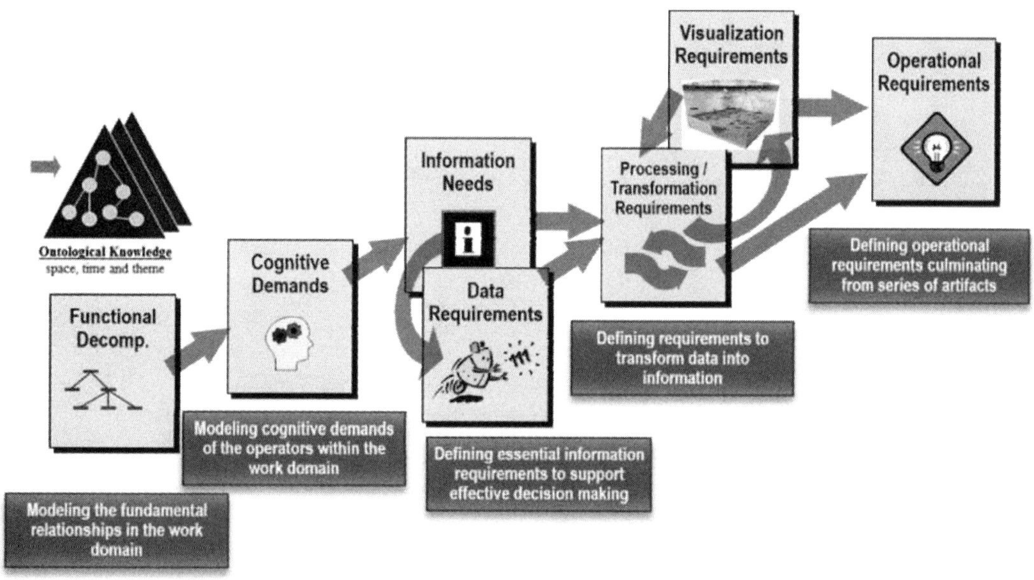

Figure 2.13 ACWA methodology: (FAN-CWR-IRR-RDR-PDC) steps. (Adapted from [81].)

- Identifying the information/relationship requirements (IRR) for successful execution of these cognitive work requirements;
- Specifying the representation design requirements (RDR) to define the shaping and processing for how the information/relationships should be represented to practitioner(s);
- Developing presentation design concepts (PDC) to explore techniques to implement these representation requirements into the syntax and dynamics of presentation forms to produce the information transfer to the practitioner(s).

2.6.2 Ontological Engineering Versus CWA

Situation awareness is usually associated with having perception, comprehension, and projection knowledge about the world. An important issue to support situation analysis is understanding the ways humans represent the world, and the development and exploitation of efficient machine representations of relevant aspects of the world. There are three domains of human reality: the physical domain (e.g., mass and energy, human bodies, infrastructures), the cognitive domain (e.g., sentiment, emotion, intention), and the symbolic domain (explicit information, media content).

Formal methods of ontological engineering can be applied with the goal of producing defensible representations of the world (e.g., situational constructs). The ambition is not to capture "the" meaning of the relevant aspects of the world, but to formulate "a" meaning of these aspects that is sufficient to engineer situation analysis support systems. There is a need to distinguish the entire world (i.e., all aspects of the world) from the world of interest to some decision makers (i.e., only aspects of interest). The spectrum of world states of interest as associated with a given task or mission is bounded by, and related to, the informational needs of users. In this regard, work domain models can be constructed based on cognitive engineering techniques to understand and formally document the information needs of decision-makers.

Ontological engineering can be applied to generate representations of the world, although cognitive engineering techniques can be used to capture the informational needs of users as illustrated in the self-explained Figure 2.14. Clearly, one could say that the cognitive engineering techniques used for developing a specification of user information needs could inform the ontologist in developing a correspondingly-bounded ontological description of those needs (scope of an ontology) as reflected in a set of ontologically described states. Concepts identified in the ontological modeling could be mapped onto concepts identified in the work domain modeling to insure consistency across and within the two modeling constructs. From this exercise would result a space over which FIAT processes can reason using an ontological representation language. The internal workings of the CSE methods and detailed descriptions of ontological engineering techniques are beyond the scope of this book. There are numerous other books available that explain these in greater details as, for instance, the following recommended reading on ontological engineering [82–85], on cognitive systems engineering [75, 76, 86], and, more specifically, on cognitive work analysis [69, 71, 73].

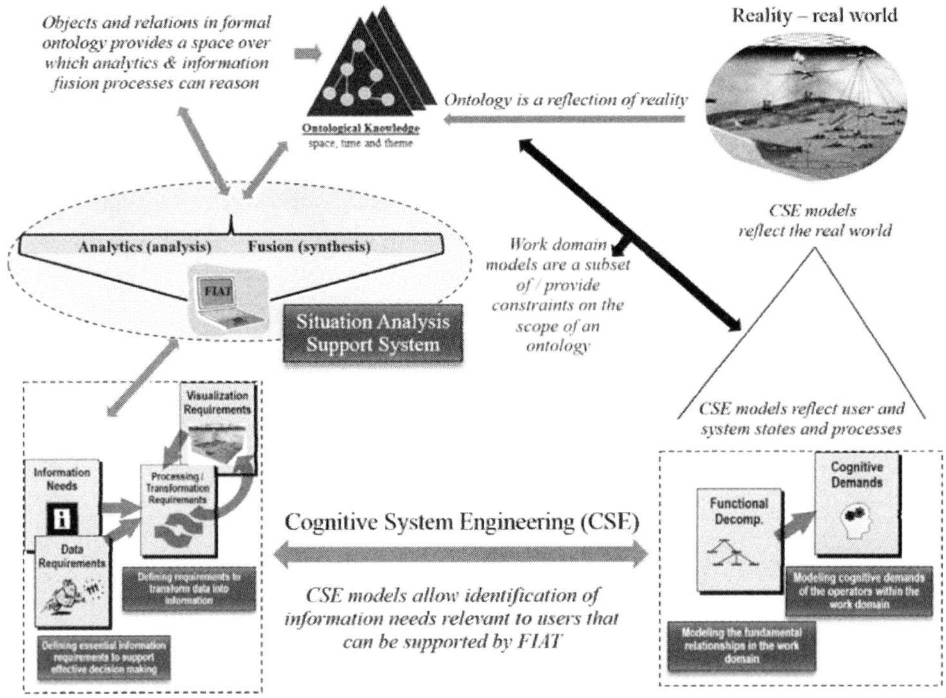

Figure 2.14 Ontological engineering versus cognitive system engineering.

2.7 Definition of Context

Data acquire meaning through context. Context establishes the basis for discerning meaning of its subjects and may occur at many levels. Exploitation of contextual knowledge is necessary for situation analysis and consequently for its support through FIAT-based systems. The most widely accepted definitions of context is from Dey [87] phrased as follows: "Context is any information that can be used to characterise the situation of an entity. An entity is a person, place, or object that is considered relevant to the interaction between the user and the application, including the user and the applications themselves."

This general definition of context has been extended by Zimmermann et al. [88], who introduced a definition comprising three canonical parts: a definition per se in general terms, a formal definition describing the appearance of context and an operational definition characterizing the use of context and its dynamic behavior. Figure 2.15 shows the five categories of the formal definition as well as the elements of the operational extension. Zimmermann et al. [88] extended the definition by the description of the following five categories of elements: individuality, activity, location, time, and relations. These five fundamental context categories determine the design space of context models. The description of the five categories of elements is presented in Table 2.1. We refer the reader to Zimmermann et al. [88] for more comprehensive details. Their underlying motivation is to provide a structure that bridges the user-developer gap from a general concept (e.g., Dey's definition) easily understandable by the user to its engineering by software devel-

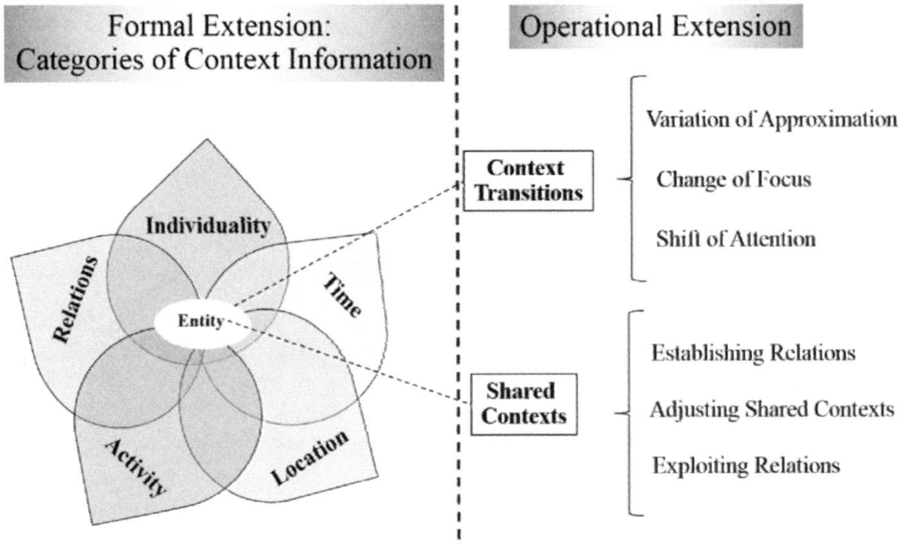

Figure 2.15 Definition of context according to Zimmermann et al. [88].

opers. This is more than required for the design of a FIAT tools suite to support situation analysis.

Dey's definition clearly states that context is always bound to an entity and that information that describes the situation of an entity is context. The activity predominantly determines the relevancy of context elements in specific situations, and the location and time primarily drive the creation of relations between entities and enable the exchange of context information among entities. The task itself is also part of the context as it characterizes the situation of the user.

Context awareness has to be distinguished from situation awareness discussed above. In part of the literature, the term has been restraint to a property of mobile devices that is defined complementarily to location awareness (e.g., users of smart phones). From Wikipedia [89], context awareness has been related to different fields as: "Context awareness originated as a term from ubiquitous computing or as so-called pervasive computing which sought to deal with linking changes in the environment with computer systems, which are otherwise static. The term has also been applied to business theory in relation to contextual application design and business process management issues."

2.8 Definitions of Information Fusion and Analytics

Situation awareness has emerged as an important concept around dynamic human decision-making in both military and civilian complex (e.g., public security and power networks) environments. Situation analysis is defined as the process that provides and maintains a state of situation awareness for the decision-maker(s), and intelligent systems based on FIAT are key enablers to meeting the demanding requirements of situation analysis in current and future complex environments.

Managing the complexity of situation understanding and of the decision space is a crucial challenge. This challenge is compounded by the wide spectrum and

Table 2.1 Description of the Five Categories of Context Elements

Individuality (comprises anything that can be observed about an entity, typically its state)	Natural entity context: comprises the characteristics of all living and nonliving things that occur naturally and are not the result of any human activity or intervention.
	Human entity context: covers the characteristics of human beings (e.g., the user's necessities, the user behavior, preferences in language, color schemes, and modality of interaction).
	Artificial entity context: denotes products or phenomena that result from human actions or technical processes. In a broad sense, this category covers descriptions for any human-built thing like buildings, computers, vehicles, books, and all sensors that measure physical or chemical properties.
	Group entity context: is a collection of entities, which share certain characteristics, interact with one another or have established certain relations between each other.
Activity	Covers the activities in which the entity is currently and in the future involved and answers the question: What does the entity want to achieve and how? It can be described by means of explicit goals, tasks, and actions.
Location	Describes location models that classify the physical or virtual (e.g., the IP address as a position within a computer network) residence of an entity, as well as other related spatial information like speed and orientation.
Time	Subsumes time information like the time zone of the client, the current time or any virtual time (intervals of time), process-oriented view of the time concept (e.g., workflows).
Relations	Social relations: describe the social aspects of the current entity context. Usually, interpersonal relations are social associations, connections, or affiliations between two or more people. For instance, social relations can contain information about friends, neutrals, enemies, neighbors, coworkers, and relatives.
	Functional relations: indicate that one entity makes use of the other entity for a certain purpose and with a certain effect (e.g., transferring a specific input into a specific output).
	Compositional relations: are the relation between a whole and its parts. In the aggregation, the parts will not exist anymore if the containing object is destroyed. For example, the human body owns arms, legs, and so forth. The association is a weaker form of the composition.

Source: [88].

diversity of data that have to be processed, fused, and eventually transformed into actionable understanding, and this, calls for considering the collaborative dimension in the quest for understanding the complexity of an actual situation. The explosion came with the advent of advanced sensing and the diversity and volume of data from multiple sources and forms (unstructured and open sources, voice records, photos, video sequences, social networks). The decision-makers/analysts cannot cope with the flow of material, with potentially severe consequences on the quality of the decision and operational processes.

Consequently, to support the military decision-makers/analysts in the making sense out of the data, a field of research called information fusion (IF) began about 25 years ago and has been slowly maturing in terms of the establishment of an underlying science and in terms of standardized engineering methodology. The overall objectives of IF technologies are then to support decision-makers/analysts in the making sense of an ever-increasing volume of data and complexity; to improve the quality of information, reduce uncertainty, and support real-time decisions and actions; and to ensure and improve dependability to face an ever-increasing complexity (networked aspect and heterogeneity).

In both defense and civilian businesses (use cases), situation analysis (SA) provides the context and knowledge for planning and acting. In the military domain, IF systems built from smart technologies have emerged as a key sense making support to SA for complex environments (hostile or cooperative). In the general business and industry domain, the term "analytics" [90, 91] is used to pursue roughly the same objective considering different focus and constraints. Constraints seem getting closer with the advancement of network technologies so there is some foreseen benefits for the convergence of these two application communities. Both information fusion and analytics are the application of computer sciences and technologies, operational research, cognitive engineering, and mathematical approaches and methods to support situations understanding.

2.8.1 Joint Directors of Laboratories (JDL) Data Fusion Model

The data fusion model maintained by the JDL Data and Information Fusion Group (JDL DIFG) is the most widely used approach for categorizing data fusion-related functions [92]. The JDL distinction, among fusion levels in Figure 2.16, provides a valuable way of differentiating between data fusion processes that relate to the refinement of objects, situations, threats, and processes. The definitions are as follows:

- *Level 0, Sub-Object Data Assessment:* Estimation and prediction of signal/object observable states on the basis of pixel/signal level data association and characterization; Level 0 assignment involves hypothesizing the presence of a signal (i.e., detection) and estimating its state.
- *Level 1, Object Assessment:* Level 1 data fusion combines data from single or multiple sensors and sources to provide the best estimate of objects and events in the battlespace in terms of their position, kinematics (e.g., tracks), identity, or identification features; Level 1 involves tracking and identification that includes reliable location, tracking, combat ID, and targeting information. Such information may be of high quality, particularly when it leverages multiple sensors to provide robustness to spoofing or deception, reliability in case of sensor malfunctions, and extended space-time coverage due to the diversity of observations.
- *Level 2, Situation Assessment:* Level 2 data fusion focuses on situation assessment. This requires objects/entities recognition in the regions of interest, as well as recognizing activities of these objects, and inferring their relationships. Issues that must be addressed, at this level, include: automated target/object recognition; automated activity recognition from multiple sensors and reports collected and stored for historical analyses; inferring relationships of objects in the scene based on their identities or coordinated behaviors and historical analyses; the capability for the automated system to estimate certainties about object identities and activities; and the capability to request human assistance or additional information from sensors or databases to resolve ambiguities.

- *Level 3, Impact Assessment:* Level 3 fusion estimates the impact of an assessed situation, that is, inferring intent of objects/entities, or groups of objects, in the regions of interest and/or the outcome of various plans as they interact with one another and with the environment. The impact estimate can include likelihood and cost/utility measures associated with potential outcomes of a player's planned actions. Issues that need to be addressed include: methods for constructing and learning a wide variety of models of threat behavior; methods for reasoning with uncertain and incomplete information for assessing threats from object activities; and methods for efficient data-mining of databases.
- *Level 4, Process Refinement:* Adaptive data acquisition and processing to support mission objectives. Level 4 processing involves planning and control, not estimation. Level 4 also involves assigning resources to tasks.

In most defense applications, data fusion processing tends to be hierarchical in nature due to the inherent hierarchies built into defense organizations and operations. As a result, the fusion process also progresses through a hierarchical series of inferences at varying levels of abstraction (data, information, knowledge as shown in Figure 2.16). The exploitation of contexts in the inferences processes targets an increase in semantics that is only obtained by what one can do with the information (i.e., actionable knowledge). There are numerous books available that present more definitions, explain concepts in details, and develop mathematical techniques and models related to Figure 2.15. For more detailed insights, use any of the following recommended reading on data and information fusion books [23, 65, 93–96].

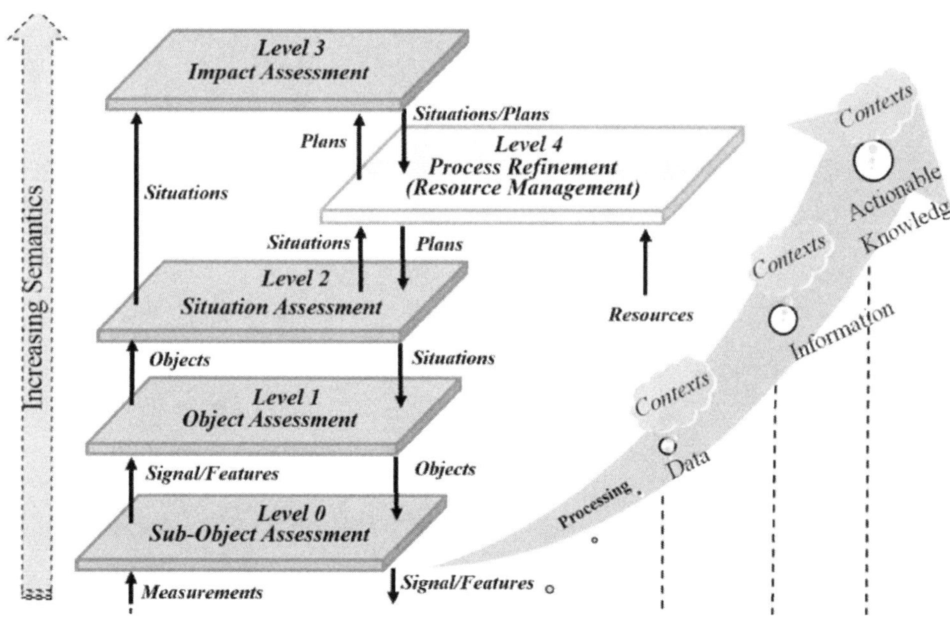

Figure 2.16 The JDL model and the data-information knowledge hierarchy.

2.8.2 Definition of Analytics

Eckerson [97] defined analytics as: "everything involved in turning data into insights into action." This is a quite broad definition that could include data and information fusion of the preceding section but does not help understanding analytics from an applied point of view. Unlike information fusion community, the analytics community did not benefit from a well-structured organization like JDL to fix the terminology in order to ease communication amongst communities. The definition of analytics associated to big data becomes more and more confusing with various vendors, consultants, and trade publications defining and offering new technologies. As Sheikh pointed out in his book [98], analytics is one of the hot topics on today's technology landscape (also referred as big data). Analytics is not new and originates from business intelligence. It has been rejuvenated with big data. Figure 2.17 illustrates Sheikh's proposal [98] to define analytics based upon a business and a technical implementation perspective.

The business value perspective looks at data in motion as it is generated through normal conduct of business. For this data, there are three variations of value, the present, the past, and the future, in the exact order as represented on Figure 2.17. When data is created, referenced, modified, and deleted during the course of normal business activities, it lives in an operational system. The operational system at any given time can tell us what we are doing now. The data from day-to-day operations (e.g., selling merchandise, reviewing applications) is accumulated for recordkeeping and starts to build history in a data warehouse. Reporting can then be done to help understanding how a business did (e.g., total sales) in the last month, quarter, or year. These analysis reports provide managers the tools to understand the performance of their departments. This leads into the question that analytics should help to answer: What should we be doing within departments and business units to improve business performance? Any tools, technologies, or systems that help with this question can qualify to be in the analytics space.

The technical implementation perspective described in Sheikh [98] is also the tangent adopted by Das [8] in his book on computational business analytics. Das' book describes the characteristics of analytics in terms of the techniques used to implement analytics solutions. They are listed in Figure 2.17. In the analytics literature, three general types of analytics make consensus even though terminology may differ. This is reflected on both Figures 2.17 and 2.18: descriptive, predictive,

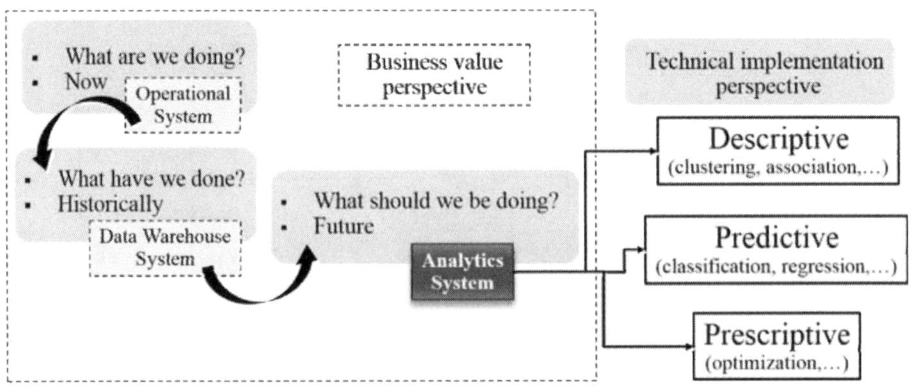

Figure 2.17 Sheikh's proposal definition of analytics.

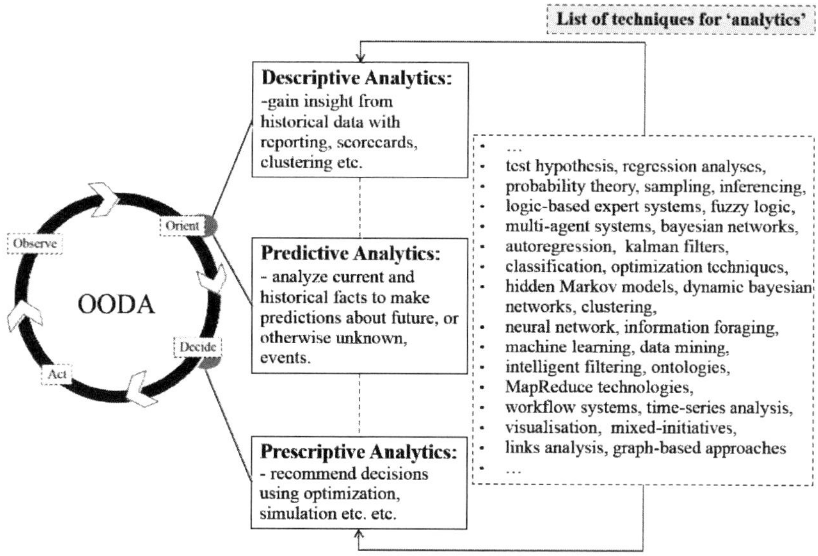

Figure 2.18 Das' views on defining analytics and relationship with the OODA loop.

and prescriptive. A lot of good reference books are available to describe these techniques particularly under data mining and machine learning fields [99–103].

As presented in Figure 2.18, descriptive and predictive analytics link with the Orient step in the OODA loop, whereas the Decide step corresponds to prescriptive analytics. While originally developed for military applications, the OODA loop has been applied frequently in the private industrial sector because of its common sense approach to making decisions. Das [8] instantiates the OODA loop in the business domain as follows: (1) observation is declining revenue figures; (2) orientation is to identify causes for declining revenue and to fully understand the company's overall financial situation and other relevant factors; (3) decision could be to enhance a marketing campaign, upgrade products or introduce new products; and (4) action is the marketing campaign or new product launch. An action in the real world generates further observations such as the increased revenue or customer base as a result of the marketing campaign.

Business teams stand in place of the "solo pilots in this metaphor," but the behaviors and benefits are the same. Deitz said [104]: "In warfare and in business, the speed at which the OODA Loop is executed becomes the largest factor in disrupting the enemy in the battle space, and the competition in the business space." Gartner VP and distinguished analyst Roy Schulte stated that disrupting your opponent often comes from using a systems approach that shortens your own OODA loop so that you can turn inside the adversary's process. The pressures of mobile, social, and big data make the need urgent to support the OODA loop steps by powerful analytics software.

Das [8] discussed a model-based approach to analytics that contributes to define and understand analytics. He described computational models as a combination of symbolic, subsymbolic, and numerical representations of a problem. This is obtained through a series of inferences (e.g., inference cycle of Figure 2.19) for description, prediction, and prescription: the tree kinds of analytics. Structured input in the form of transactional data is fed into an inference cycle for the model

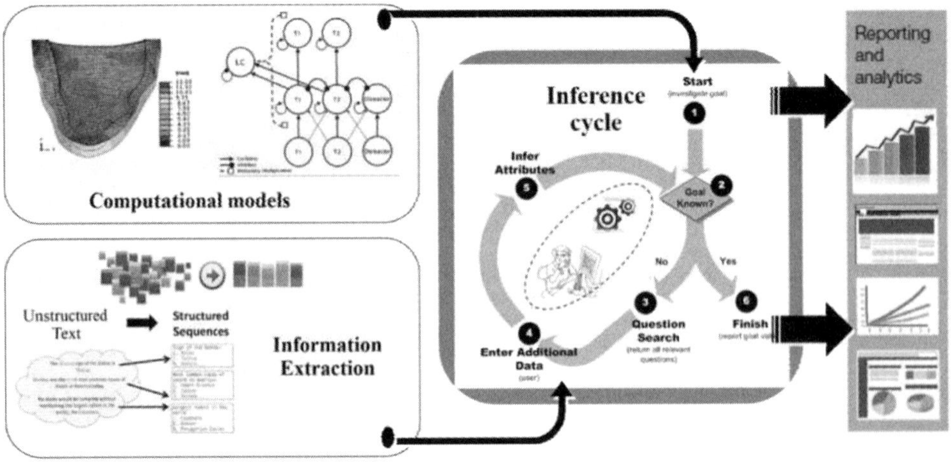

Figure 2.19 Das' model-based analytics.

to produce analytical results. If the input is unstructured text, structured sequences needs to be extracted as illustrated in Figure 2.19. The knowledge base on which implicit facts can be inferred is composed of computational models in the form of rules and structure relational data in the form of facts. Figure 2.19 illustrates this approach.

Das [8] explained how to build these computational models by mimicking human reasoning. One of the steps is to represent by expressive graphical constructs and linguistics variables and business analysts' internal mental models of things that they observe and with which they interact. Computational models can also be viewed as patterns that are embedded within huge volumes of transactional data continuously generated by many business processing systems. Such models can therefore be extracted or learned via automated learning methods.

2.9 Situation Analysis Models (Information Fusion and Analytics)

Several definitions of data fusion as well as many models and architectures have been provided through the years, most of them derived from military and remote sensing fields [3, 93, 96, 105–110]. Significant examples are information-based models such as the JDL and its revisions [92, 111–114], the DFD (Data-Feature-Decision) model of Dasarathy [115], activity-based models such as the Boyd [1] OODA loop, the intelligence cycle of Shulsky and Schmitt [116], and the Omnibus model [117]. Concepts, models, tools, trends, and comprehensive and extensive cross-analyses and surveys of high-level and low-level information fusion with respect to decision support and situation awareness can be found in books [3, 93, 96, 105–107], surveys, and outlook papers [118–122]. The major part of the literature focuses its attention on:

1. IF techniques and algorithms for integrating data from homogeneous or heterogeneous collections of potentially faulty, redundant, or inaccurate sensors and sources;

2. Process, functional, and formal models implementing IF systems; very few concern universal frameworks to address all levels of fusion;
3. Ontological-based approaches [123–125] that are very important to any definition of IF frameworks, since without context and domain representation no computer support to human sense-making is possible.

As stated earlier, the IF process is about transforming data into actionable knowledge to support the making sense of data, to reduce uncertainty and improve dependability in systems. Meaning and making sense is then intimately related to actions of any agent: human or machine. A goal-driven or activity-based approach sounds appropriate for an IF framework. Very few contributions in the IF literature have the potential to be considered as a universal IF framework. A brief description and analysis of some of them is presented next.

2.9.1 OODA-Based Models

The OODA loop [1] is a model that represents the classic decision cycle of military command and control (C2). Correspondence has been done from OODA to the JDL model in the Omnibus model [117]: Observe is level 0; Orient comprises levels 1, 2, and 3; Decide corresponds to level 4; and Act is beyond the scope of the JDL model. Act in the OODA loop has broader implication for what is meant by Level 4 that entails use of resources but only to accomplish the process refinement of Levels 1–3. Shahbazian et al. [6], with the Extended OODA, have extended Level 4 of the original JDL to C2 resource management (e.g., planning). Roy and Bossé [126] identified the presence of multiple OODA loops in the modeling of information processing schemes for a naval tactical application. More recently, Gustavsson and Planstedt [127] recognized that multiple OODA structure in a much broader and complex application, that is, the military NEC concept (equivalent to a CPSS environment). They have integrated the JDL levels in their thinking and presented an OODA fractal information fusion (FIF) modeling to support a system of systems architecture that aligns conceptually with a network-centric approach as pictured in Figure 2.20.

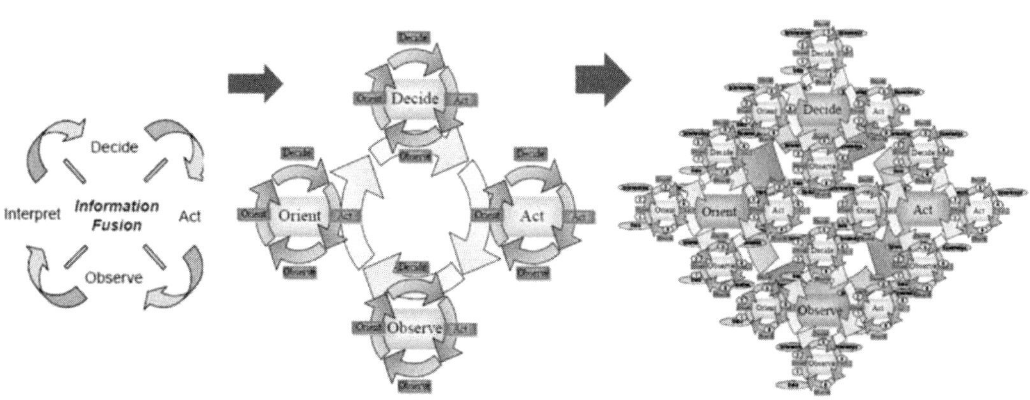

Figure 2.20 Fractal information fusion modeling. (Adapted from [128].)

Guitouni et al. [129] also used the OODA model to architect a multiagent approach for an Information Fusion and Resource Management (INFORMLab) testbed designed to develop and evaluate NEC approaches for large volume surveillance. Finally, OODA has been applied [7] to the modeling of Level 5 of the revised JDL version [114]. The Cognitive OODA [7] is a method of user and team analysis to support timely decision making. All these OODA-influenced efforts contribute towards the definition of an universal IF processing framework but still require more work to result in a framework that addresses all levels of abstraction and specificity (tasks) needed for the design of complex IF systems.

2.9.2 The Dual-Node Network (DNN) Architecture of the Revised JDL Model

The dual-node network (DNN) architecture of Bowman and Steinberg [130] is based on Department of Defense (DoD) Architecture Framework for system engineering [131]. They used the JDL model to define a dual network of data fusion (DF) and resource management (RM) processing nodes represented in Figure 2.21. A very important aspect of the DNN fusion node construct is the recognition of the ubiquity of a processing construct at any level (0–4) of the IF process. The concept of a processing node has been introduced in the framework of the JDL fusion model, as shown in Figure 2.21. According to this fusion node paradigm, the node processes the data/information provided by the sources (or other, prior fusion nodes) at the input, to produce a composite, high-quality version of some information products of interest to the users (or to other, subsequent fusion nodes) at the output. Any data/information fusion node, whatever the fusion level, contains three main subprocesses: alignment, association, and fusion. The means for implementing these functions and the data and control flow among them will vary from node to node and from system to system.

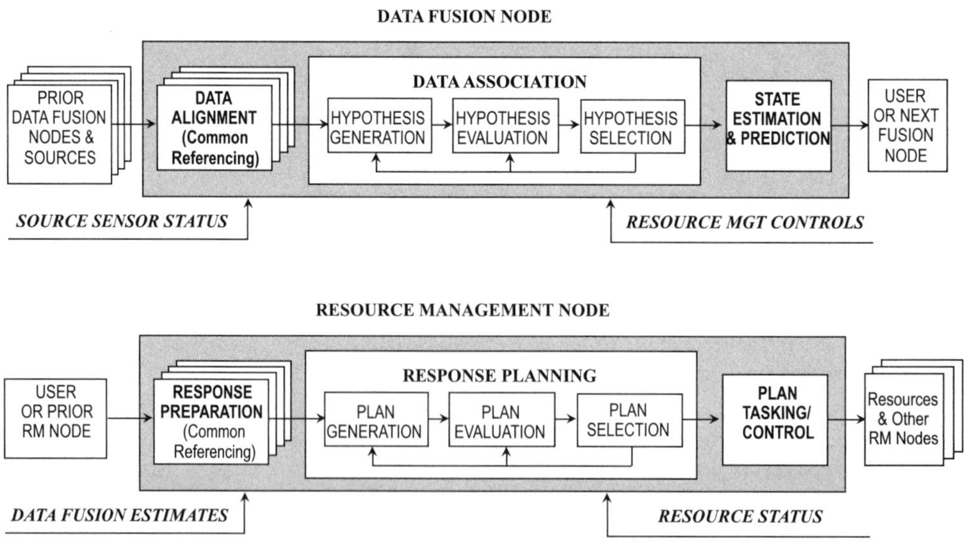

Figure 2.21 Any data/information fusion node. (Adapted from [92].)

The duality between data fusion and resource management has often been highlighted by the JDL community. This duality can be extended to include the architectures and functionality of data fusion and resource management, leading to the notion of a resource management node, as illustrated in Figure 2.21. A resource management node involves functions that directly correspond to those of a data fusion node. The common referencing has to do with normalizing performance metrics of the given problem and normalizing performance models of available resources, as well as any control format, spatio-temporal, and semantic alignment. Plan generation is about the candidate partitioning of the problem into subordinate problems and candidate assignment of resources. Plan evaluation is about evaluating the conditional net cost, and plan selection about determining a decision strategy. Finally, control (i.e., plan execution) is about generating the control commands to implement the selected resource allocation plan.

Roy, in Chapter 4 of [23], generalized this concept of a fusion node to a situation analysis node. Llinas [132] promoted that dual-node construct but identified some gaps such as adaptive internodal feedback, flow control logic between the various levels, and incorporation of contextual information to finally concludes that research is further required for higher-level IF inference and estimation processes to be considered as a universal IF framework.

2.9.3 The State Transition Data Fusion (STDF) Model

The State Transition Data Fusion (STDF) model introduced by Lambert [133] is a functional model aiming at unification of object, situation, and impact (Levels 1–3) JDL fusion. It views the world at time k as a composition of states $s(k)$. At time k, the world is understood in terms of the history of its state transitions up to time k. At time $k + 1$, different states of the world may be sensed by the sensors. This new data is fed into the observation process identifies the detections, normalizes the detections to a frame of reference, and then utilizes the prediction process to match the new observation to one or more previously predicted states of the world. An object instance of the world is represented at time t as a state vector $\mathbf{u}(t)$ of the measured values and the understanding of that object at time k is captured as a set of transitioning state vectors. A situation is represented at time t as a set of statements about the world (state of affairs) in some formal language [134] and it is understood at time k as a set of transitioning states of affairs. Finally, impact assessment in STDF presents an understanding of the world in terms of scenarios. A scenario instance at time t is expressed as a set of transitioning situations projected into the future. Scenario prediction in STDF involves assessment of intent, awareness, and capability of agents.

The STDF model captures the IF process in a broader sense. It offers well-defined notions of object, situation, and scenario and characterizes the IF process in an abstract functional model. However, STDF does not offer a computational model to represent how that functionality can be achieved. The STDF model views the underlying system in terms of a set of abstract states that are transformed in time. This appears to be the right abstraction one would need to focus on modeling the IF process, but does not provide a comprehensive formal framework to practically model and design the system. In fact, STDF explicitly leaves the choice of such a framework open to the practitioner. Details on STDF can be found in Chapter 3 of [65].

2.9.4 The Interpreted Systems Situation Analysis (ISSA) Approach

Fagin et al. [135] introduced the notion of interpreted systems as a formal semantic framework for reasoning about knowledge and uncertainty in multiagent systems. Maupin and Jousselme [136, 137] proposed using interpreted systems as a general framework for situation analysis referred here as ISSA. According to [135], a multiagent system can be conceptually divided into two components: the agents $A = \{1, ..., n\}$ and the environment e, which can be viewed as a special agent. The global state of the system with n agents is defined to be an $(n + 1)$-tuple $(s_e, s_1, ..., s_n)$, where s_e is the state of the environment and s_i is the local state of agent i. The set of all global states of the system is defined as $\delta : L_e \times L_1 \times ... \times L_n$ where L_e is the set of possible states for the environment and L_i is the set of all possible local states of agent i. To model the changes of the system's global state in time, the notion of run is introduced as a function from time to global states δ, with the assumption that time ranges over natural numbers.

A system R over δ is defined as a set of runs over δ. Agents and the environment change the global state by performing actions. A joint action is a tuple $(a_e, a_1, ..., a_n)$ of actions performed by the environment and the set of agents. Joint actions cause the system to change its global state. Agents perform actions according to some protocol, which is a rule for selecting actions. Protocols are typically described by means of programs written in some programming language. To incorporate knowledge into the framework, the notion of interpreted systems is introduced. An interpreted system I is a pair $(R; \pi)$, where R is a system over a set δ of global states, and π is an interpretation (assertion) of a set of formulas Φ over δ. Once we have a proper model of the scenario, various types of queries through model checking methods [138] is being used for situation analysis.

The ISSA approach provides a formal framework for reasoning about knowledge and uncertainty and for dealing with belief change concepts in a distributed systems context. Unlike STDF, ISSA comes with a computational underlying formal framework for modeling a distributed system. ISSA views distributed computations of concurrent and reactive systems as evolution of states, assuming multiple computational agents interacting with one another as well as with their operational environment representing the external world. The underlying computation model, in ISSA, defines the behavior of a distributed system as the set of all admissible runs originating in a distinguished set of initial system states.

The underlying view of system behavior in ISSA is geared towards theoretical aspects of system modeling, building on abstract mathematical concepts that appear to be further away from practical system design and development. Practical design and development requires a formal framework that not only offers a formal model of situation analysis concepts and processes, but also supports established system design concepts such as modularization, refinement, validation, and verification of high-level models. Controlled experiments to systematically explore the practicability of situation analysis concepts in meaningful application scenarios call for abstract executable models as a basis for rapid prototyping and experimental validation. Details on STDF can be found in Chapters 4 and 14 of [65].

2.9.5 The Archetypal Dynamics Framework

Archetypal dynamics is almost unknown to the information fusion community. Sulis [139] introduced archetypal dynamics as: "Archetypal dynamics is a formal framework for dealing with the study of meaning laden information flows within complex systems." This is a formal framework for dealing with the study of the relationships between systems, frames, and their representations and the flow of information among these different entities. The framework consists of a triad of: semantic frame (representation), realizations (system) and interpretation (agent/user). Real systems relate to semantic frames through one of the dimensions of that triad.

The following excerpt from Sulis [139] clarifies the distinction between the triad dimensions interpretation-representation-realization: "A representation is understood to be a particular situation, involving a representational frame which serves as a kind of universal object, whose realizations may be associated in some manner with the realizations and interpretations of some other semantic frame. The representation is not itself a realization or interpretation. Realizations and interpretations are ontological constructs, whereas representations are epistemological constructs like symbols and signs."

The viewpoint of archetypal dynamics is that meaning is tight with actions. The underlying dynamics is based upon the idea of a combinatorial game [139] as for the ISSA framework [140] where the interpreted systems semantics can be seen as a conjunction of the state space and game modeling approaches. A semantic frame is an organizing principle that ascribes meaning in a coherent and consistent manner to phenomena that have been parsed into distinct entities, mode of being, modes of behaving, and modes of acting and interacting. The semantic frame provides answers, in whole or in part, to the six basic epistemic questions presented on Figure 2.10.

An important aspect of archetypal dynamics is the way information is seen [139]: "not to be understood in the sense of Shannon, that is, concerned only with the quantity of information, but rather in its active sense....the act of informing or the state of being informed, which means to give a special quality or character, to imbue, to instill, to impart by gradual instruction... Information possesses content and elicits meaning." This aspect will be discussed more in the next chapter.

Archetypal dynamics is a formal approach to the modeling of information flow in complex systems used to study emergence. What is interesting for situation analysis is the mathematical framework of tapestries which is a formal representational system. Tapestries represent information flow by means of multilayered, recursive, interlinked graphical structures that express both geometry (form or sign) and logic (semantics). Sulis [141] presented a detailed mathematical description of a specific tapestry model, the causal tapestry, selected for use in describing behaving systems. Observables are represented by tapestry informons (see Chapter 5) while subjective or hidden components (for example, intellectual and emotional processes) are incorporated into a reality game that determines the tapestry dynamics.

2.10 Awareness Quality and Decision-Making: Information Quality and Quantity

SAW quality can be related to the amount of information available to an individual: no information should result in poor SAW so very low decision quality as indicated in the lower left portion of Figure 2.22. In such a case, a natural reaction would be to provide mechanisms to increase the amount of information available to the decision makers to improve SAW quality. One could even claim that a good approach to reach better SAW and decision-making would be to provide as much information as possible and provide "all information, everywhere, at all time" but more information does not automatically mean better SAW. First, all this information may exceed the human information processing capabilities, leading to cognitive overload (lower right portion in Figure 2.22). Second, it is not all of the data and information available from the environment that is relevant and useful for reaching an optimal decision. In fact, in some situations, most of the data can be seen as distracters and noise for the decision maker, and may thus reduce his/her level of SAW. The decision-maker must detect and use only a specific fraction, called "useful bandwidth" in Figure 2.22, of this information to enhance his or her SAW and decision-making processes. Such considerations lead to the concept of the right information, at the right place, at the right time, in turn, leading to the notion of information relevance where the overall goal would be to provide only useful information (actionable knowledge) to deciders (desired zone of operation in Figure 2.22).

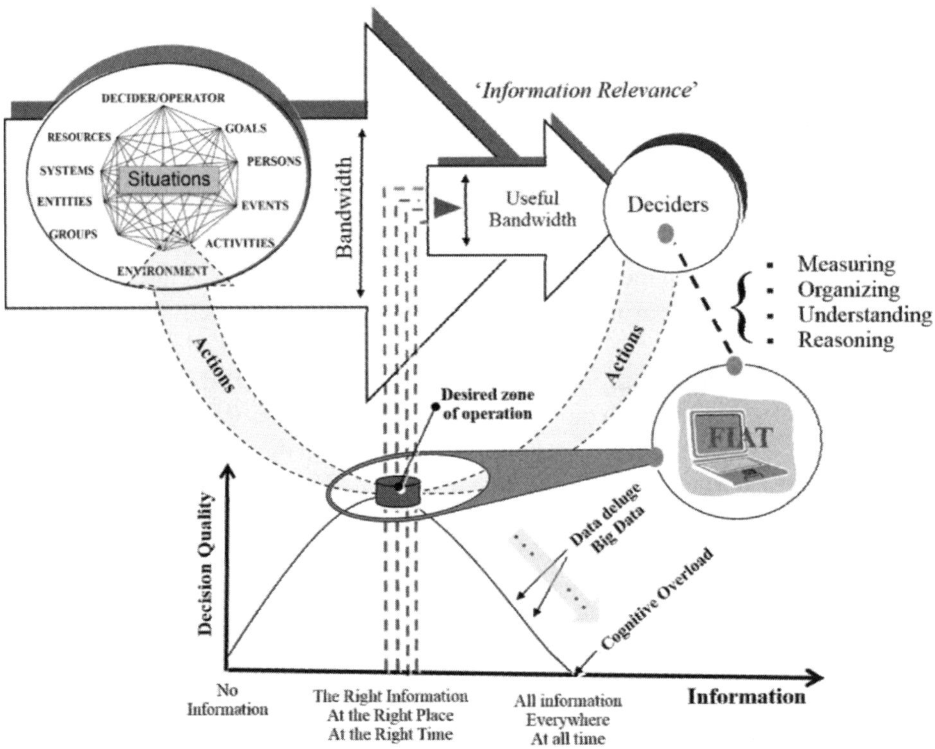

Figure 2.22 FIAT supporting deciders to be in a desired zone of operation.

A very crucial aspect of quality of information (QoI) is relevance of information: relevance shall be behind any intelligent filtering and context-aware processing concepts and impacts heavily on quality of decisions and the problem of volume of information in a big data environment. The analysis of information relevance [142–146] is one very important prerequisite to build any FIAT-based support system. In Figure 2.22, FIAT is represented by the analytics and information fusion processes to support measuring, organizing, understanding, and reasoning with data, information, and knowledge. Using an analogy borrowed from the signal processing community and pictured in Figure 2.22, one can imagine a useful information bandwidth where the overall goal would be to provide only useful information to deciders. That conceptual multidimensional useful bandwidth could be defined by assembling appropriate FIAT such as smart filtering [147–150] and metadata approaches [148, 151]. In fact, in the presence of a deluge of data, that demands classification and prioritization of information required by a given task execution [152]: "As data become abundant, the main problem is no longer finding the information as such but laying one's hands on the relevant bits easily and quickly."

A FIAT-based support system processes data, information, and knowledge and their imperfections. Three main aspects shall be considered in a holistic fashion for the design of such FIAT-based support system [153]: multi-agent systems (MAS) theories so that the distributed aspect can be adequately formalized, generalized information theory (GIT) [154] for the representation of knowledge and information imperfections, and decision theories [represented by operational research (OR) in Figure 2.23 to explicitly account for the actions and their impact on the environment. FIAT-based system design depends entirely on contexts and the domain of application. The challenge is to assemble appropriate technologies that will support: measuring, organizing, understanding, and reasoning about situations in a given complex environment (e.g., critical CPSS of Figure 2.1). There are numerous

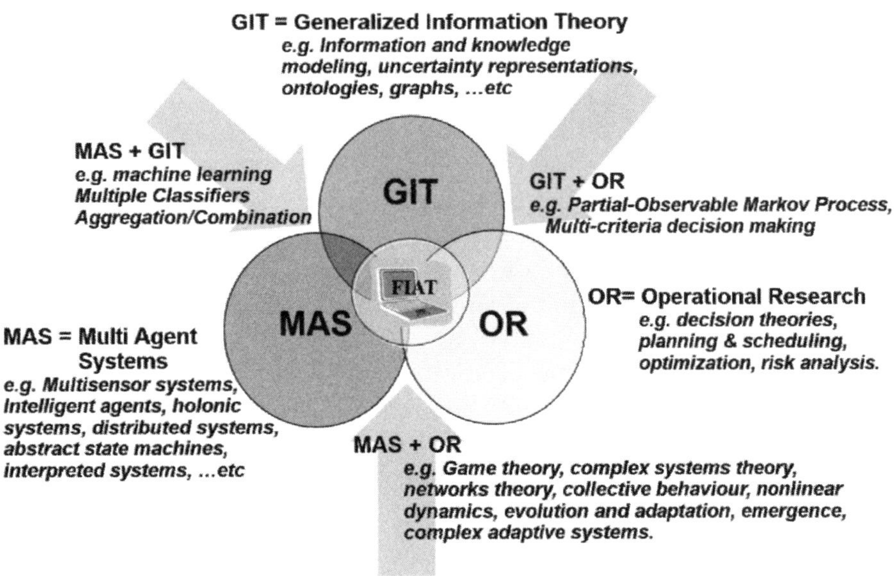

Figure 2.23 Main categories of multidisciplinary technologies for FIAT-based systems.

integrating frameworks that have been presented above, but most of them are not mastered enough and still require significant analysis efforts to benefit from them in implementing support solutions.

2.11 Conclusion

This chapter is concerned with situation awareness and the process of decision-making. Understanding the process of decision-making is a prerequisite to any design of decision support solutions. The decision-making process is a multifaceted undertaking and requires multiple perspectives to handle all its nuances and complexity. There is no universal decision-making model that might fully represent the complexity of that process. For instance, one stream of decision-making theory, based on decision theoretic paradigms, views decision making as an analytic process that corresponds closely to an estimate of the situation. Based on this estimate, the decision-maker generates several options and then identifies criteria for evaluating these options, rates the options against these criteria, and selects the best option as the basis for future plans and actions. A second stream of decision-making theory emphasizes a more inductive approach that emphasizes the acquisition of knowledge, the development of expertise, and the ability of humans to generalize from past experience. It stresses pattern recognition, creativity, experience, and initiative. Recent studies recommend having a continuum of models between these two approaches.

A great portion of decision support systems are designed to aid operators and analysts to achieve the appropriate situation awareness state for their decision-making activities, and to support the execution of the resulting actions. A technological perspective alone has led system designers to propose solutions that lack the desired cognitive fit with humans. The lack of knowledge in cognitive engineering has in the past jeopardized the design of helpful computer-based aids aimed at complementing and supporting human cognitive tasks. Moreover, this lack of knowledge has, most of the time, created new trust problems in the designed tools.

Solving the decision-making problem thus requires balancing the human factor perspective with one of the system designers and coordinating the efforts in designing a cognitively fitted system to support the decision-makers. The problem of designing a cognitively fitted decision support system is that of the cognitive systems engineering community, which is proposing methodologies to get there. Understanding the processes of how humans get situation awareness provides clues to where and how to support the underlying situation analysis processes. Let us recall that, situation analysis is a process refereeing to the examination of a situation, its elements, and their relations, to provide and maintain a product (i.e., a state of situation awareness for the decision-maker).

The multidisciplinary FIAT are key enablers to build computer-based support to situation analysis. The integration of the human element at the beginning of the analysis process is an important facet. The situation analysis process has to deal with both knowledge and uncertainty and a formalization is necessary since the support is through a computer. We review recent proposed frameworks to integrate FIAT as computer-based support systems. None of these frameworks is developed or mature enough to satisfy all requirements. We must say that those requirements are themselves not fully understood being as much multifaceted as the decision making process itself. Nevertheless, one can say that a framework in which knowledge, information and uncertainty can be represented, combined, managed, reduced, and updated is highly required. Three main aspects shall be considered: (1) the distributed aspect can be adequately formalized for instance through multi-agent systems theories; (2) decision theory in order to explicitly account for the actions and their impact on the environment; and (3) generalized information theory for the representation of knowledge and uncertainty. Information and uncertainty are the subject of the next chapter.

References

[1] Boyd, J. R., *A Discourse on Winning and Losing,* unpublished set of briefing slides available at Air University Library, Maxwell AFB, Alabama, http://www.ausairpower.net/APA-Boyd-Papers.html, 1987.

[2] Ghosh, S., *Algorithm Design for Networked Information Technology Systems*, New York: Springer, 2004.

[3] Waltz, E., *Knowledge Management in the Intelligence Enterprise*, Norwood, MA: Artech House, 2003.

[4] Richards, C., "Boyd's OODA Loop," Slideshow, http://www. dni. net/fcs/ppt/boyds_ooda_loop. ppt, 2001.

[5] Valin, P., et al., "Testbed for Distributed High–Level Information Fusion and Dynamic Resource Management," *Intl. Conf. on Information Fusion*, 2010.

[6] Shahbazian, E., D. E. Blodgett, and P. Labbé, "The Extended OODA Model for Data Fusion Systems," *Proc. of the 4th International Conference on Information Fusion (FUSION2001)*, Montreal, 2001.

[7] Blasch, E., R. Breton, and É. Bossé, "User Information Fusion Decision Making Analysis with the C-OODA Model," in *High-Level Information Fusion Management and Systems Design*, E. Blasch, É. Bossé, and D. A. Lambert, (eds.), Norwood, MA: Artech House, 2012, pp. 215–232.

[8] Das, S., *Computational Business Analytics*, New York: Taylor & Francis, 2013.

[9] Strassner, J., et al., "A Semantic Architecture for Enhanced Cyber Situational Awareness," *Secure and Resilient Cyber Architectures Conference*, 2010.

[10] Strassner, J., J. -K. Hong, and S. van der Meer, "The Design of an Autonomic Element for Managing Emerging Networks and Services," *Intl. Conf. on Ultra Modern Telecommunications & Workshops 2009 (ICUMT'09)*, 2009, pp. 1–8.

[11] Strassner, J., S. -S. Kim, and J. W. -K. Hong, "The Design of an Autonomic Communication Element to Manage Future Internet Services," in *Management Enabling the Future Internet for Changing Business and New Computing Services*, New York: Springer, 2009, pp. 122–132.

[12] Strassner, J., "Knowledge Representation, Processing, and Governance in the FOCALE Autonomic Architecture," *Autonomic Network Management Principles: From Concepts to Applications,* 2010, p. 253.

[13] Blasch, E., et al., "Implication of Culture: User Roles in Information Fusion for Enhanced Situational Understanding," *12th Intl. Conf. on Information Fusion 2009 (FUSION'09)*, 2009, pp. 1272–1279.

[14] Taylor, J., *Decision Management Systems: A Practical Guide to Using Business Rules and Predictive Analytics*, Boston, MA: Pearson Education, 2011.

[15] Minelli, M., M. Chambers, and A. Dhiraj, *Big Data, Big Analytics: Emerging Business Intelligence and Analytic Trends for Today's Businesses*, New York: John Wiley & Sons, 2012.

[16] Dittrich, D., A. Center, and M. P. Haselkorn, "Visual Analytics in Support of Secure Cyber-Physical Systems," 2009, https://staff.washington.edu/dittrich/papers/dhs-cps-ws-2009-uw.pdf.

[17] Breton, R., "The Modelling of Three Levels of Cognitive Controls with the Cognitive-OODA Loop Framework," *Def. Res. & Dev. CA-Valcartier, DRDC TR*, Vol. 111, 2008.

[18] Llinas, J., "Reexamining Information Fusion-Decision Making Inter-Dependencies," *2014 IEEE Intl. Inter-Disciplinary Conf. on Cognitive Methods in Situation Awareness and Decision Support (CogSIMA)*, 2014, pp. 1–6.

[19] Alberts, D. S., "The Agility Imperative: Precis," Unpublished white paper, http://www.dodccrp. org, 2010.

[20] Alberts, D. S., *The Agility Advantage: A Survival Guide for Complex Enterprises and Endeavors*, Washington DC: CCRP Publication Series, 2010.

[21] Alberts, D. S., "Agility, Focus, and Convergence: The Future of Command and Control," DTIC Document, 2007.

[22] Alberts, D. S., R. K. Huber, and J. Moffat, "NATO NEC C2 Maturity Model," NATO SAS-065, www.dodccrp.org, 2010.

[23] Bossé, É., J. Roy, and S. Wark, *Concepts, Models, and Tools for Information Fusion*, Norwood, MA: Artech House, 2007.

[24] Gigerenzer, G., "Bounded Rationality Models of Fast and Frugal Inference," *Revue Suisse D' Economie Politique et de Statitisique,* Vol. 133, 1997, pp. 201–218.

[25] Bryant, D. J., R. D. Webb, and C. McCann, "Synthesizing Two Approaches to Decision Making in Command and Control," *Canadian Military Journal,* Vol. 4, 2003, pp. 29–34.

[26] Guitouni, A., K. Wheaton, and D. Wood, "An Essay to Characterize Models of the Military Decision-Making Process," *11th ICCRT Symposium*, Cambridge, U.K., 2006.

[27] Von Winterfeldt, D., and W. Edwards, *Decision Analysis and Behavioral Research,* Vol. 604, Cambridge, U.K.: Cambridge University Press, 1986.

[28] Kahneman, D., and A. Tversky, "Prospect Theory: An Analysis of Decision Under Risk," *Econometrica: Journal of the Econometric Society,* 1979, pp. 263–291.

[29] Loomes, G., and R. Sugden, "Regret Theory: An Alternative Theory of Rational Choice Under Uncertainty," *The Economic Journal,* 1982, pp. 805–824.

[30] Simon, H. A., "Rational Choice and the Structure of the Environment," *Psychological Review,* Vol. 63, 1956, p. 129.

[31] Todd, P. M., and G. Gigerenzer, "Bounding Rationality to the World," *Journal of Economic Psychology,* Vol. 24, 2003, pp. 143–165.

[32] Rieskamp, J., and U. Hoffrage, "Inferences Under Time Pressure: How Opportunity Costs Affect Strategy Selection," *Acta Psychologica,* Vol. 127, 2008, pp. 258–276.

[33] Hoffrage, U., *When Do People Use Simple Heuristics, and How Can We Tell?* New York: Oxford University Press, 1999.

[34] Zsambok, C. E., and G. Klein, *Naturalistic Decision Making*, Abington, Oxford, UK: Taylor and Franics Group, 2014.

[35] Klein, G., "The Recognition-Primed Decision (RPD) Model: Looking Back, Looking Forward," *Naturalistic Decision Making,* 1997, pp. 285–292.

[36] Beach, L. R., and T. R. Mitchell, "Image Theory: Principles, Goals, and Plans in Decision Making," *Acta Psychologica,* Vol. 66, 1987, pp. 201–220.

[37] Lipshitz, R., "Converging Themes in the Study of Decision Making in Realistic Settings," in *Decision Making In Action: Models and Methods*, G.A. Klein et al. (eds.), Norwood, NJ: Ablex, 1993.

[38] Lipshitz, R., and O. Strauss, "Coping with Uncertainty: A Naturalistic Decision-Making Analysis," *Organizational Behavior and Human Decision Processes,* Vol. 69, 1997, pp. 149–163.

[39] Diaz, S. K., "Where Do I Start? Decision Making in Complex Novel Environments," DTIC Document, Naval Post-Graduate School, Monterey, CA, 2010.

[40] Budiu, R., C. Royer, and P. Pirolli, "Modeling Information Scent: A Comparison of LSA, PMI and GLSA Similarity Measures on Common Tests and Corpora," *Proc. of RIAO'07*, Pittsburgh, PA, 2007.

[41] Pirolli, P. L., *Information Foraging Theory: Adaptive Interaction with Information*, Oxford, U.K.: Oxford University Press, 2007.

[42] Llinas, J., "A Survey of Automated Methods for Sensemaking Support," *SPIE Sensing Technology+ Applications*, 2014, pp. 912206–912206-13.

[43] Roy, J., "Combining Elements of Information Fusion and Knowledge-Based Systems to Support Situation Analysis," *Defense and Security Symposium*, 2006, pp. 624202–624202-14.

[44] Roy, J., and A. Auger, "The Multi-Intelligence Tools Suite-Supporting Research and Development in Information and Knowledge Exploitation," DTIC Document, 16th International Command and Control Research and Technology Symposium, "Collective C2 in Multinational Civil-Military Operations," Quebec City, Canada, June 21–23, 2011.

[45] Roy, J., and A. B. Guyard, "Multiple Hypothesis Situation Analysis Support System Prototype," *13th Conf. on Information Fusion (FUSION)*, 2010, pp. 1–8.

[46] Giard, V. E., and B. Roy, *Méthodologie multicritère d'aide à la décision*, Paris, France: Editions Economica, 1985.

[47] Roy, J., "From Data Fusion to Situation Analysis," *4th Conf. on Information Fusion (FUSION)*, 2001, pp. 1–8.

[48] Sulis, W., "Archetypal Dynamics," in *Formal Descriptions of Developing Systems*, Vol. 121, J. Nation, et al., (eds.), Boston, MA: Kluwer Academic Publishers, 2003, pp. 180–227.

[49] Nicholson, D., "Defence Applications of Agent-Based Information Fusion," *The Computer Journal*, Vol. 54, 2011, pp. 263–273.

[50] Cugola, G., and A. Margara, "Processing Flows of Information: From Data Stream to Complex Event Processing," *ACM Computing Surveys (CSUR)*, Vol. 44, 2012, p. 15.

[51] Wang, F., et al., "Bridging Physical and Virtual Worlds: Complex Event Processing for RFID Data Streams," *Advances in Database Technology (EDBT 2006)*, 2006, pp. 588–607.

[52] Klein, M., et al., "Ontology Versioning and Change Detection on the Web," in *Knowledge Engineering and Knowledge Management: Ontologies and the Semantic Web*, New York: Springer, 2002, pp. 197–212.

[53] Krishnamurthy, B., et al., "Sketch-Based Change Detection: Methods, Evaluation, and Applications," *Proc. of 3rd ACM SIGCOMM Conf. on Internet Measurement*, 2003, pp. 234–247.

[54] Lunetta, R. S., and C. D. Elvidge, *Remote Sensing Change Detection: Environmental Monitoring Methods and Applications*, London, UK: Taylor & Francis, 1999.

[55] Pedrycz, W., *Knowledge-Based Clustering: From Data to Information Granules*, New York: John Wiley & Sons, 2005.

[56] Pedrycz, W., *Granular Computing: Analysis and Design of Intelligent Systems*, Boca, Raton, FL: CRC Press, 2013.

[57] Pedrycz, W., and S. -M. Chen, *Information Granularity, Big Data, and Computational Intelligence*, New York: Springer, 2014.

[58] Zadeh, L. A., "Toward a Theory of Fuzzy Information Granulation and Its Centrality in Human Reasoning and Fuzzy Logic," *Fuzzy Sets and Systems*, Vol. 90, 1997, pp. 111–127.

[59] Pirolli, P., and S. Card, "The Sensemaking Process and Leverage Points for Analyst Technology as Identified Through Cognitive Task Analysis," *Proc. of Intl. Conf. on Intelligence Analysis*, 2005, pp. 2–4.

[60] Roy, J., and A. B. Guyard, "A Knowledge-Based System for Multiple Hypothesis Sensemaking Support," *Proc. of 14th Intl. Conf. on Information Fusion (FUSION)*, 2011, pp. 1–8.

[61] Klein, G., B. Moon, and R. R. Hoffman, "Making Sense of Sensemaking 2: A Macrocognitive Model," *IEEE Intelligent Systems*, Vol. 21, 2006, pp. 88–92.

[62] Klein, G., et al., "A Data-Frame Theory of Sensemaking: Expertise Out of Context," *Proc. of 6th Intl. Conf. on Naturalistic Decision Making*, 2007, pp. 15–17.

[63] Endsley, M. R., and D. J. Garland, *Situation Awareness Analysis and Measurement*, Boca Raton, FL: CRC Press, 2000.

[64] Endsley, M. R., "Toward a Theory of Situation Awareness in Dynamic Systems," *Human Factors: The Journal of the Human Factors and Ergonomics Society,* Vol. 37, 1995, pp. 32–64.

[65] Blasch, E., E. Bosse, and D. A. Lambert, *High-Level Information Fusion Management and Systems Design*: Norwood, MA: Artech House, 2012.

[66] Clark, R. E., et al., "Cognitive Task Analysis," *Handbook of Research on Educational Communications and Technology,* Vol. 3, 2008, pp. 577–593.

[67] Crandall, B., G. A. Klein, and R. R. Hoffman, *Working Minds: A Practitioner's Guide to Cognitive Task Analysis*, Cambridge, MA: MIT Press, 2006.

[68] Schraagen, J. M., S. F. Chipman, and V. L. Shalin, *Cognitive Task Analysis*, New York: Taylor and Francis Group, 2000.

[69] Bisantz, A. M., and C. M. Burns, *Applications of Cognitive Work Analysis*, Boca Raton, FL: CRC Press, 2008.

[70] Elm, W. C., et al., "Applied Cognitive Work Analysis: A Pragmatic Methodology For Designing Revolutionary Cognitive Affordances," *Handbook of Cognitive Task Design,* 2003, pp. 357–382.

[71] Jenkins, D. P., *Cognitive Work Analysis: Coping with Complexity*, Surrey, UK: Ashgate Publishing, 2009.

[72] Naikar, N., A. Moylan, and B. Pearce, "Analysing Activity in Complex Systems with Cognitive Work Analysis: Concepts, Guidelines and Case Study for Control Task Analysis," *Theoretical Issues in Ergonomics Science,* Vol. 7, 2006, pp. 371–394.

[73] Vicente, K. J., *Cognitive Work Analysis: Toward Safe, Productive, and Healthy Computer-Based Work*, Boca Raton, FL: CRC Press, 1999.

[74] Vicente, K. J., "A Few Implications of an Ecological Approach to Human Factors," in *Global Perspectives on the Ecology of Human-Machine Systems, Volume 1: Resources for Ecological Psychology*, J. M. Flach et al. (eds.), Hillsdale, NJ England: Lawrence Erlbaum Associates, Inc. 1995.

[75] Rasmussen, J., A. M. Pejtersen, and L. P. Goodstein, *Cognitive Systems Engineering*, New York: Wiley, 1994.

[76] Woods, D. D., and E. Hollnagel, *Joint Cognitive Systems: Patterns in Cognitive Systems Engineering*, Boca Raton, FL: CRC Press, 2006.

[77] Rasmussen, J., "Information Processing and Human-Machine Interaction. An Approach to Cognitive Engineering," *Series in System Science and Engineering*, New York: North-Holland, Vol. 12, 1986.

[78] Burns, C. M., D. J. Bryant, and B. A. Chalmers, "Scenario Mapping with Work Domain Analysis," *Proc. of the Human Factors and Ergonomics Society Annual Meeting*, 2001, pp. 424–428.

[79] Chalmers, B., and C. Burns, "A Model-Based Approach to Decision Support Design for a Modern Frigate," *TTCP Symp. on Coordinated Maritime Battlespace Management, Space and Naval Warfare Systems Center,* San Diego, CA, 1999.

[80] Paradis, S., et al., "A Pragmatic Cognitive System Engineering Approach to Model Dynamic Human Decision-Making Activities in Intelligent and Automated Systems," DTIC Document, 2003. RTO HFM Symposium, NATO RTO-MP-088, Warsaw, Poland, 2003.

[81] Bossé, É., S. Paradis, and R. Breton, "Decision Support in Command and Control: A Balanced Human-Technological Perspective," *Data Fusion for Situation Monitoring, Incident Detection, Alert and Response Management,* Vol. 16, 2005, pp. 205–222.

[82] Corcho, O., M. Fernández-López, and A. Gómez-Pérez, "Ontological Engineering: Principles, Methods, Tools and Languages," in *Ontologies for Software Engineering and Software Technology*, New York: Springer, 2006, pp. 1–48.

[83] Fernandez-Lopez, M., and O. Corcho, *Ontological Engineering: With Examples from the Areas of Knowledge Management, E-Commerce and the Semantic Web*, New York: Springer, 2010.

[84] Gomez-Perez, A., M. Fernández-López, and O. Corcho-Garcia, "Ontological Engineering," *Computing Reviews,* Vol. 45, 2004, pp. 478–479.

[85] J. Roy and A. Auger, "Knowledge and Ontological Engineering Techniques for Use in Developing Knowledge-Based Situation Analysis Support Systems," *Defence R&D Canada–Valcartier Tech. Mem,* Vol. 757, 2006.

[86] Hollnagel, E., and D. D. Woods, *Joint Cognitive Systems: Foundations of Cognitive Systems Engineering*, Boca Raton, FL: CRC Press, 2005.

[87] Dey, A., "Understanding and Using Context," *Personal and Ubiquitous Computing,* Vol. 5, 2001, pp. 4–7.

[88] Zimmermann, A., A. Lorenz, and R. Oppermann, "An Operational Definition of Context," in *Modeling and Using Context*, New York: Springer, 2007, pp. 558–571.

[89] "Context Awareness," http://en.wikipedia.org/wiki/Context_awareness, last accessed April 22, 2015.

[90] Apte, C. V., et al., "Data-Intensive Analytics for Predictive Modeling," *IBM Journal of Research and Development,* Vol. 47, 2003, pp. 17–23.

[91] Laursen, G. H. N., and J. Thorlund, *Business Analytics for Managers: Taking Business Intelligence Beyond Reporting*, New York: Wiley, 2010.

[92] Steinberg, A. N., C. L. Bowman, and F. E. White, "Revisions to the JDL Data Fusion Model," *The Joint NATO/IRIS Conference*, Quebec City, 1998.

[93] Hall, D. L., and J. M. Jordan, *Human-Centered Information Fusion*, Norwood, MA: Artech House, 2010.

[94] Hall, D. L., and S. A. H. McMullen, *Mathematical Techniques in Multisensor Data Fusion*, Norwood, MA: Artech House, 2004.

[95] Liggins, M., D. Hall, and J. Llinas, *Handbook of Multisensor Data Fusion: Theory and Practice*, 2nd ed., New York: Taylor & Francis, 2008.

[96] Das, S., *High-Level Data Fusion*, Norwood, MA: Artech House, 2008.

[97] Eckerson, W. W., *Secrets of Analytical Leaders*, Westfield, NJ:Technics Publications, 2012.

[98] Sheikh, N., *Implementing Analytics: A Blueprint for Design, Development, and Adoption*, Walthan, MA, Morgan Kaufman, 2013.

[99] Chen, Z., *Data Mining and Uncertain Reasoning: An Integrated Approach*, New York: Wiley, 2001.

[100] Tan, P. -N., M. Steinbach, and V. Kumar, *Introduction to Data Mining, Vol. 1*, Boston, MA: Pearson Addison Wesley, 2006.

[101] Witten, I. H., and E. Frank, *Data Mining: Practical Machine Learning Tools and Techniques*, Burlington, MA: Morgan Kaufmann, 2005.

[102] Bishop, C. M., *Pattern Recognition and Machine Learning, Vol. 4*, New York: Springer, 2006.

[103] Han, J., M. Kamber, and J. Pei, *Data Mining, Southeast Asia Edition: Concepts and Techniques*, San Francisco, CA: Morgan Kaufmann, 2006.

[104] Burstein, F., and C. Holsapple, *Handbook on Decision Support Systems: Variations, Vol. 2*, New York: Springer Science & Business Media, 2008.

[105] Blasch, E., É. Bossé, and D. A. Lambert, (eds.), *High-Level Information Fusion Management and Systems Design*, Norwood, MA: Artech House, 2012.

[106] Bossé, É., J. Roy, and S. Wark, *Concepts, Models and Tools for Information Fusion*, Norwood, MA: Artech House, 2007.

[107] Liggins, M. E., D. L. Hall, and J. Llinas, (eds.), *Handbook of Multisensor Data Fusion*, Boca Raton, FL: CRC Press, 2009.

[108] Hall, D. L., *Mathematical Techniques in Multisensor Data Fusion*, Norwood, MA: Artech House, 1992.

[109] Klein, L. A., *Sensor and Data Fusion Concepts and Applications*, Bellingham, WA: SPIE Optical Engineering Press, 1993.

[110] Luo, R. C., and M. G. Kay, (eds.), *Multisensor Integration and Fusion for Intelligent Machines and Systems*, Norwood, NJ: Ablex Publishing Corporation, 1995.

[111] White, F. E., "Data Fusion Lexicon," Joint Directors of Laboratories, Technical Panel for C3, Data Fusion Sub-Panel, Naval Ocean Systems Center, San Diego, CA, 1987.

[112] Llinas, J., et al., "Revisiting the JDL Data Fusion Model II," *Proc. of 7th Intl. Conf. on Information Fusion (FUSION 2004)*, Stockholm, Sweden, 2004.

[113] Steinberg, A. N., and C. L. Bowman, "Rethinking the JDL Data Fusion Levels," *Proc. of the National Symposium on Sensor and Data Fusion, John Hopkins Applied Physics Laboratory*, 2004.

[114] Blasch, E., and S. Plano, "Level 5: User Refinement to Aid the Fusion Process," *SPIE Proceedings, Vol. 5099, Multisensor, Multisource Information Fusion: Architectures, Algorithms, and Applications*, 2003.

[115] Dasarathy, B. V., "Sensor Fusion Potential Exploitation-Innovative Architectures and Illustrative Applications," *Proc. of the IEEE*, 1997, pp. 24–38.

[116] Shulsky, A. N., and G. J. Schmitt, *Silent Warfare: Understanding the World of Intelligence*, Washington DC: Potomac Books, 2002.

[117] Bedworth, M. D., and J. C. O'Brien, "The Omnibus Model: A New Model for Data Fusion?" *Proc. of the 2nd Intl. Conf. on Information Fusion (FUSION1999)*, 1999.

[118] Khaleghi, B., et al., "Multisensor Data Fusion: A Review of the State-of-the-Art," *Information Fusion,* Vol. 14, 2013, pp. 28–44.

[119] Nakamura, N. E. F., A. A. F. Loureiro, and A. C. Frery, "Information Fusion for Wireless Sensor Networks: Methods, Models, and Classifications," *ACM Computing Surveys,* Vol. 39, 2007.

[120] Blasch, E., et al., "High Level Information Fusion Developments, Issues, and Grand Challenges," *Proc. of the 13th Intl. Conf. of Information Fusion (FUSION2010)*, Edinburg, 2010.

[121] Salerno, J. J., et al., "Issues and Challenges in Higher Level Fusion: Threat/Impact Assessment and Intent Modeling (A Panel Summary)," *Proc. of 13th Intl. Conf. of Information Fusion (FUSION2010),* Edinburg, 2010.

[122] Blasch, E., et al., "Top Ten Trends in High-Level Information Fusion," *Proc. of 15th Intl. Conf. of Information Fusion (FUSION2012)* Singapore, 2012.

[123] Kokar, M. M., J. A. Tomasik, and J. Weyman, "Formalizing Classes of Information Fusion Systems," *Information Fusion,* Vol. 5, 2004, pp. 189–202.

[124] Kokar, M. M., C. J. Matheus, and K. Baclawski, "Ontology-Based Situation Awareness," *Information Fusion,* Vol. 10, 2009, pp. 83–98.

[125] Little, E., and G. Rogova, "Designing Ontologies for Higher Level Fusion," *Information Fusion,* Vol. 10, 2009, pp. 70–82.

[126] Roy, J., and É. Bossé, "Conflict Management in the Shipboard Integration of Multiple Sensors," *Proc. of 1st Intl. Conf. on Information Fusion (FUSION1998)* Las Vegas, NV, 1998.

[127] Gustavsson, P. M., and T. Planstedt, "The Road Towards Multi-Hypothesis Intention Simulation Agents Architecture-Fractal Information Fusion Modeling," *Proc. of Winter Simulation Conf.*, 2005.

[128] Gustavsson, P. M., and T. Planstedt, "The Road Towards Multi-Hypothesis Intention Simulation Agents Architecture Fractal Information Fusion Modeling," *Proc. of 37th Conf. on Winter Simulation*, 2005, pp. 2532–2541.

[129] Guitouni, A., P. Valin, É. Bossé, and H. When, "Information Fusion and Resource Management Testbed," in *High-Level Information Fusion Management and Systems Design*, E. Blasch, É. Bossé, and D. A. Lambert, (eds.), Norwood, MA: Artech House, 2012, pp. 155–172.

[130] Bowman, C. L., and A. N. Steinberg, "Systems Engineering Approach for Implementing Data Fusion Systems," in *Handbook of Multisensor Data Fusion*, M. E. Liggins, D. L. Hall, and J. Llinas, (eds.), Boca Raton, FL: CRC Press, 2009, pp. 561–596.

[131] U.S. Department of Defense, "The DoDAF Architecture Framework Version 2.02," http://dodcio.defense.gov/Portals/0/Documents/DODAF/DoDAF_v2-02_web.pdf, 2010.

[132] Llinas, J., "A Survey and Analysis of Frameworks and Framework Issues for Information Fusion Applications," *Proc. of 5th Intl. Conf. on Hybrid Artificial Intelligence Systems, Hybrid Artificial Intelligence Systems, Lecture Notes in Computer Science Volume 6076*, 2010, pp. 14–23.

[133] Lambert, D. A., "A Blueprint for Higher-Level Fusion Systems," *Information Fusion*, Vol. 10, 2009, pp. 6–24.

[134] Lambert, D., and C. Nowak, "The Mephisto Conceptual Framework," Defence Science and Technology Organisation, DSTO-TR-2162, Australia, 2008.

[135] Fagin, R., et al., *Reasoning About Knowledge*, Cambridge, MA: MIT Press, 2003.

[136] Maupin, P., and A.-L. Jousselme, "A General Algebraic Framework for Situation Analysis," *Proc. of 8th Intl. Conf. on Information Fusion (FUSION2005)*, Philadelphia, PA, 2005.

[137] Maupin, P., and A.-L. Jousselme, "Interpreted Systems for Situation Analysis," *Proc. of 10th Intl. Conf. on Information Fusion (FUSION2007)* Quebec City, Canada, 2007.

[138] Baier, C., and J.-P. Katoen, *Principles of Model Checking*, Cambridge, MA: MIT Press, 2008.

[139] Sulis, W. H., "Archetypal Dynamics: An Approach to the Study of Emergence," in *Formal Descriptions of Developing Systems, NATO Science Series Volume 121*, 2003, pp. 185–228.

[140] Maupin, P., et al., "A Toolbox for the Evaluation of Surveillance Strategies Based on Interpreted Systems," in *High-Level Information Fusion Management and Systems Design*, E. Blasch, É. Bossé, and D. A. Lambert, (eds.), Norwood, MA: Artech House, 2012, pp. 215–232.

[141] Sulis, W., "Causal Tapestries," *Bulletin of the American Physical Society*, Vol. 56, 2011.

[142] Breton, R., et al., "Framework for the Analysis of Information Relevance (FAIR)," *IEEE Intl. Multi-Disciplinary Conf. on Cognitive Methods in Situation Awareness and Decision Support (CogSIMA)*, 2012, pp. 210–213.

[143] Hadzagic, M., et al., "Reliability and Relevance in the Thresholded Dempster-Shafer Algorithm for ESM Data Fusion," *15th Intl. Conf. on Information Fusion (FUSION)*, 2012, pp. 615–620.

[144] Pichon, F., D. Dubois, and T. Denoeux, "Relevance and Truthfulness in Information Correction and Fusion," *International Journal of Approximate Reasoning*, Vol. 53, 2012, pp. 159–175.

[145] Saracevic, T., "Relevance: A Review of the Literature and a Framework for Thinking on the Notion in Information Science. Part III: Behavior and Effects of Relevance," *Journal of the American Society for Information Science and Technology*, Vol. 58, 2007, pp. 2126–2144.

[146] White, H. D., "Relevance Theory and Citations," *Journal of Pragmatics*, Vol. 43, 2011, pp. 3345–3361.

[147] Jain, P., "Intelligent Information Retrieval," *SETIT 2005 3rd INternational Conference: Sciences of Electronic Technologies of Information and Telecommunications*, Vol. 3, 2005.

[148] Boutell, M., and J. Luo, "Bayesian Fusion of Camera Metadata Cues in Semantic Scene Classification," *Proc. of 2004 IEEE Computer Society Conf. on Computer Vision and Pattern Recognition (CVPR 2004)*, Vol. 2, 2004, pp. II-623–II-630.

[149] Lemire, D., et al., "Collaborative Filtering and Inference Rules for Context-Aware Learning Object Recommendation," *Interactive Technology and Smart Education,* Vol. 2, 2005, pp. 179–188.

[150] Yu, Z., et al., "Supporting Context-Aware Media Recommendations for Smart Phones," *IEEE Pervasive Computing,* Vol. 5, 2006, pp. 68–75.

[151] Johnson, M., and C. N. Dampney, "On Category Theory as a (Meta) Ontology for Information Systems Research," *Proc. of Intl. Conf. on Formal Ontology in Information Systems-Volume 2001*, 2001, pp. 59–69.

[152] "Data, Data Everywhere," *The Economist: A Special Report on Managing Information*, http://www.economist.com/node/15557443, February 2010.

[153] Bossé, É., A. L. Jousselme, and P. Maupin, "Situation Analysis for Decision Support: A Formal Approach," *10th Intl. Conf. on Information Fusion*, 2007, pp. 1–3.

[154] Klir, G. J., "Generalized Information Theory," *Fuzzy Sets and Systems,* Vol. 40, 1991, pp. 127–142.

CHAPTER 3

Information and Uncertainty: Definitions and Representations

The purpose of this chapter is to review fundamental concepts and prospective notions of information and uncertainty that can be exploited in the design of intelligent situation analysis and decision support systems based upon fusion of information and analytic technologies (FIAT).

3.1 Miscellaneous Views on Definition of Information

What is information? As mentioned in [1], several surveys [2–6] do not converge on a unanimous agreement on a definition of information. Floridi [1] stated: "Information is notoriously a polymorphic phenomenon and a polysemantic concept so, as an explicandum, it can be associated with several explanations, depending on the level of abstraction adopted and the cluster of requirements and desiderata orientating a theory." Floridi [1] added the reflection "It is hardly to be expected that a single concept of information would satisfactorily account for the numerous possible applications of this general field," to the following Claude E. Shannon's [7] remark on information: "The word 'information' has been given different meanings by various writers in the general field of information theory. It is likely that at least a number of these will prove sufficiently useful in certain applications to deserve further study and permanent recognition."

It is quite pertinent to replicate here from [1], the tripartite analysis of Weaver [8] on Shannon's information about: "(1) technical problems concerning the quantification of information and dealt with by Shannon's theory; (2) semantic problems relating to meaning and truth; and (3) what he called 'influential' problems concerning the impact and effectiveness of information on human behaviour, which he thought had to play an equally important role."

In the handbook *Philosophy of Information* [9], the reader is invited to a historical tour on the meaning evolution of the word "information." The word "information" is of Latin origin in the context of Plato's theory of ideas (or forms). As cited in [9], Cicero used *in-formare* to mean a representation implanted in the mind. In the fifteenth century, the French word "information" emerged with the following meanings: investigation, education, the act of informing, communicating knowledge, and intelligence. The tour drives us from abstract philosophy to quantitative science and computation to logics and linguistics. In [9], three main stances in the technical literature on information theories are presented: Information-A, knowledge, logic, what is conveyed in information answers; Information-B, probabilistic, information-theoretic, measured quantitatively; and Information-C, algorithmic, code compression, measured quantitatively. Table 3.1, entirely built from [9], gives examples and important references associated with these three stances.

The handbook [9] makes the reader aware of a plethora of approaches and methods to information that are drawing on multiple disciplines: logic, computer sciences, mathematics, cognitive sciences, and physics. All these strands have not been integrated in current computer science environments. According to the handbook editors [9], the contents have not revealed incompatibility between A-B-C, but rather complementary perspectives. Being said, a sort of unification might be possible if substantial research efforts are devoted to systems sciences or to a science of integration as mentioned in [25]. Is it through a synergetics[1] science [26–29]? Or is it through an evolution of cybernetics[2] [30–33]? The question is open and the answer is expected to provide means to benefit from the complementary perspectives of the logical A-stance about information update, the B-stance about transmission events, and the C-stance about computational activities of encoding and decoding.

This book is about prospective information technologies, labeled as FIAT in the previous chapter. The sense given to information is that of act of informing: informing who/what (a thing), about who/what, to do what (a decision, an action or a task). Transcendent in systems are tasks to be performed through dynamic processes (based on FIAT) that transform data. From [34], "Information is any type of pattern that influences the formation or transformation of other patterns." For a FIAT development prospective, information needs to be actionable to support decision-making and actions, that is, information has consequences on tasks performed by agents (machines or humans) for particular situations in the sense of Barwise and Perry [13]. Data structures are associated with the dynamic processes that manipulate them, and the tasks which these are supposed to perform. There is no information without transformation. Related to this notion of transformation is the very important idea of entropy; from its roots in the Greek "entropia" means transformation. The word was used by the German physicist Rudolph Clausius in 1850 and comes from a principle of thermodynamics dealing with energy. The idea is that everything in the universe eventually moves from order to disorder, and entropy is the measurement of that change: from ordered to disordered increases the entropy. This is under the statistical interpretation of entropy that Claude E. Shannon[3] (1940) [35] and later Weaver, Kolmogorov, and Chaitin [8, 17, 21–23, 36, 37]

1. Synergetics is the empirical study of systems in transformation, with an emphasis on total system behavior unpredicted by the behavior of any isolated components, including humanity's role as both participant and observer (http://en.wikipedia.org/wiki/Synergetics_(Fuller)).
2. Cybernetics is a transdisciplinary approach for exploring regulatory systems, their structures, constraints, and possibilities (http://en.wikipedia.org/wiki/Cybernetics).
3. Entropy effectively measures uncertainty about the actual microscopic state of matter. In the 1940s, Shannon discussed a kind of information relevant to the capacity of a telegraph line to transmit messages. The mathematical formula used to define information in terms of probabilities was the same, up to a multiplicative constant, as the formula that Boltzmann had used 50 years before to define entropy [35].

Table 3.1 Three Main Stances on Information Theories

Stances A_B_C	*Descriptive Examples (excerpts from [9, p. 11])*
Information-A A is the world of epistemic logic and linguistic semantics [10–16] A-stance about information update	The typical logic-based setting lets an agent acquire new information about what the real world is like, through acts of observation, linguistic communication, or deduction. A simple example would be an agent asking a question and learning what things are like from an answer. Thus, three features are crucial: agents that represent and use the information, dynamic events of information change, and aboutness: the information is always about some relevantly described situation or world. Here, we measure quality of information qualitatively in terms of new things agents can truly say: a quantitative measure may be handy, but it is not required. Finally, the formal paradigm for the theory is mathematical or computational logic.
Information-B B is the world of Shannon information theory, linked to entropy in physics [7, 17–20] B-stance about transmission events	The typical Shannon scenario is about a source emitting signals with certain frequencies, say, a language viewed as a global text producer, and the information that a receiver picks up from this is measured in terms of expected reduction of uncertainty. This is the sense in which seeing a particular roll of a fair die gives 3 bits of information. No specific agency seems involved here, but the scenario does analyze major features of communication which are absent on the logical approach, such as probability of signals (i.e., the long-term behavior of a source, maybe as viewed by the receiver), optimal coding, and channel capacity. Finally, mathematical paradigms for the theory are probability theory and physics.
Information-C C is the world of Kolmogorov complexity, linked to the foundations of computation [21–24] C-stance about computational activities of encoding and decoding	In the basic Kolmogorov scenario, we receive a code string and ask for its informational value. The answer is the algorithmic complexity of the string, defined as the length of the shortest program that computes it on some fixed universal Turing machine. While this looks like a totally different setting from the preceding two, there is a direct link to Scenario B. Working with the enumerable set of all prefix-free programs, we can easily find an associated probability distribution. In this way, the shortest program for a string becomes an optimal code in Shannon's sense. Thus, the following traffic arises: Information-B starts with the notion of probability as fundamental and derives an optimal code. Information-C starts with the notion of the shortest code as fundamental and derives an a priori probability from it. Further details may be found on pages 133, 171, and 281 of [9].

have made considerable advances to information theory with respect to objective views (Stances B-C).

FIAT are meant to transform data to actionable information. Examples of such data transformations are provided in [9]: from questions and answers [38], observations [39], communication [40], learning [41, 42], belief revision [43], computation [44], and inference [45] to game theoretic interaction [46]. Many other data transformations can be found in analytics, data mining and information fusion books [47–54]. For all those transformations, message container (transmission) and content (meaning) are important. However, to become actionable, the information requires meaning (Stance-A). Meaning transpires from what does one do with information, to concord with the point made in philosophy, for example, by David Lewis, who famously said that "Meaning is what meaning does." FIAT are emerging to support decision-making in action [55] and so by extension to provide meaningful information for acting that may concern a single agent (machine or human) or a multiagent process. This may lead to a more general definition of information (GDI) as in Floridi [1, 56]: information = data + meaning. It seems that even about Floridi's semantic concept of information there is no consensus [57]. Developing FIAT requires the three stances A-B-C. There are substantial mathematical tools to address objective views (B-C), but an important effort is required to address the formidable challenge presented by the subjective nature of Stance-A in a context of cyber-physical and social systems (CPSS).

Klüver [58] mentioned that the famous definition of Shannon and Weaver of the degree of information uses an objective concept of probability (i.e., a probability that is the same for all systems that receive a certain message). He stated that it makes sense in technical domains and natural sciences but makes no sense in the context of social communication between human actors. Klüver [58] continued: "Human communication as the process of social interaction is not only dependent on the meaning and degree of information of a certain message but also of the relevance or importance, respectively, the message has for the receiving system. If a message has much relevance for a receiving system, the according reactions to that message will be very different from the reactions to an unimportant message. In addition, a relevant message will be received with more attention than an unimportant one." In CPSS, humans and machines are interacting over a complex network that supports three main kinds of communications: machine-machine, human-machine, and human-human. A theory of communication that can work in such a context (e.g., social networks, IoT) must necessarily consider the important aspect of human communication processes raised by Klüver.

The following important excerpt from [59] provided an appropriate closure of this section:

> Attempts to understand how information content can be included in an accounting of the energy flux of the biosphere have led to the conclusion that, in information transmission, one component, the semantic content, or "the meaning of the message," adds no thermodynamic burden over and above costs arising from coding, transmission and translation.... The following three postulates provide the basis for discussion of a number of themes that demonstrate some important consequences. (i) Information transmission through either pathway has thermodynamic components associated with data storage and transmission. (ii) The semantic content adds no additional thermodynamic cost. (iii) For all semantic exchange, meaning is accessible only through translation and interpretation, and has a value only in context.

Finally, there is a need for more research on defining information and its attributes. For the development of FIAT, how important it is to have a general unique definition of information remains to be demonstrated. In a lot of popular writings, one can find the idea of layers that one can distinguish: data, information, knowledge, and wisdom [60–63]. Are these distinctions very clear or valuable? Again, here it still needs to be demonstrated both in the design of FIAT based systems and as contribution to information theory. From the writings of Floridi, it seems at least that there exists consensus on a general definition of information as: data + meaning [1, 56, 57]. This has been summarized on the diagram in Figure 3.1 as a conclusion of this section. We would like the reader to notify the notion of disinformation[4] shown as part of Figure 3.1 since this is quite an important aspect of the veracity dimension for the big data problem.

3.2 Semiotics

Entropy measures the uncertainty of a situation of ignorance. Ignorance and information are two opposite concepts: what measures one measures the other as well. Shannon's theory has had an immense impact on the technology of information storage, data compression, information transmission, and coding and will have an important impact in the development of FIAT. Shannon's theory was readily remarked as a syntactic theory of information but some authors claim it may address semantic issues as well [65]. Graben [65] recalls from *The Mathematical Theory of Communication* [17] three levels of communication problems (A-B-C): The technical problem A is how accurately can the symbols of communication be transmitted. The semantic problem B is how precisely do the transmitted symbols convey the desired meaning. The effectiveness problem C is how effectively does the received meaning affect conduct in the desired way. Graben [65] makes a direct link with the three semiotic dimensions discussed by Morris [66]:

- *Syntactics:* The rules that govern how signs relate to one another in formal structures. The technical problem A with noisy communication channels can be tackled by redundant codes that introduce correlations among the elements (symbols) of the messages.
- *Semantics:* Relation between signs and the concepts they denote (their denotata). (The semantic problem B addresses correlations between the transmitted symbols and their desired meanings.)

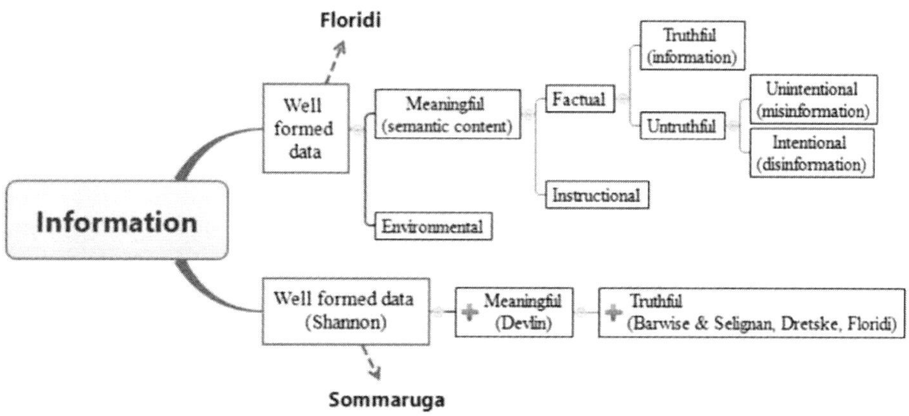

Figure 3.1 Informational map of Floridi [56] augmented from Sommaruga [64].

4. For disinformation, we include the notion of agnotology [178, 179].

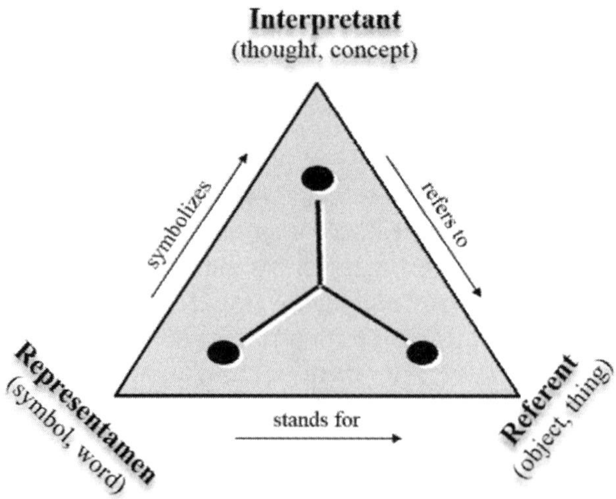

Figure 3.2 Ogden and Richards' semiotic triangle [68].

- *Pragmatics:* The systematic study of the use of signs. (The effectiveness problem C corresponds to pragmatics, addressing relations between the symbols and their impact upon the users.)

This begs a definition for the word sign which is a triadic relationship between a *representamen* [67], an *interpretant* (purely conceptual), and the referent (the actual object referred to by the *interpretant* and to which the *representamen* stands for). This trichotomy (Figure 3.2) was depicted by Ogden and Richards [68] into what is known today as the semiotic triangle.

3.3 Information and Uncertainty

First, it appears that the three words information, uncertainty, and entropy have distinct meanings for many people in different domains [69]. For instance, look at the confusion of the ordinary use of the word "information" with the word "information" in "information theory." Information is most of the time associated with the idea of knowledge through its popular use. Information in information theory, also called Shannon's information, is narrowly defined and is rather associated with uncertainty and the resolution of uncertainty. Sometimes authors confuse the two words. The same is true for the word "entropy." Authors will often use the word entropy without saying whether they are talking about thermodynamic entropy or Shannon's entropy and vice versa. The definition of the word "uncertainty" presents same kinds of discrepancies. In the economic literature [70] for instance, there exist different conceptions of uncertainty involving different types and different degrees of uncertainty as well as similar conceptions under different labels. Dequech [70] has even suggested not using the noun "uncertainty" alone but accompanied by qualifiers to specify its type and then reduce misunderstandings.

3.3.1 Information Theory and Entropy

Shannon made use of two different (but related) measures of information: entropy and mutual information. Entropy $H(X)$ is the amount of information of a discrete time discrete alphabet random process $\{X_n\}$. A detailed definition of $H(X)$ can be found in several good reference books of information theory such as [71, 72]. Shannon proved a coding theorem showing that if one wishes to code $\{X_n\}$ into a sequence of binary symbols, a receiver viewing the binary sequence can reconstruct the original process nearly perfectly. Mutual information is a measure of the information contained in one process $\{X_n\}$ about another process $\{Y_n\}$ to make decisions on it: one random process $\{X_n\}$ representing an information source and another $\{Y_n\}$ representing the output of a communication medium wherein the coded source has been corrupted by another random process called noise. Shannon introduced the notion of the average mutual information between the two processes:

$$I(X,Y) = H(X) + H(Y) - H(X,Y) \tag{3.1}$$

where $I(X, Y)$ is the sum of the two self-entropies minus the entropy of the pair. Average mutual information can also be defined in terms of conditional entropies (see Chapter 3 of [71] for details).

$$I(X,Y) = H(X) - H(X|Y) = H(Y) - H(X|Y) \tag{3.2}$$

We can read this equation in an alternative way as:

$$H(X|Y) = H(X|Y) - I(X,Y) \text{ and } H(X|Y) = H(Y) - I(X, Y)$$

For a random variable X with a probability distribution $p(.)$ defined over a finite set χ, Shannon's entropy is defined as:

$$H(X) = -\sum_{x \in \chi} p(x) \log_2 p(x) \geq 0 \tag{3.3}$$

It quantifies the unevenness of the probability distribution p. In particular, the minimum $H(X) = 0$ is reached for a constant random variable (i.e., equiprobable random variable). Entropy is also denoted as [73]:

$$S(p) = -\sum_{i=1}^{|\chi|} p(x_i) \log_2 p(x_i) \tag{3.4}$$

which underlines the fact that entropy is a feature of the probability distribution p. The source generating the inputs $x \in \chi$ is characterized by the probability distribution $p(x)$. Shannon's entropy $S(p)$ appears as the average missing information, that is, the average information required to specify the outcome x when the receiver knows the distribution p. It equivalently measures the amount of uncertainty represented by a probability distribution. In the context of communication theory, it amounts to the minimal number of bits that should be transmitted to specify x. The following excerpt from J. R. Pierce [72] shows the closed link between the amount of information conveyed by a message and the amount of uncertainty: "A message which is one out of ten possible messages conveys a smaller amount of information than a message which is one out of a million possible messages. The more we know about what message the source will produce, the less uncertainty, the less the entropy, and the less the information."

The conditional entropy, $H(X|Y)$, appears to be the average (over Y) of the entropies of the conditional probability distributions $p(X|Y = y)$:

$$H(X|Y) \equiv H(X,Y) - H(Y) = \sum_{y \in Y} p(y)\left[-\sum_{x \in \chi} p(x|y)\log_2 p(x|y)\right] \tag{3.5}$$

when the random variables X and Y have the same state space χ, with respective distributions p_x and p_y, it is possible to consider the relative entropy:

$$S_{rel}(p_X|p_Y) = -\sum_{x} p_X(x)\log_2\left[p_X(x)/p_Y(x)\right] \tag{3.6}$$

The opposite of the relative entropy defines the Kullback-Leibler divergence [74] for two probability distributions p and q on the same space χ:

$$D(p\|q) = -S_{rel}(p|q) = \sum_{x} p(x)\log_2\left[p(x)/q(x)\right] \geq 0 \tag{3.7}$$

The Kullback-Leibler divergence is an important quantity that is exploited in defining measures of information or uncertainty. Most of the measures of uncertainty so far are being developed based upon the Shannon's entropy [75–78]. However, the notion of entropy as already mentioned may not be sufficient to cover all aspects of uncertainty aggravated when more than one meaning of information (Section 2.1) creeps into a discussion. The ideas that gave rise to the entropy of physics and the entropy of Shannon are quite different, although they can be described in terms of uncertainty in similar mathematical terms. This is still a confused matter to see if one can establish significant and useful relations between both entropies [19, 79].

3.3.2 Uncertainty and Risk

Uncertainty[5] is a term used in subtly different ways in a number of fields, including philosophy, physics, statistics, economics, finance, insurance, psychology, sociology, engineering, and information science. Several uncertainty taxonomies [80–84]

5. Uncertainty applies to predictions of future events, to physical measurements that are already made, or to the unknown. Uncertainty arises in partially observable and/or stochastic environments due to ignorance and/or indolence (http://en.wikipedia.org/wiki/Uncertainty).

exist for each domain and it often happens that uncertainty is confused with the notion of risk.[6]

In the schematic approach in Figure 3.3, labeled "igloo of uncertainty," we distinguish between open and closed forms of both ignorance and knowledge and delimitate a field of uncertainty that concerns both risk and danger. Dangers are defined in terms of the possible outcomes of a given situation. A decisive difference between both is that a danger is present regardless of choice and can therefore be avoided or counteracted, whereas a risk is either optionally accepted or imposed. We refer the reader to [85, 86] for examples and a more detailed discussion on concepts presented in Figure 3.3 such as open and closed knowledge and ignorance, risks, and dangers. The authors in [85, 86] presented a discussion on the distinction between closed and open knowledge with respect to risk and closed and open ignorance with respect to danger as well as prerequisite for turning dangers to risks.

When a certain situation involves dangers and risks, it is important to reduce gaps in knowledge either by research or by learning. The development of FIAT supports this precise objective. A strategy to address gaps in knowledge first requires insight into the particular type of uncertainty. Table 3.2 presents a taxonomy of uncertainty, a description of its causes, and its related types of decision. Two layers

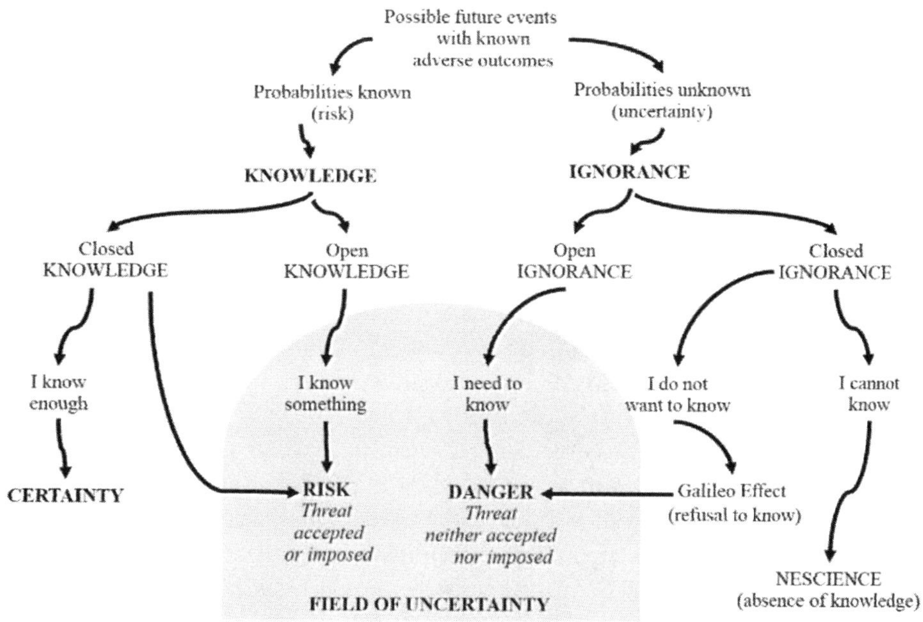

Figure 3.3 Uncertainty igloo. (Adapted from [85, 86].)

6. With regard to what is risk and how it compares to what is uncertainty, for risk we do not know what is going to happen next, but we do know what the distribution looks like. For uncertainty we do not know what is going to happen next, and we do not know what the possible distribution looks like. In other words, the future is always unknown, but that does not make it uncertain (http://www.ritholtz.com/blog/2012/12/defining-risk-versus-uncertainty/).

Table 3.2 The Taxonomy of Uncertainties and Decisions

Uncertainty	*Objective*	*Epistemical* Caused by gaps in knowledge that can be closed by research.	*Decision: Knowledge-guided* Decision-makers must both rely on existing knowledge and reflect on any remaining uncertainties. One strategy in this regard is a comparative risk assessment of similar situations.
		Ontological Caused by the stochastic features of a situation, which will usually involve complex technical, biological, and/or social systems.	*Decision: Quasi-rational* Impossible to make rational decisions and situations are largely unpredictable; past experience and probabilistic reasoning provide some guidance on how complex systems will react.
	Subjective	*Moral* Caused by a lack of applicable moral rules and we call these situations "moral uncertainties."	*Decision: Rule-guided* Decision-makers have to fall back on more general moral rules and use them to deduce guidance for the special situation in question.
		Rule Caused by uncertainty in moral rules. This means that we act on the basis of fundamental preformed moral convictions in addition to experiential and internalized moral models.	*Decision: Intuition-guided* Can make decisions only by relying on our intuition rather than knowledge, or explicit or implicit moral rules.

Source: [85].

are presented: two fundamental forms (objective, subjective), both of which are divided into two further subforms (epistemical, ontological, moral, rule).

Uncertainty has two main meanings [81] in most of the classical dictionaries. Meaning I is uncertainty as a state of mind. Meaning II is uncertainty as a physical property of information. The first meaning refers to the state of mind of an agent, which does not possess the needed information or knowledge to make a decision; the agent is in a state of uncertainty: "I'm not sure that this object is a table." The second meaning refers to a physical property, representing the limitation of perception systems: "The length of this table is uncertain." In theories of uncertain reasoning, uncertainty is often described as imperfection of information, as errors on measures for example, and does not depend on any kind of state of mind. However, an uncertain information (meaning II) can induce some uncertainty in our mind (meaning I).

Smithson [87, 88] proposed a taxonomy of ignorance where uncertainty appears as a kind of ignorance "... one of the most manageable kinds of ignorance." This taxonomy is reproduced in Figure 3.4. Smithson interprets ignorance as nonknowledge. He initially separates ignorance into two categories: the state of ignorance (error) and the act of ignoring (irrelevance). The latter corresponds to a deliberate action to ignore something irrelevant to the problem-solving situation, whereas the former is a state (of ignorance) resulting from different causes

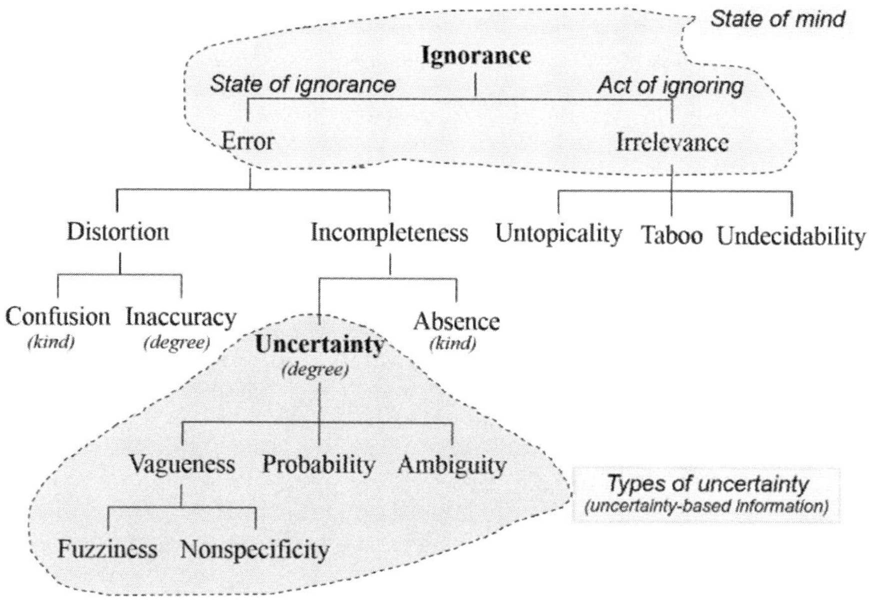

Figure 3.4 Smithson's taxonomy of ignorance.

(distorted or incomplete knowledge). For Smithson, uncertainty is incompleteness in degree as compared to absence, which is incompleteness in kind, and is subdivided into three types: probability, vagueness (being either nonspecificity or fuzziness), and ambiguity.

Uncertainty, as a state of mind (Meaning I), corresponds to ignorance in Smithson's hierarchy, while Smithson's concept of uncertainty corresponds to uncertainty-based information (Meaning II). Smithson's taxonomy could be an interesting guide for FIAT development that supports situation awareness since it accounts for different levels of processing from physical property of information (uncertainty) to mental state (ignorance).

3.3.3 Dequech's Typology of Uncertainty in Economics

Dequech [70] proposed a typology of the main concepts of uncertainty used by economists. Dequech insists upon three main distinctions presented in Table 3.3. The first distinction is between substantive and procedural uncertainty. The second distinction is that between weak and strong uncertainty. The third distinction is that between ambiguity and fundamental uncertainty. The definition of ambiguity is [89]: "ambiguity is uncertainty about probability, created by missing information that is relevant and could be known."

The basic concept of weak uncertainty is that individuals can build unique, additive (sum to 100%) and fully reliable probability distributions. This category can be subdivided into two: Knightian's risk[7] and Savage's uncertainty.[8] In Knightian's risk, individuals can act on the basis of a probability that is objective (in the sense

7. Knightian's risk after Frank Knight (1921).
8. After Leonard Savage (1954), who developed the subjective version of standard expected utility theory.

Table 3.3 Dequech's Typology of Uncertainty in Economics

Type of uncertainty	*Weak uncertainty:* Unique, additive, and fully reliable probability distribution.	*Strong uncertainty:* absence of such a distribution
Substantive uncertainty: lack of some relevant and good-quality information	*Weak uncertainty:* uncertainty about which state will obtain	*Ambiguity:* uncertainty about probability, caused by missing information that could be known; predetermined list of states
		Fundamental uncertainty: possibility of nonpredetermined structural change; nonpredetermined list of states
Procedural uncertainty: complexity in relation to limited computational and cognitive capabilities of the agents		*Procedural uncertainty*

Source: [70].

that any reasonable person would agree on it) and known. This probability may be of one of the following two kinds: a priori or statistical probability. A priori probabilities can be attributed objectively by logical reasoning, without the need to perform any experiments or trials. Statistical probabilities are relative frequencies. In *Savage's Uncertainty* [90, 91], the notion of subjective probabilities, originally developed by Ramsey and de Finetti [92], is used. Subjective theory of probability treats probability as a degree of belief and makes possible to assign precise numerical probabilities to virtually any proposition or event. Betting rates are the mechanism through which subjective probabilities can be inferred. Dequech [70] stated that Knightian's risk can be seen as a special case of Savage's uncertainty since the latter can handle subjective probabilities, with or without objective probabilities.

Decision-making under weak uncertainty, as viewed by Savage, chooses among acts as pictured in Figure 3.5. Each act, a_i, has a consequence, c_{ij}, depending on which state, s_j occurs. A state of the world is "a description of the world, leaving no relevant aspect undescribed." The set of states is exhaustive and defined independently of the set of acts. States are mutually exclusive and each state has a probability, p_j. The expected utility of each act, a_i, is the weighted average of the utilities of the several consequences, where the weights are the probabilities of the respective states:

$$u(a_i) = \sum p_j u(c_{ij}) \tag{3.8}$$

In the case of strong uncertainty associated with the notion of procedural uncertainty, the decision problem is complex and populated by individual or collective agents with limited mental and computational capabilities. In [70], based on literature in 1997 [93], two problems have been identified on procedural uncertainty: (1) large amounts of information, referred to as extensiveness; and (2) the other, called intricacy, refers "to the density of structural linkages and interactions between the

Acts	States			
	s_1	s_2	...	s_m
a_1	c_{11}	c_{12}	...	c_{1m}
a_2	c_{21}	c_{22}	...	c_{1m}
...	...	...	...	...
a_n	c_{n1}	c_{n2}	...	c_{nm}

Figure 3.5 Savage's acts-states-consequences framework.

parts of an interdependent system." One can easily relate extensiveness and intricacy to data deluge and complex networking (i.e., to the Vs dimensions of big data discussed in Chapter 1: Volume, Velocity, Variety, Veracity, and Value).

In a situation involving ambiguity, the decision-maker cannot unambiguously assign a definite probability to each and every event because some relevant information that could be known is missing. Dubois [94] explained that Savage's rational decision-making chooses according to expected utility with respect to a subjective probability but that in the presence of incomplete information: (1) decision-makers do not always choose according to a single subjective probability; (2) there are limitations with Bayesian probability for the representation of belief; and (3) a single subjective probability distribution cannot distinguish between uncertainty due to variability and uncertainty due to lack of knowledge. Dubois [94] pursued eliciting motivations for going beyond pure probability and set representations. He suggested finding uncertainty representations that combine probability and sets such as: imprecise probability theory [95, 96] (sets of probabilities), Dempster-Shafer theory [97–99] (random sets), and numerical possibility theory [100, 101] (fuzzy sets). Representations have each event with a degree of belief (certainty) and a degree of plausibility, instead of a single degree of probability.

Fundamental uncertainty in Table 3.3 is the lack of knowledge that results from the characterization of social reality as subject to nonpredetermined structural change [70]: "Future knowledge is not knowable in advance, in the sense that we do not know exactly what we are going to learn over the next years and when we are going to learn it." Innovations that cause nonpredetermined structural change in economic relationships are very illustrative of fundamental uncertainty brought by human creativity and change in knowledge in the economic reality. To illustrate further, two types of economic reality were mentioned in [102]: immutable and transmutable. An immutable reality is one in which the future path of the economy and the future conditional consequences of all possible choices are predetermined, also referred to as ontological uncertainty. A transmutable reality was described as: "the future can be permanently changed in nature and substance by the actions of individuals, groups... and/or governments, often in ways not completely foreseeable by the creators of change." There is epistemological uncertainty when "some limitation on human ability... prevents agents from using (collecting and analyzing) historical time-series data to obtain short-run reliable knowledge regarding all economic variables."

3.3.4 Dubois and Prade's Typology of Defects

A very pertinent contribution on defining information in the context of development of FIAT is from Dubois and Prade with their *Decision-Making Process: Concepts and Methods* [103]: "The term information refers to any collection of symbols or signs produced either through the observation of natural or artificial phenomena or by cognitive human activity with a view to help an agent understand the world or the current situation, making decisions, or communicating with other human or artificial agents." This definition suits very well FIAT's main objective to support decisions and actions (agents). Table 3.4 was developed from [103] where distinctions between three pairs of information qualifiers were presented: (objective, subjective), (quantitative, qualitative), and (generic, singular).

These distinctions presented in Figure 3.6 and Table 3.4 are important for engineering a FIAT solution to support decision-making. The designer can see a possibility to reduce uncertainty by gathering more data, refining models and improving processing to support the situation awareness processes represented in Figure 3.6 by the Endsley and Garland model [104]: perception, comprehension, and projection. As described in much detail in [103], an agent is supposed to be at a certain situation awareness state (epistemic) by having some information about the current world. That is made of three components (i.e., from an agent, generic knowledge, singular observations, and beliefs). Beliefs are derived from singular (i.e., related to the current situation and generic knowledge kinds of information).

Table 3.4 Distinctions of Various Types of Information

	Informal Definitions	*Representations/Examples*
Objective	Information stemming from sensor measurement and the direct perception of events	Quantitative-qualitative (e.g., balance sheet data, radar data)
Subjective	information typically uttered by individuals or conceived without resorting to direct observations	Quantitative-qualitative (e.g., testimonies, business plan)
Quantitative	Information modeled in terms of numbers, typically objective information	Numbers, intervals, functions (e.g., sensor measurements, counting processes)
Qualitative or symbolic	Subjective information	Logical or graphical (e.g., expressed in natural language)
Generic	Information refers to a collection or a population of situations or a piece of commonsense knowledge	A physical law, a statistical model built from a representative sample of observations
Singular	Information refers to a particular situation, a response to a question on the current state of affairs	An observation (a patient has fever at a given time point), or a testimony (the driver's car was blue).

Source: [103].

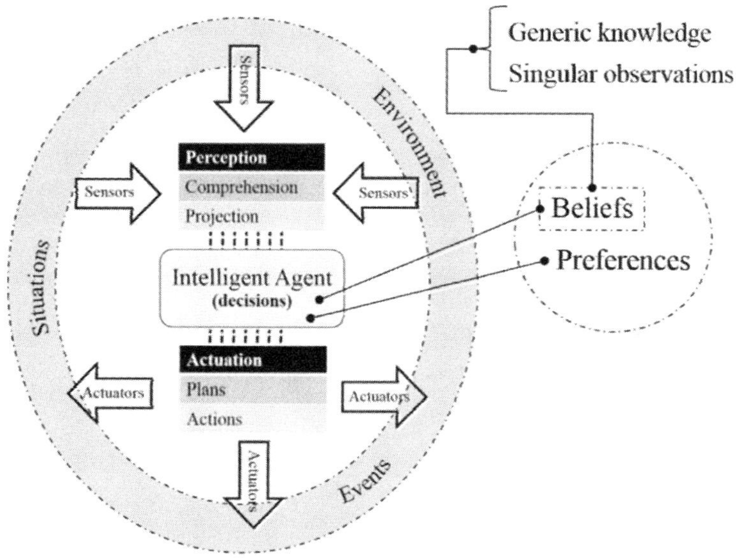

Figure 3.6 Kinds of information and epistemic state of an intelligent agent.

Decision-making requires adding another kind of information: agent preferences (Chapters 2, 7, and 16 of [105]). FIAT is supposed to instrument these kinds of information into decision support systems.

To represent the epistemic state of an agent, a representation of the states of the world is needed. Let v be the vector of attributes relevant for the agent and Ω be the domain of v. Ω is called a frame, descriptions of all states of the world. A subset A of Ω, viewed as a disjunction of possible worlds, is called an event, to be seen as a proposition that asserts $v \in \Omega$. Dubois and Prade presented four kinds of imperfection of information expressible on the frame Ω: incomplete, uncertain, gradual, and granular. Table 3.5 summarizes these four kinds of imperfection without any mathematical details but with simple examples to illustrate their meaning.

The best-known and oldest formal framework to reason under uncertainty is the probabilistic one, where uncertainty due to variability is modeled by classical probability distributions. However, a single probability distribution cannot adequately account for incomplete or imprecise information so alternative theories and frameworks have been proposed. The three main such frameworks, are, in decreasing order of generality, imprecise probability theory [95], random disjunctive sets [97, 98, 106], and possibility theory [100, 101]. In addition to [94, 103], a unified overview of various approaches to representations of uncertainty is presented in two parts by Destercke et al. [107, 108].

3.3.5 Typologies of Uncertainty

In 1993, Krause and Clark [110] proposed an alternative typology to Smithson's, centered on the concept of uncertainty. Krause and Clark distinguish two aspects: unary (i.e., uncertainty applied to individual propositions) and set theoretic (i.e., uncertainty applied to sets of propositions). Both categories lead either to conflict (conflicting knowledge) or ignorance (lack of knowledge). As subcategories,

Table 3.5 Kinds of Imperfection of Information

	Description and Examples	*Representations*
Incomplete	Not sufficient to allow an agent to answer a relevant question in a specific context (e.g., what is the current value of some quantity v?) Imprecision is a form of incompleteness, in the sense that an imprecise response provides only incomplete information (e.g., the term *minor* is imprecise, as it provides incomplete information if the question is to know the birth date of the person).	A set used for representing a piece of incomplete information is called a disjunctive set: mutually exclusive elements. A conjunctive set represents a precise piece of information: a collection of elements.
Uncertain	An agent does not know whether a piece of information is true or false. A primitive item of information being a proposition, or the statement that an event occurred or will occur. The representation of a proposition as an entity liable to being true or false (or of an event that may occur or not) is a convention.	Uncertainty qualifiers are a number (a probability) and symbolic modalities (possible, certain). Assigning to each proposition or event *A*, viewed as a subset of , a number in the unit interval to evaluate the likelihood of *A*.
Gradual	Underlies a ranking, in terms of relevance, of attribute values to which it refers. The propositions are not always Boolean and the transition to the convention of being true or false (or of an event that may occur or not) is gradual. The proposition *Pierre is young* could be neither totally true, nor totally false: the meaning of *young* will be altered by linguistic hedges expressing intensity.	Fuzzy sets [109] are used when dealing with a piece of information expressed in natural language and referring to a numerical attribute.
Granular	Changes in the level of resolution or refinement to which the set of states of affairs can represent the world; that has an impact on the possibility or not to represent relevant information. It may happen that does not contain all propositions to completely describe the problem under concern and if new propositions are added to modify the set ; this is called a *granularity change* for the representation.	The probabilistic representation of incomplete information by means of the insufficient reason principle does not resist to a change of granularity but possibilistic representation does (Section 2.4 of [103]).

we find vagueness, confidence, propensity, equivocation, ambiguity, anomaly, inconsistency, incompleteness, and irrelevance.

This model is reproduced in Figure 3.7. Compared to Smithson's taxonomy, Krause and Clark added the unary/set theoretic dichotomy, in order to introduce the concept of inconsistency and to remove the concept of incompleteness from the unary branch to the set theoretic one. Krause and Clark's classification concerns meaning II of uncertainty from a formal point of view since the distinction is based on propositions. This approach is a straightforward way to process propositional belief, a concept central to situation analysis.

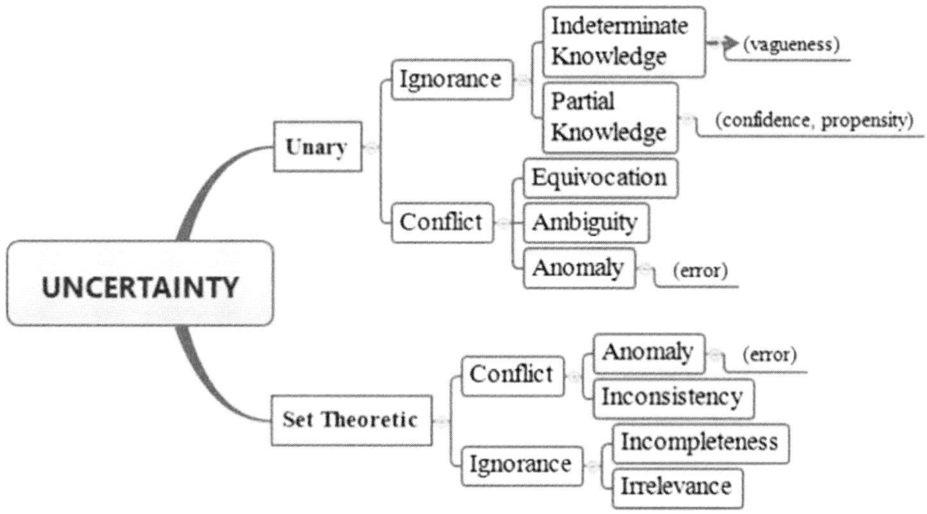

Figure 3.7 Krause and Clark taxonomy of ignorance. (Adapted from [52].)

Bouchon-Meunier and Nguyen [111] proposed a model for uncertainty (Figure 3.8). They refer to uncertainty as "imperfection on knowledge" and denote then three main types of imperfection: (1) probabilistic uncertainty, (2) incompleteness in knowledge (belief, general laws, imprecision), and (3) vague and imprecise description. The schema of Figure 3.8 is a good way to make the distinction between the two main general meanings of uncertainty. Reading the graph from right to left, uncertainty appears as a final state of mind (Meaning I) possibly caused by belief, general laws, imprecision, vagueness, or incompleteness.

Instead of a typology of uncertainty, Smets built a typology of imperfection of information [99] avoiding the confusion between the two meanings of uncertainty. The model proposed by Smets distinguishes three main categories of imperfect information (Figure 3.9): (1) imprecision: related to the content of the statement, several worlds satisfy the statement; (2) inconsistency: no world satisfies the statement; and (3) uncertainty: induced by a lack of information, by some imprecision. Smets considered imperfection as a central term, with uncertainty being a kind of imperfection. Uncertainty can be either objective (property of the information, that is, meaning II) or subjective (property of the observer, that is, meaning I).

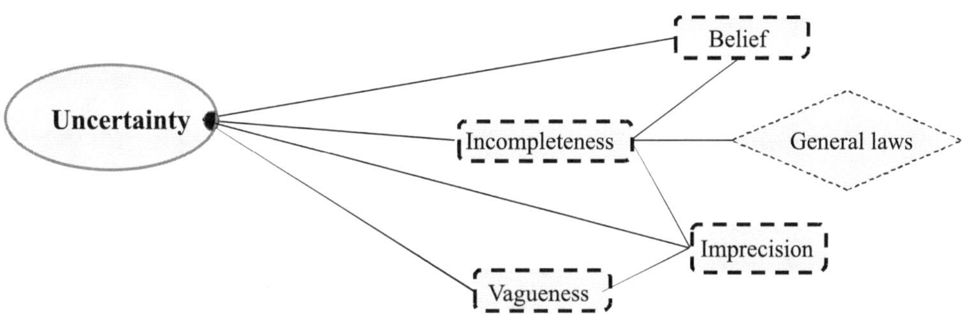

Figure 3.8 Bouchon-Meunier and Nguyen model for uncertainty. (Adapted from [52].)

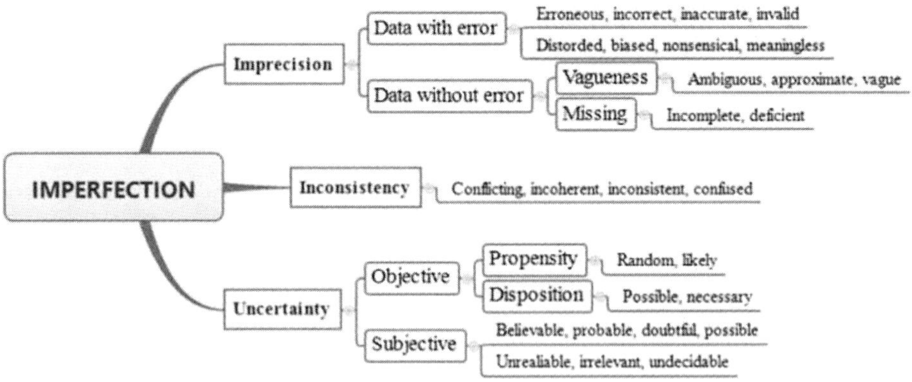

Figure 3.9 Adapted from Smets' structured thesaurus on imperfection [81].

The typology proposed by Klir et al. [112, 113] is built upon the different existing mathematical theories of uncertainty, and directly linked to measures of uncertainty. Klir conceives information in terms of uncertainty reduction and consequently introduces the term of uncertainty-based information. For Klir et al. [112, 113], uncertainty can be either fuzziness or ambiguity (two types of uncertainty). Ambiguity can itself be either nonspecificity or discord. These concepts can be related to some previously used in the other classifications: fuzziness is close to vagueness, discord is a synonym of conflict, and nonspecificity means principally imprecision or generality. In Figure 3.10 typology, they integrate the main key terms used by Smithson (fuzziness, nonspecificity, ambiguity) as well as the set theoretical aspect introduced by Krause and Clark (discord). Klir did not mention knowledge and thus stays at a lower level of processing (i.e., at the information level).

Klir's conception of uncertainty is closely related to quantitative theories of uncertainty as depicted on Figure 3.11 and briefly described in the following section

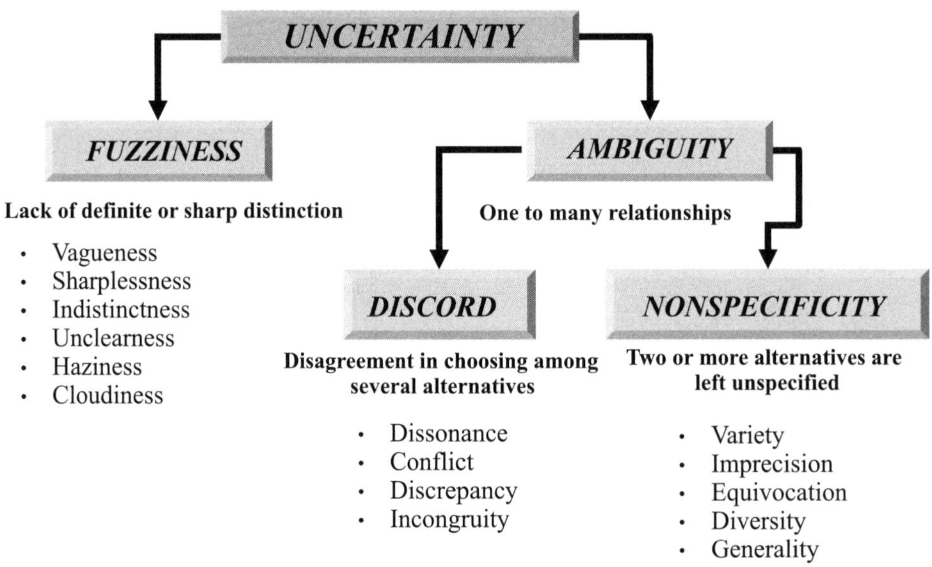

Figure 3.10 Klir and Yuan's typology. (Adapted from [77].)

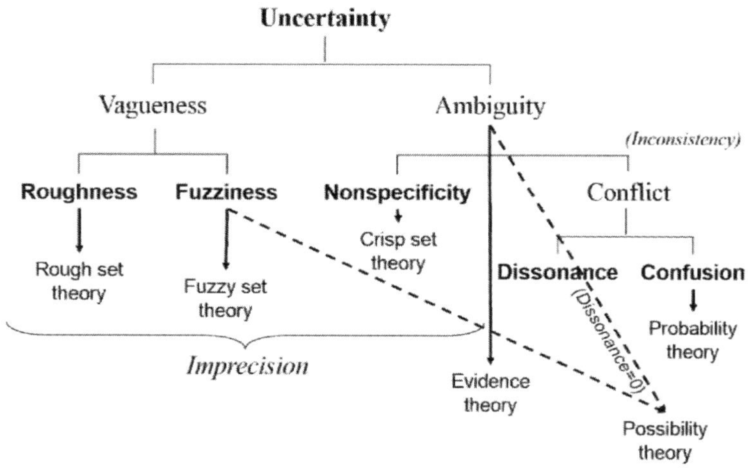

Figure 3.11 Extended Klir et al.'s typology. (Adapted from [80].)

such as probability, possibility, fuzzy sets, and evidence theories and leads to corresponding measures of uncertainty, that is, of uncertainty-based information [113]. The typology presented in Figure 3.11 is an extended version of Klir et al. [112]. It highlights four main types of uncertainty pertaining in the different theories: fuzziness, nonspecificity, dissonance, and confusion. The last two terms correspond to two aspects of conflict. The distinction imprecision-inconsistency of the model of Smets appears also: imprecision gathers fuzziness and nonspecificity and conflict corresponds to inconsistency.

3.4 Representations of Uncertainty

Material in this section has been adapted from [52, 80]. As discussed above, uncertainty and its other face, information, present quite a challenge if we look at the three semiotics dimensions: syntactics, semantics, and pragmatics. The representation of uncertainty is an important issue in many areas of science and engineering, especially in the systems design and control. To model uncertainty, many mathematical tools have been developed, being either qualitative (modal logic, nonmonotonic reasoning, and so forth) or quantitative approaches (probability theory, fuzzy sets, rough sets, random sets, belief functions, and so forth). Figure 3.12 presents a list of theories regrouped into three main categories: quantitative, qualitative and hybrid/graphical approaches. Each of the approaches in Figure 3.12 are often compared on their different strengths and weaknesses, on their better correspondence to a particular type of uncertainty, on their requirement of prior knowledge, on their computational time, on the need of independence on data. For instance, in Chapters 7 to 9 of [52], some of these approaches were discussed and compared but the discussion there was focused on information fusion applications and it is far from being complete. The unified and very instructive overview on uncertainty presented in two parts by Destercke et al. [107, 108] as well as contributions from references such as [105, 113–115] can certainly help facing the semiotics challenge, especially the treatise on decision support by Bouyssou et al. [105] where a multidisciplinary

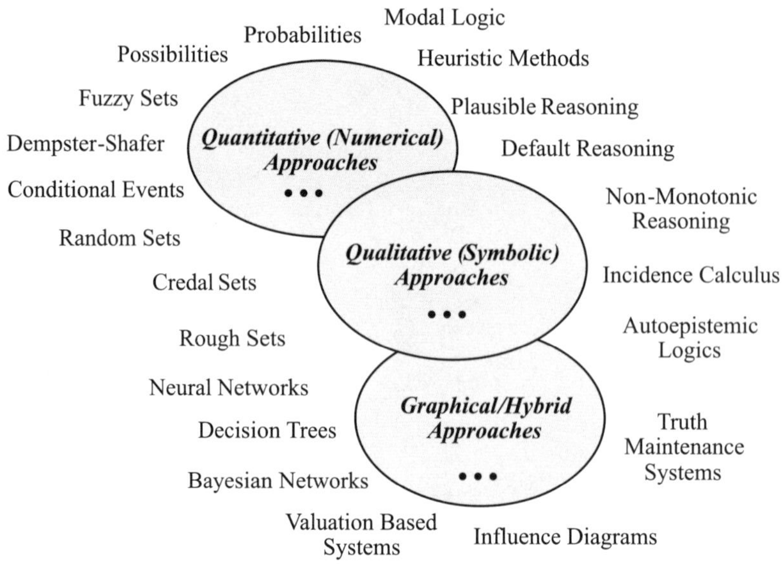

Figure 3.12 Mathematical and logical tools. (Inspired by [52].)

view (psychologists, economists, sociologists, mathematicians, computer scientists, and decision scientists) is presented. This is fundamental and instrumental for the development of FIAT as decision support systems.

3.4.1 Quantitative Approaches

Although classical set theory and probability theory have been regarded as reference frameworks for centuries, they cannot easily represent all types of uncertainty. Numerous other theories have been obtained by expanding the conceptual frameworks on which those classical theories were based. Fuzzy set theory [109] expands the classical set theory. Evidence theory (or Dempster-Shafer theory) [98] and other theories based on nonadditive measures (i.e., fuzzy measures [116]) expand probability theory. Rather than be considered as rivals, these theories should be seen as complementary to deal with different types of uncertainty and even be combined to be able to deal with a broader class of uncertainty types. In particular, to encompass the case of fuzzy events in a probabilistic framework, Zadeh proposed to generalize the classical probability theory to fuzzy probability theory defining probability measures on fuzzy events [117]. This fuzzy generalization process extends the use of crisp sets to the one of fuzzy sets.

Probability theory is the basis of numerical approaches, under Kolmogorov's axiomatization of the theory that lies on axioms imposing the use of: (1) a σ-field algebraic structure, (2) sharp values (i.e., single values) for (3) numerical evaluations taken in the (4) set of real numbers $\mathbb{R}$, (5) a normalization, and (6) countable and finite additivity conditions. However, we should keep in mind that non-Kolmogorovian approaches exist, based on the modification of the basic axioms. For example, one can obtain theories where negative, complex, or qualitative values are for example allowed, but also non-normalized probabilities. Note that dropping the additivity axiom leads to Dubois and Prade's theory of fuzzy probability, Dempster-Shafer theory, or possibility theories, among others.

Dempster-Shafer theory, also called evidence theory or belief functions theory, is a generalization of probability theory in the sense that probability functions are special cases of belief functions. Indeed, allowing belief functions to be subadditive (contrary to probability functions that are required to be additive) expands the support of the uncertainty functions to the power set of the universe of discourse. In other words, dropping the additivity axiom of probability theory allows statements such as, "I believe it will rain tomorrow with a confidence of 0.8 but I also admit that I don't know with a weight of 0.2," to be represented and processed. This high capacity for representing ignorance is one of the most attractive features of Dempster-Shafer theory.

Fuzzy sets theory aims at representing another kind of uncertainty called fuzziness (or more generally vagueness), quantified using characteristic functions (or membership functions) to assign truth degrees to events. Thus, in fuzzy sets theory the uncertainty is quantified by playing on the arity of the characteristic function (usually an infinity of values in the [0, 1] real interval) instead of juxtaposing a confidence function to a binary characteristic function as done in probability theory. Although probability and fuzzy sets theories are still nowadays considered as equivalent by a small community, they clearly address different kinds of uncertainty, and are rather complementary than rivals. Fuzziness is a fundamental concept of mathematics generalizing the notion of membership to a set allowing thus the extension of other theories of uncertainty. Indeed, a degree of confidence (probability for instance) can be attributed to the realization of an event, this event being itself more or less true with respect to a given reference. In other words, an event can be both fuzzy and random. Examples of this type are fuzzy probability, hybrid numbers, and fuzzy belief functions theories.

Built on fuzzy sets theory, but giving another meaning of the characteristic function, the theory of possibility is a theory of partial ignorance. In fact, possibility theory just as probability theory can be interpreted in the more general framework of evidence theory, possibility functions being just another special case of belief functions. The theory of possibility copes with the evaluation of confidence values (i.e., possibilities) of nested elements of the power set. This restriction on the algebraic structure allows the use of fuzzy sets theory for manipulating possibility functions, without losing the expressiveness of evidence theory when processing ignorance statements. In spite of this restriction, sets considered in this theory are crisp (a feature shared with probability theory and evidence theory), the principal concept borrowed from fuzzy sets theory being the idea of degree, here applied to the notions of possibility and necessity.

An alternative to fuzzy sets theory to represent vagueness is rough sets theory. This theory has been presented as a theory for dealing with vagueness and uncertainty, but more precisely deals with indiscernability of objects in the universe of discourse. Our limited knowledge of this universe is represented by a partition regrouping indiscernible elements. Based on this knowledge, an information (represented by a subset of the universe) is then be approximated by lower and upper bounds (i.e., two other subsets). Hence, our incapacity to class elements into one or another class is modeled by lower and upper bounds instead of membership degrees as done in fuzzy sets theory. Rough sets theory can be seen as a qualitative method for the processing of uncertainty, since its basic constituents are upper and lower limits, partitions, and sets. The numerical evaluations are not made within the theory but constrained by these qualitative notions. Indeed, on this structure, membership functions have been defined, but also belief and plausibility functions, the latter allowing rough sets to be also interpreted in the evidence theory framework.

3.4.2 Qualitative Approaches

Although it is generally accepted that classical logic is not able to deal with uncertainty (even if it could be a matter of controversy), it remains the basis of all logical approaches. Initially, logic has been conceived to reason about certain, indubitable facts. However, facing the growing and unavoidable constraint of dealing with uncertainty, this classical framework has evolved and led to logics capable of approximate reasoning. The limitations of classical logic introduce the alternative logical approaches (Figure 3.13), namely, multivalued logics, modal logics, and nonmonotonic logics. Classical logic lies on the principle of bivalence expressing the fact that a proposition is either true or false. Hence, only two truth values are allowed in classical logic. Classical logic also lies on the principle of excluded middle (EM) expressing that a proposition is either asserted or negated making it difficult to express natural language expressions with such a bivalent logic and it fails to deal with vague terms. Then, have been introduced multivalued logics.

The most common way to go beyond bivalence is to introduce supplementary truth values in the classical logic framework. With the introduction of a third truth value, a proposition can be neither true nor false. This is an important step forward for uncertainty processing based on classical logic framework. Pursuing this extension to an infinity of truth values leads to fuzzy logic, an approach considered as hybrid since it combines numerical evaluations with a logical framework. Modal logic is another important approach for dealing with uncertainty and knowledge that describe in what manner a proposition is true (possibly true, necessary true, but also knowing it is true). In modal logic, these modalities are expressed by nontruth-functional operators called modal operators. Using the appropriate combination of

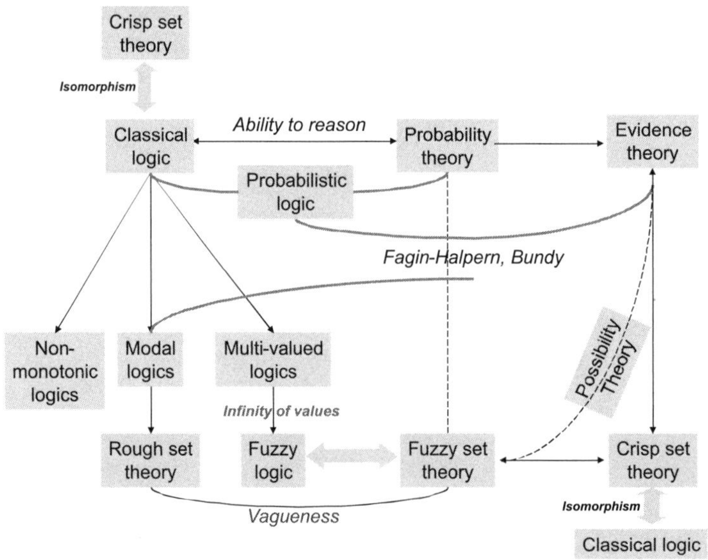

Figure 3.13 Possible links between mathematical and logical tools [80].

axioms and modalities, it is possible to build modal logic frameworks allowing to reason on the important notions of knowledge and belief.

Another basic limitation of classical logic, but also of derived logics, is their inadequacy for dealing with defeasible reasoning, the reasoning being monotonic in these frameworks. Monotonicity means, practically, that once a conclusion has been accepted as true, it cannot be defeated when confronted with new evidence. Dropping the monotonicity property leads to nonmonotonic logics. Although nonmonotonic logics are more appropriate than classical schemes for commonsense reasoning, they are much more demanding as far as reasoning schemes are concerned. Among nonmonotonic logics, default, autoepistemic, and circumscription logics are the most popular. Note that nonmonotonic reasoning is not limited to symbolic approaches. Indeed, most of numerical approaches own this property by the means of some inference or combination rules such as Bayes or Dempster.

3.4.3 Hybrid and Graphical Approaches

The three alternatives (Figure 3.13) to classical logic present different features that do not make one approach better than the other. Rather, one should select the appropriate framework for uncertain reasoning on the basis of objective criteria based on the application. However, although numerical approaches are computationally efficient, they lack clear semantics. In a tentative to resolve this conflict, the actual tendency is to reconcile both approaches. Integrating mechanisms and representations schemes from symbolic and numerical frameworks leads to hybrid approaches. Most of numerical theories have their logical counterpart. The model of construction is the same for all the theories: the event (subset of Ω) is seen as a logical formula ϕ, A referring to the concepts developed in the symbolic approaches. Figure 3.13 summarizes the various links that can be done between the numerical and symbolic approaches.

Probabilistic logic is a logic for reasoning about probabilities (i.e., the probability calculus on rules, on logical formulas). It was first introduced by Nilsson in 1986, as a "semantic generalization of logic in which the truth values of sentences are probability values" [118]. The principle of probabilistic logic lies on the fact that the truth values of propositions are their probability of occurrence. The purpose is then to deal with propositional probabilities (i.e., probabilities assigned to particular propositions or assertions). Probabilistic logic combines then logic with probability theory and reduces to ordinary logic when the probabilities of all sentences are either 0 or 1. This approach is based in the possible worlds semantic. Other approaches have been developed later by Halpern and Fagin [119–123]. The theoretical basics of probabilistic logic are those of probability theory or of classical logic.

The incidence calculus was proposed by Alan Bundy in 1985 [124, 125] as a probabilistic logic for reasoning under uncertainty. It is a method for managing uncertainty in numerical way. Different from other numerical approaches, in incidence calculus probabilities are associated with a set of possible worlds rather with formulas directly. The probability of a formula is then calculated through the incidence set assigned to the formula. Incidence calculus itself appears to be a unification of both symbolic and numerical approaches. It can therefore be regarded as a bridge between the two reasoning patterns [126].

Qualitative approaches seem to better suit reasoning on knowledge, while quantitative approaches are better candidates for uncertainty representation and management. It appears than none of the available mathematical tools is the best, and it becomes more obvious that depending on the kind of problem we are facing, one theory can force itself as a better choice. To support this argument, Smets in [127] made a formal discussion on ignorance and the need of theories well founded. Hybrid approaches (e.g., quantitative logics, incidence calculus) that mix quantified evaluations of uncertainty and high reasoning capabilities have emerged to support global situation analysis. For large space decision problems where the set of states of the world is huge, so that the explicit description of a probability or possibility distribution on the state space becomes too demanding, graphical or graph-based approaches (e.g., Bayesian networks, influence diagrams) are powerful tools for the compact representation of uncertainty or preference.

Bayesian networks were first developed to model distributed processing in reading comprehension. Bayesian networks have imposed themselves as a general representation scheme for uncertain knowledge. Pearl [128, 129] showed that the marginal distribution for individual nodes can be obtained by using only local computation. Pearl's work is the basis of graphical representation and propagation of knowledge and several architectures for exact computation of marginals using local computation in the uncertainty reasoning have emerged [130–133]. The case of singly connected Bayesian networks considered by Pearl has been expanded to multiple connected networks.

Similar techniques have been developed for propagating belief functions instead of probabilities. Gordon and Shortliffe [134] posed the problem of propagating belief functions in a diagnosis tree and present a solution that involves an approximation. Then Shafer and Logan [135] proposed an exact implementation of Dempster's rule of combination and Shenoy and Shafer [136] present a general scheme for propagating belief functions with local computation. Shenoy and Shafer architecture is valid for certain kinds of trees that admit a transformation in qualitative Markov trees and generalizes the computational scheme of Shafer and Logan for diagnosis trees as well as Pearl's scheme for Bayesian causal trees. Shenoy and Shafer [131, 136–139] generalized the idea and propose an abstract framework for computing marginals in join trees. The framework has been called Valuation-Based Systems where different formalisms can be considered such as Bayesian probabilities, belief functions, and possibilities. An overview on graphical approaches was presented by Pearl [140].

In a valuation-based system (VBS), knowledge is represented by the graph itself, with its nodes (variables of interest and their possible values) and its links corresponding to dependencies between variables. Moreover, the knowledge is represented by the valuations associated to the links. Two kinds of knowledge can be distinguished: generic and singular knowledge as in Table 3.4. VBS is a general framework allowing different formalisms dealing with uncertainty to be used as in probability theory, valuations are called probability potentials; in Dempster-Shafer theory, valuations are called basic probability assignment, potentials; in possibility theory, valuations are called possibility potentials. According to Shenoy, using VBS for decision problems is a more efficient method than that of decision trees and of influence diagrams.

3.5 General Uncertainty Principles in Data and Information Transformations

The concept of system composition defined in Figure 1.8 (Chapter 1) as composability and compositionality is very important to deal with complexity through system scalability, modularity, and abstraction. System composition can exploit homomorphism in constructing networked information technology (NIT) systems [141]: layered systems and services as anticipated for cyber physical and social systems (CPSS) environment. The development of FIAT is sine qua non for CPSS or NIT systems as mentioned in Chapter 1. In analyzing and designing such FIAT-based systems, the common practice is that the entire system be decomposed into subsystems of manageable size, and these subsystems are later composed [142]. Fundamental principles must guide the process of decomposition and composition and it is inevitable that information is lost or gained in the process. The three uncertainty principles presented by Klir et al. in [113] and reproduced on Figure 3.14 become essential: maximum uncertainty, minimum uncertainty, and uncertainty invariance.

The principle of minimum uncertainty is basically an arbitration principle that is applied mainly for simplification problems. When a system is simplified, it is usually unavoidable to lose some information contained in the system. The amount of information that is lost in this process results in the increase of an equal amount of relevant uncertainty. A sound simplification of a given system should minimize the loss of relevant information while achieving the required reduction of complexity. When properly applied, the principle of minimum uncertainty guarantees that no information is wasted in the process of simplification. There are many simplification strategies, which can perhaps be classified into three main classes: (1) simplifications made by eliminating some entities from the system (variables, subsystems, and so forth); (2) simplifications made by aggregating some entities of the system

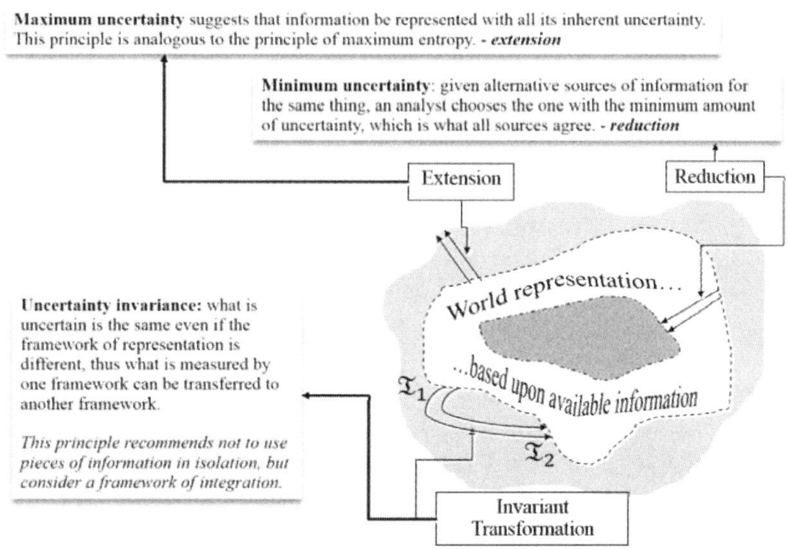

Figure 3.14 Klir's three uncertainty principles.

(variables, states, and so forth); and (3) simplifications made by breaking overall systems into appropriate subsystems ([113, p. 108]).

The principle of maximum uncertainty is essential for inductive reasoning and requires in any inductive inference using all information available but make sure that no additional information is unwittingly added. That is, that conclusions resulting from any inductive inference maximize the relevant uncertainty within the constraints representing the premises. This principle guarantees that our ignorance be fully recognized and that our conclusions are maximally noncommittal with regard to information not contained in the premises [113, p. 110].

To transform the representation of a problem solving situation in one theory, $\mathfrak{T}_1$, into an equivalent representation in another theory, $\mathfrak{T}_2$, the principle of uncertainty invariance requires that: (1) the amount of uncertainty associated with the situation be preserved when we move from $\mathfrak{T}_1$ into $\mathfrak{T}_2$ to guarantee that no uncertainty is added or eliminated solely by changing the mathematical theory by which a particular phenomenon is formalized, and (2) the degrees of belief in $\mathfrak{T}_1$ be converted to their counterparts in $\mathfrak{T}_2$ by an appropriate scale, at least ordinal to guarantee that certain properties, which are considered essential in a given context (such as ordering or proportionality of relevant values), be preserved under the transformation [113, p. 120]. Finally, as an introduction to next section, the following citation from [113] is quite pertinent: "The principle of uncertainty invariance can be made operational only when the theories involved possess well-justified and unique measures of uncertainty."

3.6 Measures of Uncertainty

Klir [112] posed that uncertainty can be quantified in an equivalent manner that information can be as proposed by Shannon with his known theory of information. Such measures of uncertainty allow, for example, "extrapolation of evidence, assessing the strength of relationship between groups of variables, assessing the influence of given input variables on given output variables, measuring the loss of information when a system is simplified and the like." In the classical theory of probability, Shannon's entropy is the tool used for quantifying uncertainty in a sense that it verifies a set of desirable properties for probability distributions. In the presence of incomplete and vague information the probabilistic representation is inadequate so that an imprecise probability theory can be used. The following list of imprecise theories is mainly from Abellán [143]: imprecise probabilities [95], Dempster-Shafer's theory (DST) [97, 98], interval-valued probabilities [144], order-two capacities [145], upper-lower probabilities [146–149], or general convex sets of probability distributions also called credal sets [146, 150–154].

Events in probability theory, as well as focal elements in evidence theory, are crisp subsets of the universe of discourse or the frame of discernment. Evidence theory can thus be generalized to fuzzy evidence theory. Zadeh first proposed such an extension [155] and it is worth mentioning other contributions, for example, Smets [156], Yager [157], Yen [158], and Lucas and Araabi [159]. In the area of Generalized Information Theory (GIT), gathering among other all the theories mentioned above, three main types of uncertainty (Figure 3.10) are usually mentioned: fuzziness, nonspecificity, and discord (or randomness). In an equivalent way that information is quantified by Shannon's entropy in the probability framework, or by the Hartley measure [160] in classical set theory, information or preferably uncertainty-based information [113] can be quantified by different measures commonly called measures of uncertainty.

In the framework of evidence theory, both nonspecificity and discord coexist leading to a multitude of uncertainty measures. They are called nonspecificity [105, 161], discord, entropy or conflict [162–165], total or global uncertainty [164–169], or aggregate uncertainty [143, 150, 152, 170–173], provided that they capture more or less one of the two kinds of uncertainty or both. Fuzziness is the type of uncertainty represented in fuzzy sets theory, which is clearly distinct from discord. Two main approaches exist for measuring fuzziness, namely either entropy-like measures when the membership function is related to a probability distribution [29, 30, 31, 32] or nonspecificity-like measures when an extension to the classical measure of cardinality is involved [33, 34]. Measuring uncertainty in fuzzy probability theory (i.e., discord and fuzziness) has been studied by many authors: Zadeh [5], DeLuca and Termini [32], and Xie and Bedrosian [35].

Because a system might deal with all three types of uncertainty [16, 36], the fuzzy evidence framework is then adequate for this general case. In particular, fuzzy evidence theory has been recently applied to solve different problems in knowledge discovery [37], images segmentation [38], or pattern classification [39]. Although most of the uncertainty measures of evidence theory can be extended to fuzzy evidence theory [40], to our knowledge, not much work has been done that addresses the problem of quantifying total uncertainty of a fuzzy body of evidence. In [12], the authors introduce mean, mode, and entropy of a fuzzy body of evidence. Indices of fuzziness, nonspecificity, and discord are proposed in [41] and an equivalent approach is also proposed in [42]. However, as argued in [36]: "The consideration of each facet of uncertainty separately, followed by injudicious combination of measures of them, leads to serious problems in interpretation and validation"; rather, the same authors suggested looking at more meaningful axiomatic approaches, a hard but important task.

To quantify the uncertainty representations, Shannon's entropy has been used as the starting point and the most common approach to justification is the axiomatic approach, that is, by assuming a set of necessary basic properties that a measure must verify [113]. Based on Klir and Wierman's classification, Liu et al. [76] developed a circular typology of uncertainty that is reproduced in Figure 3.15(a). Initially, a new term, "fuzzy confusion," was introduced but replaced here by nondisguishability that refers to the total uncertainty in the fuzzy probability theory. Additionally, three general terms are used to designate the combinations: (1) nonspecificity and discord (i.e., ambiguity), (2) nonspecificity and fuzziness (i.e., imprecision), and (3) fuzziness and confusion (i.e., nondisguishability). The term uncertainty refers to the combination of these three basic kinds. The dotted circle on Figure 3.15(a), which encircles the notion of fuzziness and vagueness, is plotted to illustrate that the typology of the various kinds of uncertainty may be classified in different ways. For instance, beyond the concept of fuzziness is the broader concept of vagueness, which simply means that borderline cases arise when representing set elements. Modeling borderline cases by degrees of truth or membership is only one of many solutions. Vagueness is thus a kind of uncertainty that is detected when borderline cases arise (i.e., objects for which we cannot decide if they belong or not

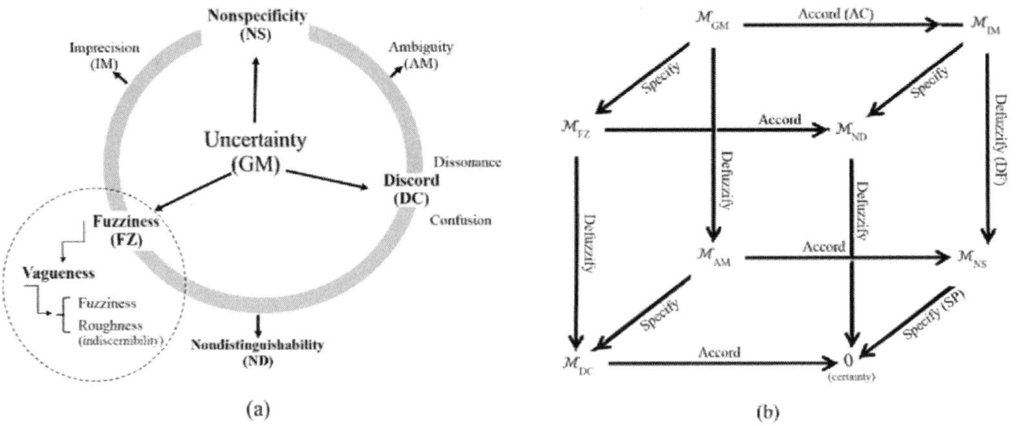

Figure 3.15 (a, b) Circular uncertainty typology. (Adapted from [76, 77].)

to a given concept). Just like for uncertainty, different kinds of vagueness can be identified.

In the framework of evidence theory (Dempster-Shafer), the belief function can model both nonspecificity and discord. The fuzzy sets theory, representing and managing vague information, deals with fuzziness and nonspecificity as main kinds of uncertainty. The most adequate framework for representing uncertainty when dealing with all three kinds of uncertainty is the combination of the evidence and fuzzy sets theory (i.e., fuzzy evidence theory) [77]. Each type of uncertainty can be quantified in fuzzy evidence theory and a general measure of uncertainty named M_{GM} is being proposed [76, 77] along with a new measure of total uncertainty in fuzzy sets theory, named M_{IM}. The general measure of uncertainty is an aggregate measure including all kinds of uncertainty. Figure 3.15 illustrates the approach used in [76, 77] to show the consistency with respect to existing uncertainty measures in classical set, fuzzy sets, probability, fuzzy probability and evidence theories. The approach is based on the reduction of the uncertainty of a fuzzy basic probability assignment, from M_{GM} which quantifies fuzziness, discord, and nonspecificity to 0 where no uncertainty can be measured [Figure 3.15(b)].

Three basic operations aiming at artificially reducing the uncertainty of a fuzzy basic probability assignment are proposed: (1) defuzzification, (2) specification, and (3) accordance. This scheme leads then to six different ways, as pictured in Figure 3.15(b), to make M_{GM} decreasing to 0, through the different quantities of uncertainty according to the circular typology [Figure 3.15(a)] such as: M_{AM}, M_{IM}, M_{FZ}, M_{NS}, M_{DC}, M_{ND}. (See [76, 77] for additional details on M_{GM}.)

1. Defuzzification transforms a fuzzy basic probabilistic assignment into a crisp one. When applied to a fuzzy set, defuzzification gives a crisp set, and while applied to a fuzzy probability distribution, defuzzification gives a classical probability distribution.
2. Specification that transforms a fuzzy basic probabilistic assignment into a fuzzy probability distribution. When applied to a fuzzy set, specification gives a nonspecific fuzzy set (pure fuzzy set), and while applied to a crisp set, specification gives a singleton.

3. Accordance that transforms a fuzzy basic probabilistic assignment into a fuzzy set. When applied to a fuzzy probability distribution, accordance gives a nonspecific fuzzy set, and while applied to a classical probability distribution, accordance gives a singleton.

3.6.1 Requirements for a Measure of Total Uncertainty

In the classical theory of probability, Shannon's entropy is the tool used for quantifying uncertainty. The approach is to verify a set of desirable properties for probability distributions. In situations where the probabilistic representation is inadequate, the approach is an axiomatic one, by assuming a set of necessary basic properties that a measure must verify. Abellán et al. [143, 173] extended the set of required properties originally defined by Klir et al. [113] for a total uncertainty measure on the Dempster-Shafer theory as reported in Table 3.6. Their extension concerns a monotonicity property that, in fact, is quite important.

Abellán et al. [143] and Liu et al. [77] examined a certain number of measures associated with Figure 3.15 (M_{AM}, M_{IM}, M_{FZ}, M_{NS}, M_{DC}, M_{ND}, M_{GM}) with respect to the set or a subset of properties listed in Table 3.6. Abellán et al. [143] proposed a measure of total uncertainty based on maximum entropy that separates the nonspecificity and conflict parts. That measure, according to [143], meets all P1 to P6 requirements. The analysis on other measures like M_{AM} needs to be revisited since it was done prior to recent results of [174, 175] where they modified M_{AM} to overcome the remarks of Klir et al. [176] about M_{AM} on subadditivity property (P4). This recent result may have as well an impact on the (P6) property for M_{AM} and consequently to M_{GM}.

3.6.2 General Remarks on Measures of Uncertainty

Although there have been significant contributions [75, 76, 143, 150, 151, 173] on measures of uncertainty recently, great challenges are still unresolved:

1. More investigations are required to attach meaning to those measures. The pragmatics has not been established yet. What those measures are really measuring?
2. Meeting a set of mathematical properties may not mean a meaningful translation to real-world systems.
3. Empirical studies such as that by Burkov et al. [177] need to be conducted on numerous scenarios in order to generalize results.

3.7 Conclusion

The three words information, uncertainty, and entropy have distinct meanings for many people in different domains. The sources and characterization of uncertainties in engineering situation analysis and decision support systems are quite important. The sources of uncertainty are generally categorized as either aleatory or epistemic.

Table 3.6 Requirements for a Measure of Total Uncertainty

Total Uncertainty (TU) requirements from Abellán et al. [143]	
P1	*Probabilistic consistency:* When all the focal elements of a basic probability assignment m are singletons, then a total uncertainty measure must be equal to the Shannon entropy: $TU(m)=\sum_{x\in X} m(x)\log m(x)$
P2	*Set consistency:* when a set A exists such that $m(A) = 1$, then a TU must collapse to the Hartley measure: $TU(m) = \log\lvert A\rvert$
P3	*Range:* The range of $TU(m)$ is $[0, \log\lvert X_i\rvert]$.
P4	*Subadditivity:* Let m be a basic probabilistic assignment on the space $X \times Y$, m_x, and m_y be its marginal basic probabilistic assignments on X and Y, respectively; then a TU must satisfy the following inequality: $TU(m)\leq TU(m_X)+TU(m_Y)$
P5	*Additivity:* Let m be a basic probabilistic assignment on the space $X \times Y$, m_x, and m_y be its marginal basic probabilistic assignments on X and Y, respectively, such that these marginals are not interactive $(m(A\times B)=m_X(A)\ \ m_X(B)$, with $A\subseteq X, B\subseteq Y$ and $m(C)=0$ if $C\neq A\times B)$; then a TU must satisfy the equality: $TU(m)=TU(m_X)+TU(m_Y)$
P6	*Monotonicity:* A total uncertainty measure must not decrease the total quantity of uncertainty in a situation where a clear decrease in information (increment of uncertainty) is produced. Formally, in evidence theory with belief *Bel* and plausibility *Pl*, $Pl_1(A)-Bel_1(A)<Pl_2(A)-Bel_2(A), \forall A\subseteq X \Rightarrow TU(m_1)<TU(m_2)$
Behavior requirements from Abellán et al. [143]	
RB1	The calculation of a TU should not be too complex.
RB2	A TU must not conceal the two types of uncertainty (conflict and non-specificity) coexisting in the evidence theory.
RB3	A TU must be sensitive to changes in evidence either directly or via its parts of conflict and nonspecificity.
RB4	The extension of a TU in DST on more general theories must be possible.

Uncertainties are characterized as epistemic, if one sees a possibility to reduce them by gathering more data or by refining models. Uncertainties are categorized as aleatory if one does not foresee the possibility of reducing them. Identifying epistemic uncertainties that have the potential of being reduced has a pragmatic value for the design of FIAT-based situation analysis and decision support systems.

References

[1] Floridi, L., "Semantic Conceptions of Information," *The Stanford Encyclopedia of Philosophy*, E. N. Zalta (ed.), http://plato.stanford.edu/archives/Spr2011/entries/information-semantic.

[2] Braman, S., "Defining Information: An Approach for Policymakers," *Telecommunications Policy,* Vol. 13, 1989, pp. 233–242.

[3] Losee, R. M., "A Discipline Independent Definition of Information," *Journal of the American Society for Information Science,* Vol. 48, 1997, pp. 254–269.

[4] Machlup, F., and U. Mansfield, *The Study of Information: Interdisciplinary Messages*, New York: John Wiley and Sons, 1984.

[5] Debons, A., and W. J. Cameron, *Perspectives in Information Science*, Dordrecht: Springer International Publishing, 1975.

[6] Hoffmann, E., "Defining Information: An Analysis of the Information Content of Documents," *Information Processing & Management,* Vol. 16, 1980, pp. 291–304.

[7] Shannon, C. E., N. Sloane, and A. D. Wyner, *Claude E. Shannon: Collected Papers*, New York: John Wiley & Sons, 1993.

[8] Shannon, C. E., and W. Weaver, "The Mathematical Theory of Information," University of Illinois Press, 1971.

[9] Adriaans, P., and J. van Benthem, *Philosophy of Information*, New York: North Holland, 2008.

[10] Lewis, D., *General Semantics*, New York: Springer, 1972.

[11] Hintikka, J., "Knowledge and Belief," Ithaca, NY: Cornell University Press, 1962.

[12] R. Fagin, et al., *Reasoning About Knowledge*, Cambridge, MA: MIT Press, 2003.

[13] Barwise, J., and J. Perry, "Situations and Attitudes," MIT Press, 1983.

[14] Dretske, F., "Knowledge and the Flow of Information," MIT Press, 1981.

[15] Bar-Hillel, Y., and R. Carnap, "Semantic Information," *The British Journal for the Philosophy of Science,* Vol. 4, 1953, pp. 147–157.

[16] Tarski, A., "The Semantic Conception of Truth and the Foundations of Semantics," *Philosophy and Phenomenological Research,* Vol. 4, 1944, pp. 341–376.

[17] Shannon, C. E., *The Mathematical Theory of Communications*, Urbana, IL: The University of Illinois Press, 1949.

[18] Barnum, H., "Entropy and Information Causality in General Probabilistic Theories," *New J. Phys.,* Vol. 12, 2010, p. 033024.

[19] Short, A. J., and S. Wehner, "Entropy in General Physical Theories," *New J. Phys.,* Vol. 12, 2010, p. 033023.

[20] Fisher, R. A., "Theory of Statistical Estimation," *Mathematical Proceedings of the Cambridge Philosophical Society*, 1925, pp. 700–725.

[21] Kolmogorov, A. N., *Foundations of the Theory of Probability*, 2nd ed. New York: Chelsea Publishing, 1956.

[22] Kolmogorov, A. N., "Three Approaches to the Quantitative Definition of Information," *Problems of Information Transmission,* Vol. 1, 1965, pp. 1–7.

[23] Li, M., and P. M. Vitányi, *An Introduction to Kolmogorov Complexity and Its Applications*, New York: Springer Science & Business Media, 2009.

[24] Parzen, E., *Modern Probability Theory and Its Applications*, New York: John Wiley & Sons, 1960.

[25] Poovendran, R., "Cyber-Physical Systems: Close Encounters Between Two Parallel Worlds [Point of View]," *Proc. of the IEEE,* Vol. 98, 2010, pp. 1363–1366.

[26] Bushev, M., *Synergetics: Chaos, Order, Self-Organization*, New York: World Scientific, 1994.

[27] Fuller, R. B., *Synergetics: Explorations in the Geometry of Thinking*, New York: Macmillan, 1975.

[28] Fuller, R. B., *Synergetics*, Washington DC: Pacific Tape Library, 1975.

[29] Haken, H., *Advanced Synergetics*, New York: Springer, 1983.

[30] Machado, J. T., B. Pátkai, and I. J. Rudas, *Intelligent Engineering Systems and Computational Cybernetics*, New York: Springer, 2009.

[31] Paritsis, N., "Social Systems, Their Observation and Description," 6th Congress Symposium on Sociocybernetics in European Systems Science Union, 2005.

[32] Schwaninger, M., and J. P. Ríos, "System Dynamics and Cybernetics: A Synergetic Pair," *System Dynamics Review,* Vol. 24, 2008, pp. 145–174.

[33] Tharumarajah, A., A. J. Wells, and L. Nemes, "Comparison of Emerging Manufacturing Concepts," *1998 IEEE International Conference on Systems, Man, and Cybernetics*, 1998.

[34] Vigo, R., "Representational Information: A New General Notion and Measure of Information," *Information Sciences,* Vol. 181, 2011, pp. 4847–4859.

[35] Gell Mann, M., and S. Lloyd, "Information Measures, Effective Complexity, and Total Information," *Complexity*, Vol. 2, 1996, pp. 44–52.

[36] Delahaye, J. -P., and H. Zenil, "Towards a Stable Definition of Kolmogorov-Chaitin Complexity," *Fundamenta Informaticae XXI* (2008), 1–15, IOS Press, 2008.

[37] Delahaye, J. -P., and H. Zenil, "Randomness and Complexity: Fron Leibniz to Chaitin," *World Scientific*, 2007

[38] Kamp, H., and M. Stokhof, "Information in Natural Language," *Handbook of the Philosophy of Information*, Vol. 8, P. Adriaans and J. van Bentham (eds.), New York: North-Holland, 2008, pp. 49–112.

[39] Baltag, A., H. van Ditmarsch, and L. Moss, "Epistemic Logic and Information Update," in *Philosophy of Information*, P. Adriaans and J. van Benthem, (eds.): New York: North-Holland, 2008.

[40] Devlin, K., and D. Rosenberg, "Information in the Study of Human Interaction," in *Philosophy of Information*, Vol. 8, P. Adriaans and J. van Benthem, (eds.), New York: North-Holland, 2008, pp. 685–710.

[41] Adriaans, P., "Learning and the Cooperative Computational Universe," in *Philosophy of Information*, Vol. 8, P. Adriaans and J. Van Benthem, (eds.), New York: North-Holland, 2008, pp. 133–167.

[42] Kelly, K., "Ockhams Razor, Truth, and Information," in *Philosophy of Information*, Vol. 8, P. Adriaans and J. Van Benthem, (eds.), New York: North-Holland, 2008, pp. 321–360.

[43] Rott, H., "Information Structures in Belief Revision," in *Philosophy of Information*, Vol. 8, P. Adriaans and J. Van Benthem, (eds.), New York: North-Holland, 2008, pp. 457–482.

[44] Abramsky, S., "Information, Processes and Games," in *Philosophy of Information*, Vol. 8, P. Adriaans and J. Van Benthem, (eds.), New York: North-Holland, 2008, pp. 483–549.

[45] van Benthem, J., and M. Martinez, "The Stories of Logic and Information," in *Philosophy of Information*, Vol. 8, P. Adriaans and J. Van Benthem, (eds.), New York: North-Holland, 2008, pp. 217–280.

[46] Walliser, B., "Information and Beliefs in Game Theory," in *Philosophy of Information*, Vol. 8, P. Adriaans and J. Van Benthem, (eds.), New York: North-Holland, 2008, pp. 551–580.

[47] Das, S., *Computational Business Analytics*, London, U.K.: Taylor & Francis, 2013.

[48] Schroeck, M., et al., "Analytics: The Real-World Use of Big Data," 2012. http:/www-935.ibm.com/services/us/gbs/thoughtleadership/ibv-big-data-at-work.html.

[49] Zikopoulos, P., and C. Eaton, *Understanding Big Data: Analytics For Enterprise Class Hadoop and Streaming Data*, New York: McGraw-Hill Osborne Media, 2011.

[50] Laursen, G. H. N., and J. Thorlund, *Business Analytics for Managers: Taking Business Intelligence Beyond Reporting*, New York: Wiley, 2010.

[51] Blasch, E., E. Bossé, and D. A. Lambert, *High-Level Information Fusion Management and Systems Design*: Norwood, MA: Artech House, 2012.

[52] Bossé, É., J. Roy, and S. Wark, *Concepts, Models, and Tools for Information Fusion*, Norwood, MA: Artech House, 2007.

[53] Waltz, E., *Knowledge Management in the Intelligence Enterprise*, Norwood, MA: Artech House, 2003.

[54] Liggins, M., D. Hall, and J. Llinas, *Handbook of Multisensor Data Fusion: Theory and Practice*, 2nd ed., New York: Taylor & Francis, 2008.

[55] Klein, G. A., et al., *Decision Making in Action: Models and Methods*, Norwood, NJ: Ablex Publishing Corporation, 1993.

[56] Floridi, L., *Philosophical Conceptions of Information*, New York: Springer, 2009.

[57] Adriaans, P., "A Critical Analysis of Floridi's Theory of Semantic Information," *Knowledge, Technology & Policy,* Vol. 23, 2010, pp. 41–56.

[58] Klüver, J., "A Mathematical Theory of Communication: Meaning, Information, and Topology," *Complexity,* Vol. 16, 2011, pp. 10–26.

[59] Crofts, A. R., "Life, Information, Entropy, and Time: Vehicles for Semantic Inheritance," *Complexity,* Vol. 13, 2007, pp. 14–50.

[60] Ahsan, S., and A. Shah, "Data, Information, Knowledge, Wisdom: A Doubly Linked Chain," *Proc. of 2006 Intl. Conf. on Information Knowledge Engineering*, 2006, pp. 270–278.

[61] Bernstein, J. H., "The Data-Information-Knowledge-Wisdom Hierarchy and Its Antithesis," *NASKO,* Vol. 2, 2011, pp. 68–75.

[62] Hey, J., "The Data, Information, Knowledge, Wisdom Chain: The Metaphorical Link," *Intergovernmental Oceanographic Commission,* 2004.

[63] DIKW Hierarchy, http://en.wikipedia.org/wiki/DIKW_Pyramid, last accessed: March 5, 2015

[64] Sommaruga, G., "Formal Theories of Information," *Lecture Notes in Computer Science,* Vol. 5363, 2009.

[65] Beim Graben, P., "Pragmatic Information in Dynamic Semantics," *Mind and Matter,* Vol. 4, 2006, pp. 169–193.

[66] Morris, C. W., *Foundations of the Theory of Signs, Vol. 1*, Chicago, IL: University of Chicago Press, 1938.

[67] Peirce, C. S., *Peirce on Signs: Writings on Semiotic*, Durham, NC: UNC Press Books, 1991.

[68] Ogden, C. K., et al., *The Meaning of Meaning*, New York: Harcourt, Brace & World, 1946.

[69] Uminsky, S., *Information Theory Demystified*, http://www.ideacenter.org/contentmgr/showdetails.php/id/1236, March 10, 2015.

[70] Dequech, D., "Uncertainty: A Typology and Refinements of Existing Concepts," *Journal of Economic Issues,* Vol. 45, 2011, pp. 621–640.

[71] Gray, R. M., *Entropy and Information Theory*, New York: Springer Science & Business Media, 2011.

[72] Pierce, J. R., *An Introduction To Information Theory: Symbols, Signals and Noise, Second Edition*, New York: Dover Publications, 1980.

[73] Lesne, A., "Shannon Entropy: A Rigorous Notion at the Crossroads Between Probability, Information Theory, Dynamical Systems and Statistical Physics," *Mathematical Structures in Computer Science,* Vol. 24, 2014, p. e240311.

[74] Kullback, S., and R. A. Leibler, "On Information and Sufficiency," *The Annals of Mathematical Statistics,* 1951, pp. 79–86.

[75] Liu, C., et al., "Reducing Algorithm Complexity for Computing an Aggregate Uncertainty Measure," *IEEE Trans. on Systems, Man and Cybernetics, Part A: Systems and Humans,* Vol. 37, 2007, pp. 669–679.

[76] Liu, C., et al., "Measures of Uncertainty for Fuzzy Evidence Theory," Technical Report DRDC-Valcartier TR2010-223, 2010.

[77] Liu, C., "A General Measure of Uncertainty-Based Information," Ph.D, Electrical and Computer Engineering, Université Laval, Québec, 2004.

[78] Garrido, A., "Classifying Entropy Measures," *Symmetry,* Vol. 3, 2011, pp. 487–502.

[79] Ben-Naim, A., *Entropy Demystified: The Second Law Reduced to Plain Common Sense*, New York: World Scientific, 2008.

[80] Bossé, É., A. Jousselme, and P. Maupin, "Knowledge, Uncertainty and Belief in Information Fusion and Situation Analysis," *NATO Science Series III Computer and Systems Sciences,* Vol. 198, 2005, pp. 61–81.

[81] Jousselme, A. -L., P. Maupin, and É. Bossé, "Uncertainty in a Situation Analysis Perspective," *Proc. of 6th Intl. Conf. of Information Fusion*, 2003.

[82] Han, P. K., W. M. Klein, and N. K. Arora, "Varieties of Uncertainty in Health Care A Conceptual Taxonomy," *Medical Decision Making,* Vol. 31, 2011, pp. 828–838.

[83] Ramirez, A. J., A. C. Jensen, and B. H. Cheng, "A Taxonomy of Uncertainty for Dynamically Adaptive Systems," *2012 ICSE Workshop on Software Engineering for Adaptive and Self-Managing Systems (SEAMS)*, 2012, pp. 99–108.

[84] Regan, H. M., M. Colyvan, and M. A. Burgman, "A Taxonomy and Treatment of Uncertainty for Ecology and Conservation Biology," *Ecological Applications,* Vol. 12, 2002, pp. 618–628.

[85] Tannert, C., H. D. Elvers, and B. Jandrig, "The Ethics of Uncertainty," *EMBO Reports,* Vol. 8, 2007, pp. 892–896.

[86] Faber, M., R. Manstetten, and J. L. Proops, "Humankind and the Environment: An Anatomy of Surprise and Ignorance," *Environmental Values,* Vol. 1, 1992, pp. 217–241.

[87] Smithson, M., *Ignorance and Uncertainty: Emerging Paradigms*, New York: Springer-Verlag Publishing, 1989.

[88] Bammer, G., and M. Smithson, *Uncertainty and Risk: Multidisciplinary Perspectives*, London: Earthscan, 2012.

[89] Camerer, C., and M. Weber, "Recent Developments in Modeling Preferences: Uncertainty and Ambiguity," *Journal of Risk and Uncertainty,* Vol. 5, 1992, pp. 325–370.

[90] Savage, L. J., The Foundations of Statistics, New York: Courier Corporation, 1972.

[91] Friedman, M., and L. J. Savage, "The Expected-Utility Hypothesis and the Measurability of Utility," *The Journal of Political Economy,* 1952, pp. 463–474.

[92] Galavotti, M. C., "The Notion of Subjective Probability in the Work of Ramsey and de Finetti," *Theoria,* Vol. 57, 1991, pp. 239–259.

[93] Hodgson, G. M., "The Ubiquity of Habits and Rules," *Cambridge Journal of Economics,* Vol. 21, 1997, pp. 663–684.

[94] Dubois, D., "Uncertainty Theories: A Unified View," in SIPTA school 08 - UEE 08, Montpellier, France, 2008. http://www.cost-ic0702.org/summercourse/files/uncertainty.pdf.

[95] Walley, P., *Statistical Reasoning with Imprecise Probabilities*, London, U.K.: Chapman and Hall, 1991.

[96] Walley, P., "Towards a Unified Theory of Imprecise Probability," *International Journal of Approximate Reasoning,* Vol. 24, 2000, pp. 125–148.

[97] Shafer, G., *A Mathematical Theory of Evidence*, Princeton, NJ: Princeton University Press, 1976.

[98] Dempster, A. P., "Upper and Lower Probabilities Induced by a Multivalued Mapping," *The Annals of Mathematical Statistics,* 1967, pp. 325–339.

[99] Smets, P., "Imperfect Information: Imprecision and Uncertainty," in *Uncertainty Management in Information Systems*, New York: Springer, 1997, pp. 225–254.

[100] Dubois, D., and H. Prade, *Possibility Theory: An Approach to Computerized Processing of Uncertainty*, New York: Plenum Press, 1988.

[101] Zadeh, L., "Fuzzy Sets as the Basis for a Theory of Possibility," *Fuzzy Sets and Systems,* Vol. 1, 1978, pp. 3–28.

[102] Dequech, D., "Uncertainty: Individuals, Institutions and Technology," *Cambridge Journal of Economics,* Vol. 28, 2004, pp. 365–378.

[103] Dubois, D., and H. Prade, "Formal Representations of Uncertainty," *Decision-Making Process: Concepts and Methods,* 2009, pp. 85–156.

[104] Endsley, M. R., and D. J. Garland, *Situation Awareness Analysis and Measurement*, Boca Raton, FL: CRC Press, 2000.

[105] Bouyssou, D., et al., *Decision Making Process: Concepts and Methods*, New York: John Wiley & Sons, 2013.

[106] Molchanov, I., *Theory of Random Sets*, New York: Springer Science & Business Media, 2006.

[107] Destercke, S., D. Dubois, and E. Chojnacki, "Unifying Practical Uncertainty Representations–I: Generalized P-Boxes," *International Journal of Approximate Reasoning,* Vol. 49, 2008, pp. 649–663.

[108] Destercke, S., D. Dubois, and E. Chojnacki, "Unifying Practical Uncertainty Representations: II. Clouds," *Int. J. of Approximate Reasoning*, 49 (2008) 664–677.

[109] Zadeh, L. A., "Fuzzy Sets," *Information and Control,* Vol. 8, 1965, pp. 338–353.

[110] Krause, P., and D. Clark, *Representing Uncertain Knowledge: An Artificial Intelligence Approach*, Boston, MA: Kluwer Academic Publishers, 1993.

[111] Bouchon-Meunier, B., and H. T. Nguyen, *Les incertitudes dans les systèmes intelligents*, Paris, France: Presses Universitaires de France, 1996.

[112] Klir, G., and B. Yuan, *Fuzzy Sets and Fuzzy Logic,* Vol. 4, Upper Saddle River, NJ: Prentice Hall, 1995.

[113] Klir. G. J., and M. J. Wierman, *Uncertainty-Based Information: Elements of Generalized Information Theory*, New York: Physica-Verlag HD, 1999.

[114] Ayyub, B. M., and G. J. Klir, *Uncertainty Modeling and Analysis in Engineering and the Sciences*, Boca Raton, FL: CRC Press, 2006.

[115] Klir, G. J., *Uncertainty and Information: Foundations of Generalized Information Theory*, New York: John Wiley & Sons, 2005.

[116] Grabisch, M., M. Sugeno, and T. Murofushi, *Fuzzy Measures and Integrals: Theory and Applications*, New York: Springer-Verlag, 2000.

[117] Zadeh, L. A., "Probability Measures of Fuzzy Events," *Journal of Mathematical Analysis and Applications,* Vol. 23, 1968, pp. 421–427.

[118] Nilsson, N. J., "Probabilistic Logic," *Artificial Intelligence,* Vol. 28, 1986, pp. 71–87.

[119] Fagin, R., and J. Y. Halpern, "Reasoning About Knowledge and Probability," *Journal of the ACM (JACM),* Vol. 41, March 1994, pp. 340–367.

[120] Fagin, R., et al., *Reasoning About Knowledge,* Vol. 4, Cambridge, MA: MIT Press, 1995.

[121] Halpern, J. Y., *Reasoning About Uncertainty*, Cambridge, MA: MIT Press, 2003.

[122] Halpern, J. Y., and R. Fagin, "Two Views of Belief: Belief as Generalized Probability and Belief as Evidence," *Artificial Intelligence,* Vol. 54, 1992, pp. 275–317.

[123] Halpern, J. Y., "An Analysis of First-Order Logics of Probability," *Artificial Intelligence,* Vol. 46, 1990, pp. 311–350.

[124] Bundy, A., "Incidence Calculus: A Mechanism for Probabilistic Reasoning," *Journal of Automated Reasoning,* Vol. 1, 1985, pp. 263–283.

[125] Bundy, A., *Incidence Calculus*, New York: John Wiley & Sons, 1992.

[126] Liu, W., A. Bundy, and D. Robertson, "On the Relations Between Incidence Calculus and ATMS," in *Symbolic and Quantitative Approaches to Reasoning and Uncertainty*, New York: Springer, 1993, pp. 249–256.

[127] Smets, P., "Varieties of Ignorance and the Need for Well-Founded Theories," *Information Sciences,* Vol. 57, 1991, pp. 135–144.

[128] Pearl, J., *Probabilistic Reasoning in Intelligent Systems Networks of Plausible Inference*, New York: Morgan Kaufmann Publishers, 1988.

[129] Pearl, J., "Fusion, Propagation, and Structuring in Belief Networks," *Artificial Intelligence,* Vol. 29, 1986, pp. 241–288.

[130] Lauritzen, S. L., and D. J. Spiegelhalter, "Local Computations with Probabilities on Graphical Structures and Their Application to Expert Systems," *Journal of the Royal Statistical Society. Series B (Methodological)*, 1988, pp. 157–224.

[131] Lepar, V., and P. P. Shenoy, "A Comparison of Lauritzen-Spiegelhalter, Hugin, and Shenoy-Shafer Architectures for Computing Marginals of Probability Distributions," *Proc. of 14th Conf. on Uncertainty in Artificial Intelligence*, 1998, pp. 328–337.

[132] Jensen, F. V., *An Introduction to Bayesian Networks* Vol. 210, London, U.K.: UCL Press, 1996.

[133] Nielsen, T. D., and F. V. Jensen, *Bayesian Networks and Decision Graphs*, New York: Springer Science & Business Media, 2009.

[134] Gordon, J., and E. H. Shortliffe, "A Method for Managing Evidential Reasoning in a Hierarchical Hypothesis Space," *Artificial Intelligence,* Vol. 26, 1985, pp. 323–357.

[135] Shafer, G., and R. Logan, "Implementing Dempster's Rule for Hierarchical Evidence," *Artificial Intelligence,* Vol. 33, 1987, pp. 271–298.

[136] Shenoy, P. P., and G. Shafer, "Axioms for Probability and Belief-Function Propagation," in *Classic Works of the Dempster-Shafer Theory of Belief Functions*, New York: Springer, 2008, pp. 499–528.

[137] Shenoy, P. P., "Valuation-Based Systems for Bayesian Decision Analysis," *Operations Research,* Vol. 40, 1992, pp. 463–484.

[138] Shenoy, P. P., "Binary Join Trees for Computing Marginals in the Shenoy-Shafer Architecture," *International Journal of Approximate Reasoning,* Vol. 17, 1997, pp. 239–263.

[139] Shenoy, P. P., and G. Shafer, "Propagating Belief Functions with Local Computations," *IEEE Expert,* Vol. 1, 1986, pp. 43–52.

[140] Pearl, J., *Causality: Models, Reasoning and Inference,* Vol. 29, Cambridge, U.K.: Cambridge University Press, 2000.

[141] Ghosh, S., *Algorithm Design for Networked Information Technology Systems*, New York: Springer, 2004.

[142] Kikuchi, S., and V. Perincherry, "Handling Uncertainty in Large Scale Systems with Certainty and Integrity," 2004. http://esd.mit.edu/symposium/pdfs/papers/kikuchi.pdf.

[143] Abellán, J., and A. Masegosa, "Requirements for Total Uncertainty Measures in Dempster-Shafer Theory of Evidence," *International Journal of General Systems,* Vol. 37, 2008, pp. 733–747.

[144] De Campos, L. M., J. F. Huete, and S. Moral, "Probability Intervals: A Tool for Uncertain Reasoning," *International Journal of Uncertainty, Fuzziness and Knowledge-Based Systems,* Vol. 2, 1994, pp. 167–196.

[145] Grabisch, M., and C. Labreuche, "A Decade of Application of the Choquet and Sugeno Integrals in Multi-Criteria Decision Aid," *Annals of Operations Research,* Vol. 175, 2010, pp. 247–286.

[146] Abellán, J., and S. Moral, "Upper Entropy of Credal Sets. Applications to Credal Classification," *International Journal of Approximate Reasoning,* Vol. 39, 2005, pp. 235–255.

[147] Couso, I., and L. Sánchez, "Upper and Lower Probabilities Induced by a Fuzzy Random Variable," *Fuzzy Sets and Systems,* Vol. 165, 2011, pp. 1–23.

[148] Jaffray, J. -Y., and F. Philippe, "On the Existence of Subjective Upper And Lower Probabilities," *Mathematics of Operations Research,* Vol. 22, 1997, pp. 165–185.

[149] Suppes, P., and M. Zanotti, "On Using Random Relations to Generate Upper and Lower Probabilities," *Synthese,* Vol. 36, 1977, pp. 427–440.

[150] Abellán, J., G. Klir, and S. Moral, "Disaggregated Total Uncertainty Measure for Credal Sets," *International Journal of General Systems,* Vol. 35, 2006, pp. 29–44.

[151] Abellan, J., and S. Moral, "Maximum of Entropy for Credal Sets," *International Journal of Uncertainty, Fuzziness and Knowledge-Based Systems,* Vol. 11, 2003, pp. 587–597.

[152] Abellán, J., and S. Moral, "Difference of Entropies as a Non-Specificity Function on Credal Sets," *International Journal of General Systems,* Vol. 34, 2005, pp. 201–214.

[153] Karlsson, A., "Evaluating Credal Set Theory as a Belief Framework in High-Level Information Fusion for Automated Decision-Making," 2008. Universit of Skovde Technical Report 2010. http:/www.diva-portal.org/smash/get/diva2:345982/FULLTEXT03.

[154] Liu, Z. -G., et al., "Credal Classification Rule for Uncertain Data Based on Belief Functions," *Pattern Recognition,* Vol. 47, 2014, pp. 2532–2541.

[155] Zadeh, L. A., "Toward a Theory of Fuzzy Information Granulation and Its Centrality in Human Reasoning and Fuzzy Logic," *Fuzzy Sets and Systems,* Vol. 90, 1997, pp. 111–127.

[156] Smets, P., "The Degree of Belief in a Fuzzy Event," *Information Sciences,* Vol. 25, 1981, pp. 1–19.

[157] Yager, R. R., "Generalized Probabilities of Fuzzy Events from Fuzzy Belief Structures," *Information Sciences,* Vol. 28, 1982, pp. 45–62.

[158] Yen, J., "Generalizing the Dempster-Schafer Theory to Fuzzy Sets," *IEEE Trans. on Systems, Man and Cybernetics,* Vol. 20, 1990, pp. 559–570.

[159] Lucas, C., and B. N. Araabi, "Generalization of the Dempster-Shafer Theory: A Fuzzy-Valued Measure," *IEEE Trans. on Fuzzy Systems,* Vol. 7, 1999, pp. 255–270.

[160] Hartley, R. V., "Transmission of Information," *Bell System Technical Journal,* Vol. 7, 1928, pp. 535–563.

[161] Dubois, D., and H. Prade, "A Note on Measures of Specificity for Fuzzy Sets," *International Journal of General System,* Vol. 10, 1985, pp. 279–283.

[162] Yager, R. R., "Entropy and Specificity in a Mathematical Theory of Evidence," *International Journal of General System,* Vol. 9, 1983, pp. 249–260.

[163] Höhle, U., "Entropy with respect to plausibility measures," *Proc. of 12th IEEE Intl. Symp. on Multiple-Valued Logic*, 1982, pp. 167–169.

[164] Klir, G. J., and A. Ramer, "Uncertainty in the Dempster-Shafer Theory: A Critical Re-Examination," *International Journal of General System,* Vol. 18, 1990, pp. 155–166.

[165] Klir, G. J., and B. Parviz, "A Note on the Measure of Discord," *Proc. of 8th Intl. Conf. on Uncertainty in Artificial Intelligence*, 1992, pp. 138–141.

[166] Lamata, M. T., and S. Moral, "Measures of Entropy in the Theory of Evidence," *International Journal of General System,* Vol. 14, 1988, pp. 297–305.

[167] Klir, G. J., and R. M. Smith, "On Measuring Uncertainty and Uncertainty-Based Information: Recent Developments," *Annals of Mathematics and Artificial Intelligence,* Vol. 32, 2001, pp. 5–33.

[168] Klir, G. J., and R. M. Smith, "Recent Developments in Generalized Information Theory," *International Journal of Fuzzy Systems,* Vol. 1, 1999, pp. 1–13.

[169] Pal, N. R., J. C. Bezdek, and R. Hemasinha, "Uncertainty Measures for Evidential Reasoning II: A New Measure of Total Uncertainty," *International Journal of Approximate Reasoning,* Vol. 8, 1993, pp. 1–16.

[170] Maeda, Y., H. T. Nguyen, and H. Ichihashi, "Maximum Entropy Algorithms for Uncertainty Measures," *International Journal of Uncertainty, Fuzziness and Knowledge-Based Systems,* Vol. 1, 1993, pp. 69–93.

[171] Harmanec, D., and G. J. Klir, "Measuring Total Uncertainty in Dempster-Shafer Theory: A Novel Approach," *International Journal of General System,* Vol. 22, 1994, pp. 405–419.

[172] Jousselme, A. -L., et al., "Measuring Ambiguity in the Evidence Theory," *IEEE Trans. on Systems, Man and Cybernetics, Part A: Systems and Humans,* Vol. 36, 2006, pp. 890–903.

[173] Abellán, J., and S. Moral, "Measuring Total Uncertainty in Dempster-Shafer Theory of Evidence: Properties and Behaviors," *Annual Meeting of the North American Fuzzy Information Processing Society 2008 (NAFIPS 2008)*, 2008, pp. 1–6.

[174] Shahpari, A., and S. A. Seyedin, "Measuring Mutual Aggregate Uncertainty in Evidence Theory," *7th Intl. Symp. on Telecommunications (IST)*, 2014, pp. 12–19.

[175] Shahpari, A., and S. A. Seyedin, "Using Mutual Aggregate Uncertainty Measures in a Threat Assessment Problem Constructed by Dempster-Shafer Network, in a Threat Asses-

ment Problem Constructed by Dempster-Shafer Network," *Systems, Man, and Cybernetics: Systems, IEEE Transactions On*, Vol. 45, 2015, pp. 877–886.

[176] Klir, G. J., and H. W. Lewis III, "Remarks on 'Measuring Ambiguity in the Evidence Theory,'" *IEEE Trans. on Systems, Man and Cybernetics, Part A: Systems and Humans*, Vol. 38, 2008, pp. 995–999.

[177] Burkov, A., et al., "An Empirical Study of Uncertainty Measures in the Fuzzy Evidence Theory," *Proc. of 14th Intl. Conf. on Information Fusion (FUSION)*, 2011, pp. 1–8.

[178] Croissant, J. L., "Agnotology: Ignorance and Absence or Towards a Sociology of Things That Aren't There," *Social Epistemology*, Vol. 28, 2014, pp. 4–25.

[179] Proctor, R., and L. L. Schiebinger, *Agnotology: The Making and Unmaking of Ignorance*, Stanford, CA: Stanford University Press, 2008.

CHAPTER 4

Information Characterization and Representations

The purpose of this chapter is to review fundamental concepts and define new ones on information characterization and uncertainty representations in view of being exploited in the design of intelligent situation analysis and decision support systems based upon fusion of information and analytics technologies (FIAT). This chapter gives a framework for assessing the quality and reliability of information available and a base for assessing the resulting decisions. It parallels the experimental framework of physics, with equipment calibration and uncertainties in observations and measurements.

4.1 Introduction

The perspective taken in this chapter is that of processing pieces of information with analytics and information fusion methods. That perspective imposes a need to characterize and represent those pieces of information. For instance, fusion is the process of combining information from different sources to provide robust and complete description of environment or process of interest. Fusion is of special significance in applications where large data sets must be combined, fused, and distilled to obtain information of appropriate quality and integrity on which decisions can be made. It is not restricted to data set management and ought to be considered to perform inferences, handle redundancies, and exploit complementarities. Fusion is generally envisioned within complex systems that generally dispose of knowledge sources (e.g., data, a priori knowledge) grouped within associated knowledge bases. Enabling a fusion capability implies combining or fusing data from multiple sources to achieve a predefined objective. The basic concepts of data/information fusion first appeared in the literature is the 1960s when mathematical models for data manipulation like the Bucy-Kalman filter [1] were developed and implemented in different application fields: defense, robotics, pattern recognition, remote sensing, aerospace systems, and medical.

In the literature, several definitions related to information fusion can be found. The most accepted definition is the JDL one discussed in the previous chapter. It can be simply stated as [2]: “Data fusion is the process of combining data or information to estimate or predict entity states.” Thus, fusion consists on the development of methods that allow the extraction and conciliation, from different knowledge sources, of meaningful elements for decision-making. A more general formulation from Wald [3] presents information fusion as a "formal framework that expresses the means and tools for the alliance of data originating from different sources; it aims at obtaining information of greater quality, the exact significance of 'greater quality' depends upon the considered application." Notice that in Wald's definition the focus is shifted towards the quality of the fusion result including all aspects of interest as the technical parameters, the performance, the effectiveness, and the user benefit. The major part of information fusion literature and efforts focuses its attention on:

1. Information fusion techniques and algorithms for integrating data from homogeneous or heterogeneous collections of potentially faulty, redundant, or inaccurate sensors;
2. Structural and functional models implementing information fusion systems.

Nevertheless, and besides the huge amount of the conducted research activities, information fusion is still a relatively new research field where the basic methodology is still under construction. Since fusion is interdisciplinary and independently growing out of various research fields, many basic definitions and terminology lack higher-level coherence. For instance and generally speaking, the terms sensor fusion, data fusion, multisensor data fusion, information integration, and information fusion have been used in various publications without much discrimination. This is possibly due to various meanings associated to the word "information" as discussed in Chapter 3. Studies have been discussing the differences between data, information, and knowledge for years, but the results are still controversial. These studies tried to find the differences among concepts properties/characteristics. Input can take different meanings according to the considered situation. The statement that "Tokyo is the capital of Japan" can be data when stored in a database. It can be information when told to a person. It can be knowledge of a person who knows it. This suggests that an input, in itself, has no unique meaning. Data, information, and knowledge are hence most likely context-dependent representations.

This chapter proposes to lift the fog around input characterization by distinguishing their intrinsic representation and their positioning within higher-level structures/processes (e.g., actions: to transform, to inform, and to know). Internal structure includes attributes, objects definitions, relationships, and behaviors. The higher-level positioning considers a system's input contextual role attribution for data, information, and knowledge. This chapter is organized into sections addressing the following topics on information element structural representation, information functional objectives and its knowledge processing aspects, information elements characterization and imperfections modeling, the concept of fusion cells, information fusion system's components building blocks, and finally on a global and homogeneous fusion framework.

4.2 Definition of an Information Element

The success of information fusion systems is strongly related to the way its basic components are defined, the quality of their associated knowledge as well as the information or knowledge produced by the fusion process itself. Unfortunately, there is a relatively small body of literature in the fusion community that addresses this topic. Moreover, there is no agreement on what information is and what defines its characteristics as discussed in the previous chapter.

An information definition should capture its nature and allows frameworks, theories and results to be transferred across disciplinary boundaries [4]. Whitaker [5] defined information as: "data that is relevant to the considered application" and stated the same definition and principles can automatically be repeated for information fusion. According to Losee [6], information can be defined in terms of a process or a function: "Information is the value currently attached or instantiated to a characteristic or variable returned by a function, or a process. The value returned by a function is informative about the function's argument, or about the function, or about both." This definition has the merit of positioning information as a relational or functional concept linking data sets [7].

For instance, consider the following statements that are potential system's inputs. First, state a given value of $x = 41$. There is no more to understand from x than it is a positive integer. Second, consider the statement $x = 41°C$. It denotes an element from a set of meaningful semantics. The value is related to temperature. Given that 41°C is the temperature of somebody, one can start reasoning about it and conclude a fever and a need to be examined by a medical doctor.

This example highlights that observing data from a given set is not enough to make it an informative entity. Information hence requires a content set, how its outcome is obtained, and what it refers to. In this way, Stonier [8] highlighted that the full information extent cannot be grabbed when working only on the contents set. This leads to the following definition of information [7, 9], which extends the definition proposed by Losee [6] by including a formal structure associated with the previous information characteristics. Definition, an information element, is: "an entity composed of a definition set and a content set linked by a functional relationship called informative relation, associated with internal & external contexts."

Therefore, as shown in Figure 4.1, the main components of an information element are:

1. A definition set, giving the potential information input elements (what the information refers to);
2. A contents set encoding the possible knowledge produced by the information (e.g., measurements or estimations of physical parameters, decisions, hypothesis);
3. An input-output relationship representing the functional link model (e.g., mathematical, physical) that associates the input elements with the produced information contents;
4. An internal context gathering intrinsic characteristics, constraints, or controls about the informative relation itself;
5. An external context containing data, information, or knowledge useful to the elaboration of the meaning or the interpretation of the information element.

The information element (J) is denoted as: J = (*Information definition set, Informative relation, Information content set, Internal Context, External Context*).

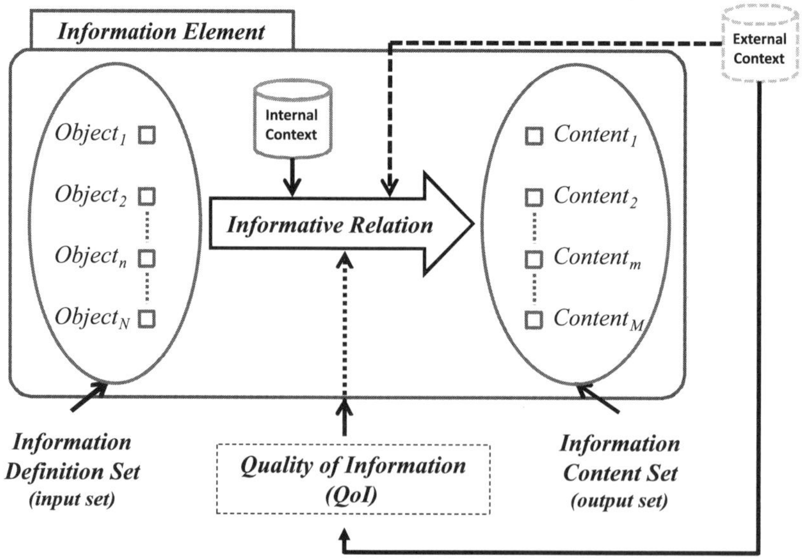

Figure 4.1 A basic information element (*J*) structure.

Based on this definition, the information element's sets could include as well data than knowledge. Its structure is thus generic enough to be considered as information-base building block.

4.2.1 Objects, Contents, and Informative Relation

The meaning of these (information element) components is rather straightforward. Objects and contents represent entities linked through the informative relation. However, the nature of these entities may be either hard (quantitatively defined with numbers, individuals, and so forth) or soft (qualitatively defined with words, opinions, predictions, and so forth) as described in Table 4.1. While hard entities appear to be rather self-contained, soft entities obviously require a context in which the qualitative descriptors are defined. An informative relation may be purely impersonal (or hard) when it does not depend on external conditions to link objects to contents. However, it may also behave in a softer way by using cognitive factors such as subjective judgments, opinions, and perception [e.g., (*J*) representing human experts outputs]. To deal with both kinds of entities or relations, (*J*) must embed context to be fully defined [10].

4.2.2 Internal/External Contexts

A simple example may help making the difference between internal and external contexts. For instance, when the informative relation stands for a sensor making an acquisition, the actual setting values of the sensor are part of the internal context (these settings intrinsically control the content outputs). However, the reasons (conditions) why the sensor has been set up with these settings belong to the external context, including the context of observation. Indeed, the context of observation is an important part of the perception process as it is all that can have an influence on the perception of an event and all that is needed to understand an observation (or

Table 4.1 Hard and Soft Data Characteristics

	Hard Data	*Soft Data*
Nature	Hard information is quantitative: numbers (in finance, these are balance sheet data, asset returns); it is also rather backward-looking (e.g., balance sheet data).	Soft information is qualitative: words (opinions, ideas, projects, comments); it is rather forward-looking (e.g., business plan).
Collecting method	Collection of hard information is impersonal, and it does not depend upon the context of its production; hard information is therefore exhaustive and explicit.	Collecting soft information is personal and includes its production and treatment context.
Cognitive factors	Absent	Subjective judgment, opinions, and perception are integral components of soft information.

Source: [10].

a message). Therefore, different situations may be perceived, relying on the same set of sensory information items, if they are interpreted within different contexts of observation. To let computers be somehow intelligent, one has to provide them with this context for a specific domain and a specific aim. Part of this context is the domain knowledge that every human uses to interpret and understand any perception.

4.2.3 Illustrations of (*J*) at Signal and Data Levels of Abstraction

Consider a digital image as a two-dimensional array of gray levels. What is considered is not information but a kind of abstract data (i.e., a geometrically organized set of pixels with associated values). This abstract data fits the information element paradigm when associated with the basic objects that are observed (or imaged). This association is instantiated by the physical model expressing the process used by the imaging sensor. In Figure 4.2, two image information elements of an area north of Quebec City are given using two imaging modalities (i.e., two distinct informative physical relations: radar imaging and optical panchromatic imaging). The two informative relations operate on the same definition set (i.e., resolution cells) as the two sensors have almost the same resolutions when acquiring images (one pixel covers a 30m by 30m square on the ground). Using different sensors' resolution, separate definition sets would have been required to deal with the variable size of the observed ground pieces.

To be able to link objects (input set) to contents (output set), an informative relation embeds knowledge, allowing the links to be built. The extent and complexity of the involved knowledge depend on the information element. However, the existence of this embedded knowledge makes the difference between data itself and information elements. For example, imagine a basic resolution cell (i.e., an input object) is observed as having a zero gray level. The interpretation of this cell in terms of its thematic contents (i.e., giving a semantic meaning to the observed gray level) cannot be conducted if the sensor's physical model is not known. If the sensor is an imaging radar (e.g., active system), then a resolution cell zero-value

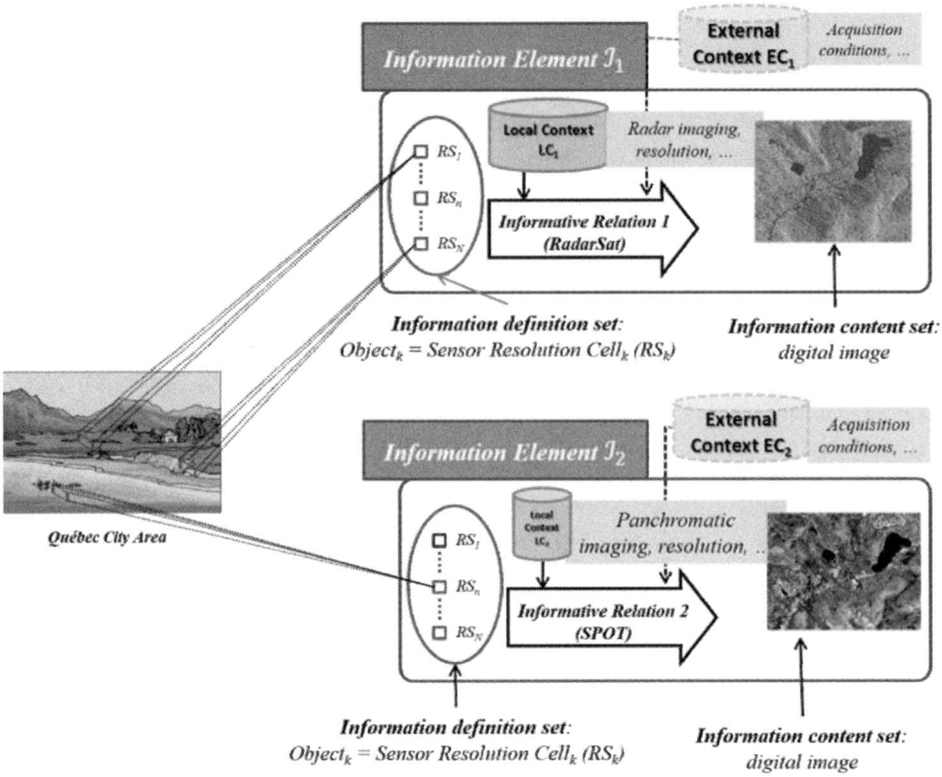

Figure 4.2 Remote sensing example giving two *J*s and two informative relations.

content can be interpreted as being a flat surface or may correspond to a shadow area. However, if the imaging sensor corresponds to a given spectral band in multi-spectral imaging (passive system, for example, visible, infrared), then the resolution cell thematic content corresponds to the real-world specific content absorbing the emitted electromagnetic signals within the considered spectral band. From this example, one can see that the physical model (i.e., informative relation) linking inputs (e.g., gray values) and outputs (e.g., content interpretation) makes the essential difference between data, herein the inputs only, and information.

The informative relation embedded knowledge extent and complexity depend on the information element. For instance, in image processing it can be illustrated by the use of different processing techniques. To this effect, consider the difference between the classical image filtering and the filtering adapted to speckled images (e.g., radar, sonar, ultrasound imaging). The last requires knowledge with more complexity, in this case corresponding to the mathematical modeling of the physical process and measure parameters exploited by the imaging sensors (i.e., a multiplicative Rayleigh noise). A higher level of knowledge complexity might thus be associated to references on systems intelligence. This taxonomy is widespread in the fields of scene interpretation, information fusion, and data/knowledge mining. Within the information element framework, intelligent system terminology shall thus be interpreted as an indicator of the informative function refinement/complexity.

Figure 4.3 shows two *J*s based on an imaging example. The two *J*s clearly do not belong to the same level of abstraction. The top one (signal level) links scene parts to pixels while the bottom one (data level) quantifies those pixels. It is fundamental to note that as the level of abstraction increases, the extent of *J* contexts widens as well. Thus, *J* at the data level encompasses within its contexts all the contexts of the underlying signal level *J*: the informative relation that extracts features from the image (the definition set) probably needs to know the resolution of the sensor that produced the image pixels. That information belongs to the signal level context (LC). This example illustrates the growing extent of *J* contexts as *J* level of abstraction increases.

4.2.4 Illustrations of (*J*) at Higher Levels of Abstraction

Most researchers agree that there is a difference between data and information even though it is not precise where the boundaries are within a specific context (see the lengthy discussion in [11]). Higher levels of information abstraction may origin from processed sensor data, surveillance and monitoring reports, human intelligence reports, operational information, and information obtained from open sources (Internet, papers, radio, TV). High-level information, which is also called soft data, has recently been reported by the information fusion community [12]. Soft data items are observations that are generated by humans. They may be as unconstrained natural language [13, 14] and expressed through textual data or speech signal, but can also made of semiconstrained data items such as XML files

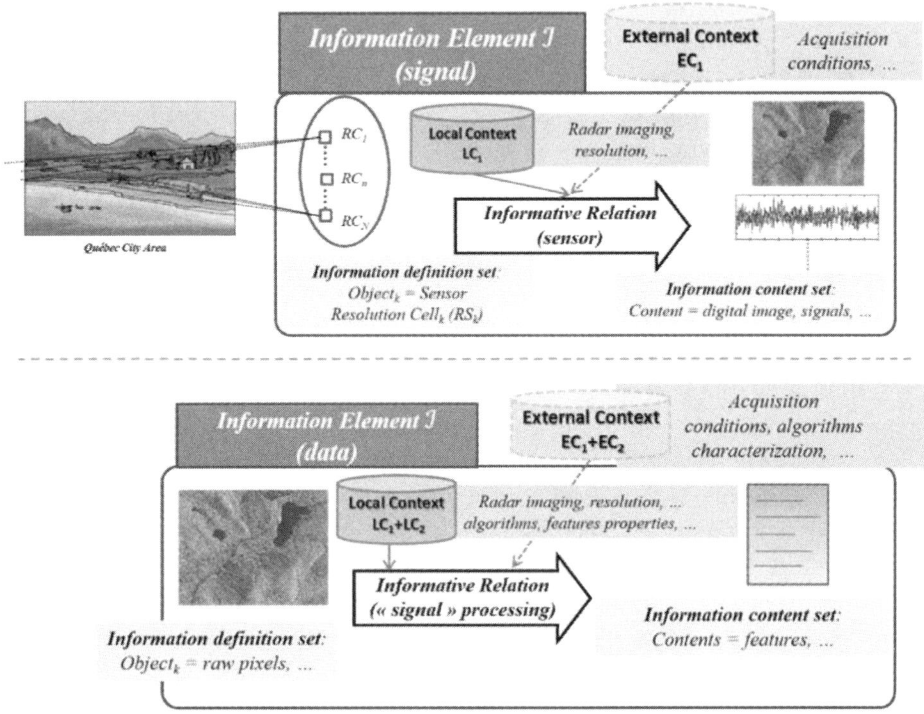

Figure 4.3 Generic information elemnts for two levels of abstraction (signal, data).

or databases. Graph structures are often used to store high-level information as well. Most electronic sensors are feature-based and do not generally provide information on relationships. Laudy [10, 15] used the graph-matching (conceptual graphs) approach to fuse hard and soft information. Najgebauer et al. [16] used a semantic graph as storage for facts and events to evaluate a terrorist threat. Even if such algorithms and representations seem heterogeneous, they reflect various possible implementations for information elements components (informative relation, internal and external contexts) for different levels of abstractions and applications. The information element structure itself remains unchanged and supports the various levels of abstraction.

4.2.5 Basic General Properties of an Information Element

The presented information element definition and structure are extremely general and can be applied to all types of encountered information elements: sensor-issued information, data transformation information, feature extraction information, and decisional information. However, some basic characteristics can be defined:

1. *Exhaustivity:* An information element is called exhaustive if and only if the content set contains all possible outcomes produced by the informative relation. This property, related to the information content set, is also called the closed-world assumption. Otherwise, the information is said to be operating under the open-world assumption.
2. *Exclusivity:* Exclusivity is another important property used to characterize an information element. In fact, an information element is called exclusive if and only if two different information contents cannot be simultaneously produced as the outcome of the informative relation.
3. *Incompleteness:* Positioning explicitly the informative relation within the basic information element structure clarifies the concept of information incompleteness at all levels of abstraction (data, information, knowledge or physical, informative, cognitive). Single information does not capture all relevant aspects of a phenomenon, an entity, or an input object. This is obvious for a sensor-based information element because a sensor specifically makes use of a specific physical process to extract one or several facets of the observed objects. This clearly means that the huge amount of informative aspects of an observed object will be restricted to those acquired through the physical window of the considered sensor. Examples can be brought for the informative and cognitive levels. For instance, at the knowledge level as defined by Plato as being "justified truth belief," incompleteness manifests as degrees of truth, belief, membership, relevance, and so forth.

4.3 Quality of Information (QoI)

The subject of information and data quality has been receiving significant attention in the recent years in many areas including communications, business processes, personal computing, health care, and databases. At the same time, the problem of information quality in the fusion-based human-machine systems for decision-making has attracted less attention. The main body of the literature on information fusion concerns with building an adequate uncertainty model without paying much attention to the problem of representing and incorporating other quality characteristics into fusion processes. Many relevant research questions, related to the information quality problem in the designing fusion-based systems, are raised including [17]:

- What is the ontology of quality characteristics?
- How do we assess the information quality of incoming heterogeneous data as well as the results of processes and information produced by users?
- How do we combine quality characteristics into a single quality measure?
- How do we evaluate the quality of the quality assessment procedures and results?
- How do we compensate for various information deficiencies?
- How do quality and its characteristics depend on the context?
- How does subjectivity (i.e., user biases) affect information quality?

Nevertheless, there is no unique definition of information quality. In fact, several definitions of information quality are available in the literature:

1. "Quality is the totality of characteristics of an entity that bear on its ability to satisfy stated and implied needs" [18].
2. "Quality is the degree to which information has content, form, and time characteristics, which give it value to specific end users" [19].
3. "Quality is the degree to which information is meeting user needs according to external, subjective user perceptions" [20].
4. "Quality is fitness for use" [21].

While having different emphases, all of these definitions point to the fact that information quality is a user-centric notion and needs to be measured in terms of the potential or actual value for the users. In the human system context, users can be either humans or automated agents and models. QoI is information about information, or meta-information, and the best way of representing and measuring the value of this meta-information is through its attributes since "without clearly defined attributes and their relationships, we are not just unable to assess QoI; we may be unaware of the problem" [22].

These attributes have to be considered in relations to specific user objectives, goals, and functions in a specific context (internal/external contexts in Figure 4.1). Due to the fact that all users, whether human, semiautomated, or automatic processes, have different data and information requirements, the set of attributes considered and the level of quality considered satisfactory vary with the user's perspective, the type of the models, algorithms, and processes comprising the system. Therefore, the general ontology designed to identify possible attributes and relations between them for a human-machine integrated system will require instantiation in every particular case. Considerable research targeting to study and to classify various aspects of quality of information has been conducted in both civilian and military domains. The following sections are devoted to detail this particular issue.

4.3.1 The NATO SAS Command and Control (C2) Conceptual Reference Model

In military domain, a very good representative of that body of research is the efforts conducted under the NATO Research and Technology Office Research Group SAS-050 to define a conceptual reference model of command and control (C2) [23–25]. The SAS-050 conceptual reference C2 model consists of a set of key variables and relationships to understanding C2. Among those sets of variables is a set of attributes related to information quality, pictured in Figure 4.4, which is believed to impact directly the quality of decisions. It consists of twelve attributes or variables including information accuracy, information completeness, information consistency, information correctness, information currency, information precision, information relevance, information timeliness, information uncertainty, information service characteristics, information sharability, and information source characteristics.

Key to decisions is information. Within the NATO C2 reference model, quality of information is directly linked with the quality of decisions as shown in Figure 4.4. The quality of information and decisions impact directly trustworthiness, survivability, and dependability in systems. Three concepts that are roughly equivalent in goals and addressing similar threats [26, 27]. Dependability [26] in systems, especially in cyber physical systems, is a huge and very challenging domain of application. The detailed description and system implementation workings of these concepts are beyond the scope of this book. There are numerous other books available that explain these concepts in greater details down to the mathematical description by means of equations, statistical formulae, and methods [28–30].

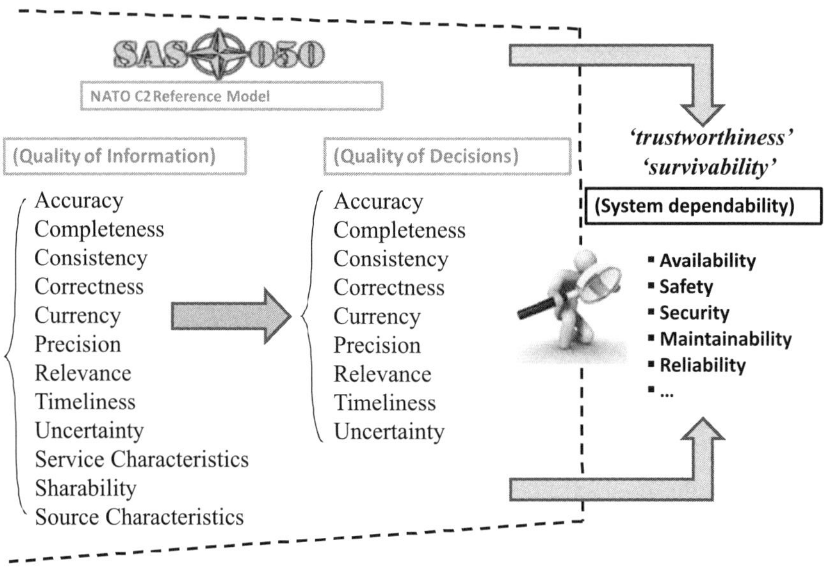

Figure 4.4 QoI within the NATO C2 reference model (SAS-050).

4.3.2 QoI Framework of Wang and Strong

In the civilian domain, one of the most cited frameworks for QoI is, from [20], reproduced in Figure 4.5. QoI is defined in terms of certain characteristics of the data itself called data quality dimensions. The authors have classified QoI into four major categories illustrated in Figure 4.5: intrinsic, contextual, representational, and accessibility. The intrinsic level characterizes an information element in terms of its nature, interpretation, and modeling while considering it as independent entity, out of any global fusion context. Conversely, the contextual level characterizes an information element in terms of its impact, completeness, relevance, conflict, and redundancy within a global fusion context. The representational level addresses the formal representation of the problem (and its context). It encompasses both the inner element representation (e.g., discrete versus continuous output set) and the semantic alignment that allows an information element to be considered as either input or output for the fusion process (i.e., to be involved in a processing elements flow). The accessibility level characterizes information on that accessibility. Accessibility might be physical (e.g., with regard to a sensor's availability) as well as organizational (e.g., holding the right to access and process specific information elements). The intrinsic facet concerns information characteristics that are independent of the application context in which this information is exploited. On the other side, the contextual facet gathers characteristics allowing the understanding of the information usefulness and its informative contribution (its impact, completeness, relevance, conflict, redundancy) when integrated into a larger information processing context. Representational and accessibility emphasize the importance of the role of systems. There is almost a direct concordance with the framework of [20] (Figure 4.5) and

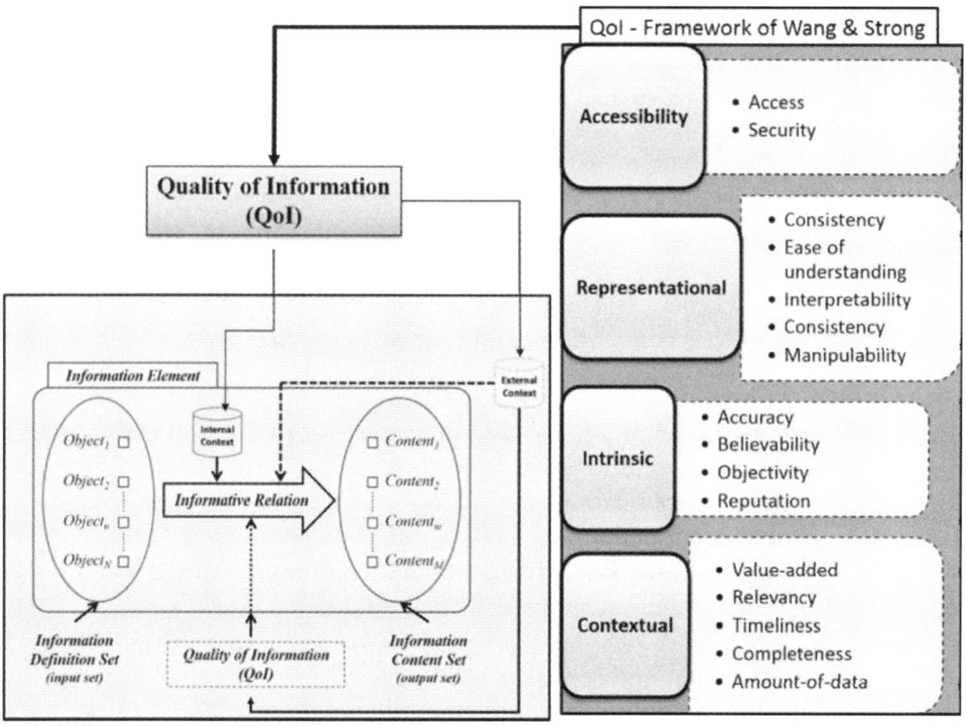

Figure 4.5 QoI framework of Wang and Strong versus an information element.

the NATO SAS-050 (Figure 4.4). QoI criteria, identified through contextual system analysis, impact the informative relation of an information element from the management functions of its internal and external contexts.

4.3.3 On Defining Ontologies of QoI

There have been multiple views on information quality ontologies, identifying quality attributes and classifying them into broad categories and relations. In [20], data quality was classified into four categories: intrinsic, contextual, representational, and accessibility. In [31], three categories were enumerated: pragmatic, semantic, and syntax while in [22], four sets were identified: integrity, accessibility, interpretability, and relevance. In [32], they used relevance, reliability, completeness, and uncertainty. In other words, there is no clear understanding of what dimensions define information quality from the perspective of information fusion process designers and how different dimensions defining information quality are interrelated. The information quality ontology presented in [17, 33] was one of the first attempt to fill this gap.

More recently, significant efforts [34–38] have been pursued to address the evaluation of techniques of uncertainty reasoning and to define an ontology called URREF (Uncertainty Representation and Reasoning Evaluation Framework) under the International Society of Information Fusion (ISIF) Evaluation of Technologies for Uncertainty Representation Working Group (ETURWG). These efforts contribute towards the understanding of QoI and the formalization of some criteria that could be used for the integration in the information element structure of Figure 4.1. However, research is still ongoing and requires additional efforts to get a QoI ontology. In fact, from all the efforts just mentioned, there are mainly four main aspects of information quality that offer formalizations exploitable in computer-based support systems, namely on, uncertainty, reliability, completeness, and relevance as summarized in Figure 4.6. Some of these formalizations will be described in the following sections.

4.4 Information Element Characterization: Intrinsic Dimensions

Information element (J) characterization is crucial for any further use. The characterization is done considering the following fundamental axes:

1. Information inner sets scope (e.g., resolution and level of abstractions): how the definition and the content sets of J are built (level of details, problem representation, and so forth);
2. Informative relation arity (in logic, mathematics, and computer science, the arity of a function or a relation is the number of arguments or operands that the function takes): how the definition and the content sets of J are considered through the informative relation;
3. Informative relation objectivity: on which ground the informative relation links input and output sets;
4. Information imperfections (e.g., the four aspects of QoI of Figure 4.6).

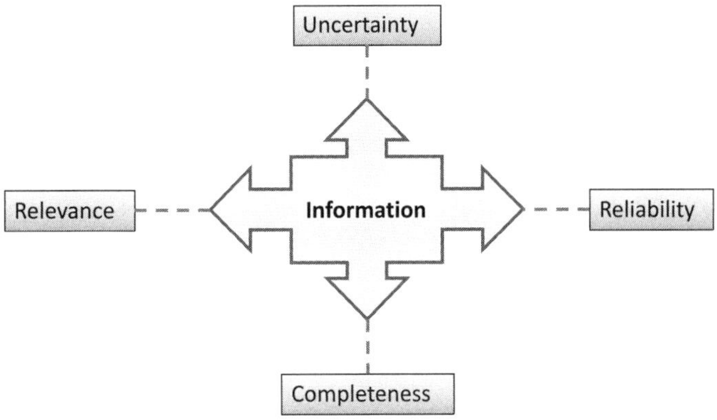

Figure 4.6 Four aspects of quality of information that offer exploitable formalizations. (From [33].)

4.4.1 Information Inner Sets Scope

The first characterization issue of an information element is mainly intrinsic as it concerns the information input/output levels of granularity or scope. The information input scope characterizes the set of elementary objects from the definition set, concerned by the information. The information output scope characterizes the set of elementary outcomes from the contents set, concerned by the information. This granularity resolution defines on which level of details the informative relation operates.

4.4.2 Informative Relation Arity

The information relation links input and output sets. However, this relation may consider input and output sets as: single-arity, plural-arity, and fuzzy-arity. Figure 4.7 illustrates such relations from input sets pointing toward output sets. Two of the frequently encountered relations are Figure 4.7(a) (e.g., person's size) or Figure 4.7(b) (e.g., group of observed patterns class). However, the participation of such sets to the informative relation may also be distributed or fuzzy [Figure 4.7(c)]. For instance, information produced by a fuzzy classification process typically associates one object from an input single element set to a series of elements (or classes) from an output fuzzy elements set.

The ideal situation that information represents a pure input object corresponding to one and only one high-level category or class from the contents set seems to be utopia in practice. For instance, in sensing imaging, input objects consist of resolution cells that generally include several thematic classes (i.e., information content) at once. These cells are referred to as mixels (i.e., mixed pixels). In this case, one object from the input set can be considered as a mixture of multiple output set's objects defining a relationship seeing the input set as a single-arity and the output set as a fuzzy-arity. In any case, the information elements characterization process is crucial for any information fusion system since several information elements may be cascaded. Therefore, one information element outcome constitutes the input of another one.

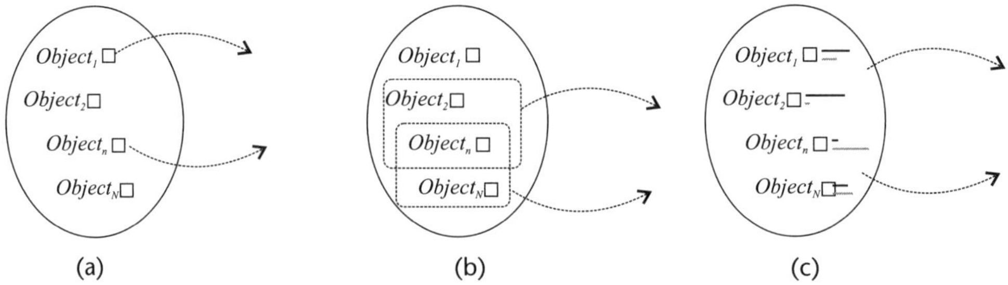

Figure 4.7 How an informative relation may consider an input set: (a) single-arity, (b) plural-arity, and (c) fuzzy-arity.

4.4.3 Informative Relation Objectivity

When the mixture inside an input object relies on the nature of a content set, the characterization is purely intrinsic. The imaging example is a perfect illustration where information links together mixels (input definition set) and thematic classes (output content set), with thematic classes defining the mixels breaking down. However, if the cutting is performed according to extra knowledge or a global system purpose, the characterization becomes more extrinsic, for instance, from the other three dimensions of the Wang and Strong QoI framework.

4.4.4 Information Imperfections

From an intrinsic point of view, all the information element components (input definition, content outcome, informative relation) may be sources of imperfection, especially the informative relation according to its nature and behavior (relying on the available external sources of knowledge used to perform the link between input and output sets). Figure 4.8 describes the basic structure of an information element $J = (\Theta, X, \Omega)$, where Θ (resp. Ω) denotes the definition (content) set and X the informative relation.

4.4.4.1 Uncertainty

Most data fusion efforts are concerned with adequate modeling of information uncertainty which is the result of noisy, imprecise, ambiguous, erroneous, or ill-suited for the problem, observations, and incomplete or poorly defined prior knowledge. To describe uncertainty, we assume that the two following conditions hold:

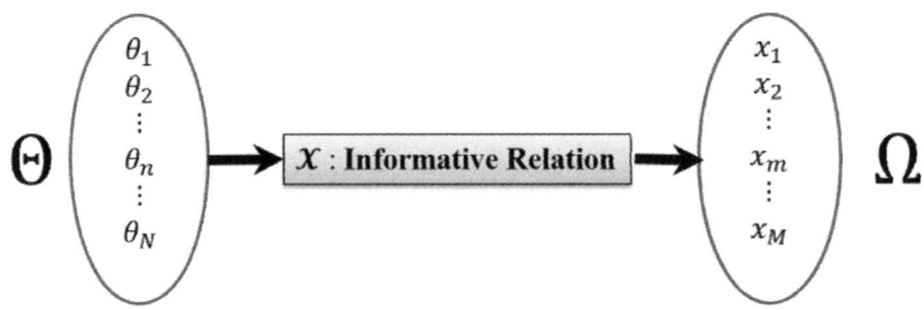

Figure 4.8 Generic structure of an information element J.

1. The informative relation is a punctual outcome relation: one information content outcome x is produced by X.
2. The set of information contents Ω is assumed to be exhaustive and exclusive: the information content produced by X is unique and certainly belongs to Ω.

Uncertainty affects information if and only if the true information content x_{True} is unknown with certainty. The main objective of uncertainty imperfection modeling and processing is to model and to process this lack of knowledge to determine with total certainty the unknown true content outcome of the considered information (for instance, the class of an observed object and the decision to consider). In other words, only one content x in Ω is the true outcome, but we do not know which one.

In the framework of information processing and fusion, two major approaches are used to model and to process the uncertainty type of imperfection: probabilistic and evidential approaches. Given that uncertainty affects punctual outcome informative functions, both probabilistic and evidential approaches consider the total certainty as having a unit belief or certainty measure. Depending on the available knowledge concerning the true information content x_{True}, both approaches differ in the way this unit total certainty measure is distributed over the different events from Ω.

The probabilistic approach makes a punctual certainty distribution of the unit total certainty onto different information contents, that is, each information content $x \in \Omega$, called a singleton, captures a partial amount of certainty Pr{x} and all Pr{x} summing up to unit. Pr{x} is interpreted as the probability or the basic belief for x to be the true information content. An information element, $J = (\Theta, X, \Omega)$, affected by uncertainty and for which a probability distribution of uncertainty is available, is called a probabilistic information. As an example, Pr{temperature of Paul = 38°C} = 0.25, Pr{temperature of Paul = 39°C} = 0.7, defines a typical probabilistic uncertain information.

The evidential approach (based on evidence theory [39]) constitutes an excellent alternative to the probabilistic approach when the available knowledge about true information content x_{True} does not allow making a punctual certainty distribution. In fact, the evidential approach spread out the unity total certainty over different subsets of Ω, building a subsets certainty distribution. Each information content subset, or event, $A \subseteq \Omega$ captures a partial amount of certainty, noted as $m(A)$. Summing up all the partial amounts of certainty over Ω gives one. $m(A)$ is interpreted as the mass of belief that the true information content x_{True} is in A.

An information element $J = (\Theta, X, \Omega)$, affected by uncertainty, and for which a mass distribution of uncertainty is available, is called an evidential information. For instance, m(Paul temperature ∈ {38°C, 39°C}) = 0.25, m(Paul temperature ∈ {39°C, 40°C}) = 0.75, defines a typical evidential information.

Notice that the probabilistic approach can be considered as a special case of the evidential approach. Indeed, relaxing the punctual probabilistic certainty distribution constraint using an evidential approach brings a practical and convenient tool for knowledge representation. However, this relaxation reduces the precision and the quality of information processing results when compared to the probabilistic approach.

4.4.4.2 Imprecision

Imprecision type of imperfection is an issue pertaining to the intrinsic quality of information. It refers to the case where the available knowledge about the true information content is expressed as a subset Ω_1 of the information content set Ω (i.e., we only know that the true information content belongs to Ω_1: $x_{True} \in \Omega_1 \subseteq \Omega$. In this case, the information $J = (\Theta, X, \Omega)$ is called an imprecise information. Therefore, imprecise information involves the lack of precise knowledge about the information content and thus should not be considered as erroneous.

In the decision-making domain, imprecision represents the uncertainty as the state of mind of an agent who does not possess the needed information or knowledge to make a precise decision: the agent is in a state of uncertainty ("I'm not sure that this object is a table" [40]). Special kinds of imprecise information include:

- Disjunctive information content subset (e.g., John's age is either 31 or 32, the class of the object is either C_1 or C_2...);
- Negative information content subset (e.g., John's age is not 30...);
- Range information content subset (e.g., John's age is between 30 and 35, or John's age is over 30);
- Error margins content subset (e.g., measured missile range is 100 ± 5 km).

When $\Omega_1 = \{x_{True}\}$, the information J is a precise information, whereas when $\Omega_1 = \Omega$, the Ω_1 encompasses the entire set of possible information contents Ω, the information J becomes a null (also called missing data or total ignorance) information. Imprecise information is generally associated with an accuracy measurement quantifying the closeness of agreement between the produced information outcome (i.e., Ω_1) and the true information content x_{True}. From an information processing point of view, an imprecise information is considered as a special case of evidential information where the total certainty is attributed to the subset Ω_1 (i.e., $m(\Omega_1) = 1$, and $m(A) = 0$ for all $A \neq \Omega_1$).

4.4.4.3 Remarks on Imprecision Versus Uncertainty

The aim of information processing and fusion approaches, when dealing with both forms of imperfection (imprecision and/or uncertainty), is mainly to determine the unique true information content x_{True} with the highest precision and certainty degree.

Imprecision is often confused with uncertainty because both imperfection types are related to the same root. They come from punctual informative functions where the unique true content is unknown precisely (case of imprecision) or certainly (case of uncertainty). Moreover, both imprecision and uncertainty can be present at the same time and one can cause the other. It is important to be able to tell the difference between these two antagonistic concepts, even if they can be included in a broader meaning for uncertainty (knowing that $x_{True} \in \Omega_1 \subseteq \Omega$ does not imply the precise and certain knowledge of x_{True}). To illustrate the difference and potential mixture between imprecision and uncertainty, consider the following two situations:

1. Paul has at least two children and I am sure about it.
2. Paul has three children but I am not sure about it.

In the first piece of information, the number of children is imprecise but certain, whereas, in the second piece of information, the number of children is precise but uncertain [41].

4.4.4.4 Ambiguity

Information is said to be ambiguous if it is unclear to what the information refers. This corresponds to the situation where the information can be interpreted in several ways, or when its truth or validity is not totally verified. From an informational point of view and assuming that the information content set Ω is exhaustive, two types of ambiguity are encountered:

- *Nonspecificity:* Multiple content outcomes are produced simultaneously by the informative function.
- *Partial truth*: The information content is partially produced.

Zadeh [42] proposed modeling this imperfection type as a fuzzy set defined on the information content set, where each content outcome x is associated with a membership value $\mu(x) \in [0,1]$ representing the strength or the truth of production of the outcome x by the informative function. $\mu(x) = 0$ means that x is not produced (not obtained or not concerned) and $\mu(x) = 1$ means that x is fully produced.

An information element $J = (\Theta, X, \Omega)$, affected by ambiguity and for which a membership function $\mu(.)$ is available, is called an ambiguous (or fuzzy) information. In this case, it is clearly nonsense to specify a single information content. Therefore, the major objective of the application of fuzzy concepts is to conduct the fusion of multiple ambiguous information elements.

4.4.4.5 Remark on Hybrid Imperfections

Hybrid forms of imperfections are frequently encountered, especially the following situation. The information element J is affected by uncertainty (J has a punctual informative function where the true output x_{True} is unknown with certainty), but the available knowledge about x_{True} is ambiguous, that is, each content outcome x is associated with a possibility value $\pi(x) \in [0,1]$ representing the possibility strength for the outcome x to be the unique true information content. This type of information elements is referred to as possibilistic information [43].

4.5 Contextual Information Characterization

When performing information fusion tasks, humans naturally start by giving personalized quality evaluations about the gathered information. The process of establishing such evaluations involves the nature of information sources as well as their respective informative contribution tackling a considered problem. When building his own representation of the world, the human uses his idea of information quality to obtain a plausible model, given his knowledge. It may not be the optimal result, but it is a good starting point for a more refined subsequent model. From an information fusion perspective, successfully and efficiently characterizing information quality is a key point as it impacts the production of reliable results and decisions, which is the ultimate goal of all information processing and fusion systems.

QoI is addressed in this chapter, at only two levels: intrinsic and contextual levels (according to Wang and Strong's framework). The intrinsic level consists in considering the information element as a conceptual entity out of the global fusion context, leading to an information characterization in terms of scope and perfection/imperfection issues. This level was addressed in the previous section. The contextual level concerns information characterization from a higher level of knowledge that derives from the information impact (i.e., completeness, relevance, reliability, conflict, redundancy) within the global fusion environment. Thus, contextual characterization is a fundamental part of the fusion process as it influences how the process evolves. Indeed, knowing that a source is producing wrong or unreliable information in a particular situation is critical to modify the behavior of the fusion algorithm: the measurements can be pruned away or their importance in the merging step can be lowered. Two major contextual characterizations are considered in this study: information reliability and information consistency.

4.5.1 Information Reliability

Traditional approaches of information and decision theory assume information elements to be fully reliable. Nevertheless, the concept of information reliability allows for various interpretations and representations [44]. Two approaches are generally considered for modeling information reliability.

4.5.1.1 Weighting Information Reliability

The first reliability modeling approach considers the information element $J = (\Theta, X, \Omega)$ as an input/output link and assumes the existence of a general information quality measure, intended to capture how frequently a source of information agrees with a ground truth, as well as prior available knowledge, allowing the estimation of the quality measure. In this case, reliability is measured by a performance measure of the information element using (for instance, recognition or false alarm rates). This corresponds to the conventional commonness reliability concept understanding as the relative stability of the first-order uncertainty. Therefore, this approach associates the information element J with a weight $R(J)$ measuring its reliability. This reliability weight $R(J)$ is drawn from a scale with either ordinal or numeric values, ordering information reliability. A common case assigns values from an interval (e.g., $[0,1]$) to $R(J)$. Giving the upper bound of the scale to $R(J)$ indicates that the information J should be given the highest reliability, whereas the assignment of the lowest value means that the information has no or little reliability. An example of reliability weight estimation is given in [45].

4.5.1.2 Modeling Reliability from the Belief Model Behavior

Most part of information fusion literature is about a second reliability modeling approach, which is defined as a higher-order uncertainty as it deals with uncertainty about the evaluation of uncertainty [46]. In fact, given an imperfect information $J = (\Theta, X, \Omega)$ within the selected framework of modeling, the available knowledge related to intrinsic imperfections, generally leads to the establishment of a belief model on the information outcome (e.g., a probability distribution, a mass function, a membership function, or a possibility distribution). Thus, this second reliability modeling approach focuses its attention on modeling and representing reliability effects on the computed beliefs and not on the information itself (i.e., how reliable is the computed probability distribution?).

4.5.1.3 Constraints and Reality

Several constraints strongly limit the complete formalization of information reliability:

- Information may be very reliable with a given context and not reliable with another context.
- The determination of reliability weights is application-dependent and is mostly determined through empirical manners.

Within the general framework of information fusion systems, both approaches of reliability modeling are generally merged into a two-step procedure. Using the first approach, and based on the available knowledge, a reliability weight $R(J)$ is estimated; then the second approach tunes the computed belief model using this reliability weight. Notice that in some applications $R(J)$ is considered as a vector $R(J) = [R(x_1), R(x_2), \ldots, R(x_{\mathrm{m}})]$ where $R(x_{\mathrm{m}})$ denotes the reliability weight of the specific information content outcome x_m with M referring to the amount of considered information contents.

4.5.2 Information Consistency

Information consistency is an important information contextual characteristic as it describes how much a given information element is in agreement with related or prior information. This agreement has two facets:

- A temporal agreement is related to the age of the information. For instance, several blood test information produce results with limited time validity. The extent to which the current information is suitable for use is called information timeliness.
- A completeness (or trustworthiness) agreement is related to the proportion of information outcome relevance to the global task at hand. In this context, information having all relevant key features of all considered information input objects means that the information is complete.

From a mathematical point of view, information consistency modeling is certainly the most difficult contextual element to characterize. Nevertheless, some information fusion approaches express information consistency by assigning to the information J a value $Cons(J)$ called consistency degree drawn from a well-defined scale having upper and lower bounds (e.g., unit interval [0,1]). For instance, within the possibility theory framework, the consistency degree $Cons(J)$ of an information is expressed as the highest possibility value as in (4.1) from the outcome sets. In this case, if $Cons(J) < 1$, then no information content has the total unit possibility value and the information element is called inconsistent.

$$Cons(J) = \max_{m=1,M} \pi(x_m) \tag{4.1}$$

The next section will describe further on the different belief models that are currently used to deal with both intrinsic and contextual information characterization, emphasizing on how they can represent the different uncertain information element types.

4.5.3 Information Relevance

A three-part critical review [47–49] on relevance over the past 30 years or so has been published and provides an updated framework within which the still widely dissonant ideas and works about relevance might be interpreted and related. Reference [49] revealed the importance of relevance in information sciences: "Relevance is a, if not even the, key notion in information science in general and information retrieval in particular." Information relevance is thus a critical aspect of the design of situation analysis support systems based on analytics and information fusion technologies. The aspect is so important for an environment such as big data that it deserves a book by itself. That will be the subject of a companion book to address information relevance for analytics and fusion of information. Pichon et al. [50, 51] described relevance under these terms: "It is assumed here that the reliability of a source of information involves two dimensions: its relevance and its truthfulness. A source is said to be relevant if it provides useful information regarding a given question of interest. If the source is a human agent, irrelevance means that the provided information does not pertain to the question it answers, for instance because the agent is actually ignorant.... There are various forms of lack of truthfulness. A source may declare the contrary of what it knows, or just say less, or something different, even if consistent with its knowledge. Lack of truthfulness for a sensor may take the form of a systematic bias."

Our world can be considered today as more chaotic, complex, and unpredictable as any period in history. The opportunity to face complex situations stimulates the challenge of developing adequate decision support systems as discussed in Chapter 2. Emerging cyber physical and social environments, supported by important technological advances, are evolving very fast. While more information is made available to provide better, clearer, and more certain operating pictures, the time window to process the information may shrink considerably. This raises the need to qualify the information based on their degree of relevance with respect to managing that situation. The relevance of the information can be defined as the degree to which the information can meet decision-makers' needs. In the presence of a deluge of information and interactions, decision-makers need decision support systems (e.g., intelligent filtering) based upon information relevance concepts and criteria [52].

It is beyond the scope of the current book to analyze the recent contributions [53–58] on the subject of information relevance but to stress the importance of that aspect of information quality within the structure of the information element in any design of situation analysis support systems within a context of data deluge.

4.6 Belief Model Representation

In the framework of information fusion systems, several theories and approaches are used to represent different belief models and to apply reasoning mechanisms on these models. The most widely uncertainty management theories are probability, Dempster-Shafer, and possibility theories. The fuzzy set theory is mainly concerned by ambiguous information management.

All these approaches mainly focus on modeling different types of intrinsic information imperfections for these models to be processed, fused, and interpreted through appropriately designed reasoning mechanisms. This section briefly presents the general mathematical frameworks of these approaches in terms of their context of application and the belief models and reasoning mechanisms that they introduce.

4.6.1 Probability Theory

The most studied belief model is clearly the theory of probabilities [59]. Probabilistic belief models provide a powerful and consistent tool for describing uncertainty in a broad range of situations, leading naturally into the areas of information fusion and decision-making.

4.6.1.1 Definition

The application context of probability theory concerns information elements $J = (\Theta, X, \Omega)$ assumed to be punctual (i.e., producing a single content outcome), under uncertainty constraint (i.e., the unique true content outcome x_{True} is unknown) and where the available knowledge about true content outcome x_{True} is given in terms of a probabilistic belief model. Obviously, within this probability framework, both concepts of information elements and random variables are identical. Indeed, an information element $J = (\Theta, X, \Omega)$ corresponds to a random variable X representing a random experiment over a definition set Θ and where Ω denotes the set the experiment's potential outcomes.

The probabilistic approach defines a punctual certainty distribution dividing a total certainty of 1 among the different information contents: each information content $x \in \Omega$, called a singleton, captures a partial amount of certainty $\Pr\{x\}$ and summing all $\Pr\{x\}$ from Ω gives "1." Thus, $\Pr\{x\}$ is interpreted as the probability or belief that "x" is the true information content.

Two equivalent methods are used to characterize a probabilistic belief model. Close to human reasoning, the first method directly comes from the previous definition and defines Pr as a probability density function (PDF):

$$\begin{aligned} &\Pr : \Omega \to [0,1] \\ &x_m \to \Pr\{x_m\} \end{aligned} \tag{4.2}$$

such that:

- $\Pr(\varnothing) = 0$: no belief is attributed to the empty set.
- $\sum_{m=1,M} \Pr\{x_m\} = 1$: summing the probabilities given to all different singleton x_m gives "1."

The occurrence probability $\Pr(A)$ of an event A is obtained as the addition of the occurrence probabilities of singletons forming the subset A.

The second method considers the probabilistic belief model as a probability measure defined over the power set 2^Ω of Ω, the set of all subsets of Ω including Ω and $\varnothing$:

$$\begin{aligned} &\Pr : 2^\Omega \to [0,1] \\ &A \in 2^\Omega (A \subseteq \Omega) \to \Pr\{A\} = \sum_{x_m \in A} \Pr\{x_m\} \end{aligned} \tag{4.3}$$

such that:

- $\Pr(\varnothing) = 0$ and $\Pr(\Omega) = 1$
- If $A \cap B = \varnothing$ *then* $\Pr(A \cup B) = \Pr(A) + \Pr(B)$.

Notice that both methods, defining the probabilistic belief model, are strictly equivalent from the mathematical point of view. Also, the basic exhaustivity and exclusivity characteristics of the information content set can be straightforward deduced from these basic definitions. A probabilistic information element $[J, \Pr] = [(\Theta, X, \Omega), \Pr]$ is thus the name given to an information element affected by the uncertainty type of imperfection and for which a probabilistic belief model of uncertainty is available.

4.6.1.2 Fusion Mechanisms

Even with limitations, probabilistic modeling techniques play an essential role in developing information fusion methods as they provide a robust and complete description of a probabilistic environment or process of interest. Consider the two following probabilistic information elements leading to the same information content set Ω:

$$\begin{aligned} &[J_1, \Pr_1] = [(\Theta_1, X_1, \Omega_1), \Pr_1] \\ &[J_2, \Pr_2] = [(\Theta_2, X_2, \Omega_2), \Pr_2] \end{aligned}$$

The aim of probabilistic information fusion is to obtain a global joint probability distribution:

$$\left[\mathcal{I}_{1\cap 2},\mathrm{Pr}_{1\cap 2}\right]=\left[\left(\Theta_1\times\Theta_2,X_{1\cap 2},\Omega\right),\mathrm{Pr}_{1\cap 2}\right] \tag{4.4}$$

where $\mathrm{Pr}_{1\cap 2}$ is the joint probabilistic belief model measuring the probability of occurrence of events A (subset of content outcomes through $\left[J_1,\mathrm{Pr}_1\right]$ and B (subset of contents outcomes through $\left[J_2,\mathrm{Pr}_2\right]$). The Bayesian chain-rule of conditional probability distributions is used to expand the joint probability distribution in terms of conditional and marginal distributions:

$$\mathrm{Pr}_{1\cap 2}(A\cap B)=Pr_1(A/B)\cdot Pr_2(B)=Pr_2(B/A)\cdot Pr_1(A) \tag{4.5}$$

When information elements J_1 and J_2 are statistically independent, $\mathrm{Pr}_{1\cap 2}$ is equal to:

$$\mathrm{Pr}_{1\cap 2}(A\cap B)=Pr_2(B)\cdot Pr_1(A) \tag{4.6}$$

When J_1 and J_2 are not statistically independent, the joint probability distribution is slightly difficult to determine. Nevertheless, within the framework of a decision-making problem, probabilistic information fusion uses the Bayesian formalism and transforms the decision-making process into a decision likelihood ratio computation issue.

4.6.2 Dempster-Shafer Theory

Dempster-Shafer theory, also called the theory of evidence [39], has become a popular tool in artificial intelligence for knowledge representation and decision making problems.

4.6.2.1 Definition

Let $J=(\Theta,X,\Omega)$ denote a punctual information element (i.e., producing a single content outcome) and assume the contents set Ω to be exhaustive and exclusive (i.e., J produces a unique true content outcome $x_{True}\in\Omega$). This assumption explains why Ω is also called the frame of discernment. The application context of the Dempster-Shafer theory can be formulated as follows:

1. $J=(\Theta,X,\Omega)$ is an uncertain information element (x_{True} is unknown).
2. Given some evidential corpus, the agent's available knowledge about the true content outcome x_{True} is modeled by a function m, called basic probability assignment defined as a mapping from 2^{Ω} into [0, 1] such that:

$$\begin{aligned}&m(\varnothing)=0\\&\sum_{A\in 2^{\Omega}}m(A)=1\end{aligned} \tag{4.7}$$

The mass $m(A)$ associated with a subset is $A \subseteq \Omega$ interpreted as the part of belief allocated exactly to the hypothesis that x_{True} is in A [39]. No further information is provided about x_{True} belonging to a specific subset of A. Indeed, each of these subsets has its own mass. Subsets $A \subseteq \Omega$ such $m(A) > 0$ are called focal elements of m. In other words, the Dempster-Shafer theory allows to split up the total unit truth into all subsets of Ω, whereas this distribution is limited, within the probability theory framework, to singletons only.

The *belief* (*Bel*) and *plausibility* (*Pl*) set functions are equivalent representations of the mass function $m(A)$ for an event $\subseteq \Omega$:

$$Bel(A) = \sum_{B/B \subseteq A} m(A) \tag{4.8}$$

$$Pl(A) = \sum_{B/B \cap A \neq \phi} m(B) \tag{4.9}$$

Bel(A) measures the total belief that x_{True} is in A (i.e., the minimal belief that the subset A occurs): it represents the confidence that x_{True} lies in A or any subset of A, whereas *Pl*(A) measures the total belief that x_{True} might be in A (i.e., the maximal belief that the subset A occurs): any subset B having nonempty intersection with A support this belief. Starting from these definitions, it is straightforward to show that $Pl(A) = 1 - Bel(A^C)$ where A^C is the complement of $A\left(A \cup A^C = \Omega\right)$ *and* $A \cap A^C = \varnothing$. As the functions m, *Bel,* and *Pl* can be computed from each other, it is equivalent to talk about one of them or also about the corresponding body of evidence (subsets from Ω).

An evidential information element $[J,m] = [(\Theta,X,\Omega),m]$ is thus the name given to an information element affected by the uncertainty type of imperfection and for which an evidential belief model of uncertainty is available.

4.6.2.2 Fusion Mechanisms

Evidential combination rules defines specific aggregation methods for data obtained from multiple information elements that provide different assessments for the same frame of discernment Ω. However, the Dempster-Shafer theory is based on the assumption that these information elements are independent. Notice that the independence concept in the framework of Dempster-Shafer theory is not an important constraint, as in probability theory. It simply means that different information elements are established separately on the basis of the observation of different informational facets.

Dempster-Shafer theory includes many possible ways for combining evidences. The original combination rule of multiple basic belief assignments known as the Dempster rule can be argued to be a generalization of Bayes' rule [39]. Recent works [60, 61] highly question that statement. This rule strongly emphasizes the agreement between multiple information elements and ignores all the conflicting evidence through a normalization factor (i.e., a strict and/or conjunctive behavior).

For instance, let $[J_n] = [(\Theta,X,\Omega),m_n]_{n \in \{1,2\}}$, denote two evidential information elements having the following basic belief assignments:

$$m_n = 2^{\Omega} \rightarrow [0,1] \tag{4.10}$$

$$m_n(\varnothing) = 0 \text{ and} \sum_{A/A \subseteq \Omega} m_n(A) = 1 \tag{4.11}$$

where $m_n, n \in \{1,2\}$, characterizes the information element Jn belief that the unique occurring singleton x_{True} belongs to a specific contents subset A of Ω. Combining J_1 and J_2 through the Dempster rule produces a new information element with the following basic probability assignment:

$$m_{1\oplus 2}(A) = \frac{1}{1-K} \sum_{A_i \cap A_j = A} m_1(A_i) m_2(A_j) \tag{4.12}$$

where

$$K = \sum_{A_i \cap A_j = \varnothing} m_1(A_i) m_2(A_j) \tag{4.13}$$

is the normalization factor that can also be considered as a conflict measure between the two individual information elements. $K = 0$ corresponds to the absence of conflict, whereas $K = 1$ indicates a complete contradiction between m_1 and m_2. Indeed, when $K = 0$, no nonintersecting subsets of Ω are given masses by both basic probability assignments m_1 and m_2, through the combination. However, $K = 1$ is obtained when only nonintersecting subsets (or events) of are given masses by both basic probability assignments m_1 and m_2, through the combination process.

Notice that the normalization operation is always conducted in order to guarantee the fact that the joint information content set is exhaustive: the new basic probability assignment must split up the unit total belief over all subsets of Ω. This is not related to the consistency degrees of the considered information elements. However, the conflict measure K may be considered as a reliability weight of the joint information element, that is, $R_{1\oplus 2} =$ 'k'.

The use of the Dempster rule has come under serious criticism when encountering significant conflict in the information to be merged. Consequently, researchers have developed modified Dempster rules that attempt to represent the degree of conflict in the final result [62–64]. This conflict issue and the construction of the basic probability assignment associated with it is the critical distinction between all of the tailored Dempster-type rules. To employ any of these combination rules in an actual application, it is essential to understand how conflict should be treated in that particular application context.

4.6.3 Fuzzy Sets Theory

Introduced in 1965 by Zadeh [42], fuzzy sets theory has a number of properties that make it suitable for formalizing and representing vagueness in everyday life and when describing a system characteristics. It is based on the premise that key information elements in human thinking are not just crisp numbers but more of fuzzy sets or classes of phenomena or objects in which the transition from membership to nonmembership is gradual rather than abrupt. This theory is especially suited to provide methods to deal with and fuse ambiguous information.

4.6.3.1 Definition

Let $\Omega = \{x_m, m = 1, \ldots, M\}$ denote a universal set. A fuzzy set A over Ω is defined by a membership function:

$$\begin{aligned} &\mu_A : \Omega \rightarrow [0,1] \\ &x_m \rightarrow \mu_A(x_m) \end{aligned} \tag{4.14}$$

where $\mu_A(x_m)$ refers to the degree of membership of the element $x_m \in \Omega$ to the fuzzy set A. A membership value of 1, $\mu_A(x_m) = 1$, involves that the corresponding element x_m definitely belongs to the fuzzy set A. On the contrary, if the membership value is 0, $\mu_A(x_m) = 0$, the corresponding element x_m definitely does not belong to the fuzzy set A.

Thus, modeling ambiguous information elements with fuzzy sets theory is straightforward. Let $J = (\Theta, X, \Omega)$ denote an information element for which the content set Ω is "only" assumed to be exhaustive (the potential output(s) are within Ω). Moreover, let us consider the information output scope as a fuzzy set A over Ω, where $\mu_A(x_m)$ measures the membership degree of the content x_m to the output scope. In other words, an ambiguous information element is modeled by means of a fuzzy set defined over the information content set. This type of representation is extremely powerful and allows to explicit ambiguous information elements with various kinds of ambiguity:

1. Nonspecificity corresponds to information elements with multiple elements scope where several information contents have full membership values. In this case, Ω is not an exclusive contents set.
2. Mixture of multiple elements scope is when several information contents have strictly positive membership values such that

$$\sum_{m=1}^{M} \mu_A(x_m) = 1$$

3. Fuzzy singleton ambiguity is when only one content has a strictly positive membership value (not necessarily equal to 1).
4. A fuzzy set of elements scope is when all contents have positive membership degrees [i.e., $\mu_A(x_m)$, $\in [0,1]$, for all $m = 1 \ldots M$].

A fuzzy information element $[J, \mu] = [(\Theta, X, \Omega), \mu]$ is thus the name given to an information element affected by ambiguity and for which a fuzzy belief model is available.

4.6.3.2 Fusion Mechanisms

Fuzzy sets theory provides a host of attractive aggregation connectives for integrating membership values representing different ambiguous information elements. These connectives can be categorized into the following three classes:

- *Disjunctive fusion:* Union produces a high membership degree whenever any one of the input membership degrees, representing degrees of satisfaction of individual information elements, is high.
- *Conjunctive fusion:* Intersection connectives produce a high output only when all of the input membership degrees have high values.
- *Compromise fusion:* Compensative connectives have the property that a higher degree of satisfaction of one of the individual information elements can compensate for a lower degree of membership of another information element to a certain extent.

4.6.4 Possibility Theory

Possibility theory [65] is based on fuzzy sets theory [43]. At the semantic level, the application context of possibility theory concerns information elements $J=(\Theta,X,\Omega)$ assumed to be punctual (i.e., producing a single content outcome), under uncertainty constraint (i.e., the unique true content outcome x_{True} is unknown), and where the available knowledge about the true content outcome x_{True} is ambiguous (summing all the pieces of knowledge does not give 1).

4.6.4.1 Definition

A basic belief model is a possibility distribution π when it assigns each potential information content x_m from Ω, a value $\pi(x_m)\in[0,1]$ (or a set of graded values) representing the possibility degree for $x_m \in \Omega$ to be the true occurring content outcome x_{True}:

$$\begin{aligned} \pi &: \Omega \rightarrow [0,1] \\ x_m &\rightarrow \pi(x_m) \end{aligned} \tag{4.15}$$

From a possibility distribution π, a possibility measure Π, and a necessity measure N can be derived for every subset $A\subseteq\Omega$ as follows:

$$\begin{aligned} \Pi(\mathrm{A}) &= \max_m\left(\left\{\pi\left(x_m\right)/\,x_m\in A\right\}\right) \\ \mathcal{N}(A) &= 1-\Pi(A^C) \end{aligned} \tag{4.16}$$

The possibility measure $\Pi(A)$ estimates to what extent the true content outcome x_{True} is believed to be in the subset A. When $\Pi(A) = 0$, it means that x_{True}

being in A is an impossible situation; however, when $\Pi(A) = 1$, it does not mean that x_{True} is necessary in A, but this fact is considered as usual or totally possible.

The necessity measure $N(A)$ evaluates the degree of necessity that x_{True} belongs to A by measuring the possibility of x_{True} not belonging to A. For instance, $N(A)$ is equal to 1 only when A^C is impossible. When $N(A) = 0$, it only means that it would not be surprising that x_{True} does not belong to A. Notice that the exhaustive and exclusive characteristics of the information content set cannot be derived from the definition of the possibility distribution itself, but are assumed to be implicitly verified.

The possibility theory also offers an indicator *Cons*(J) measuring the consistency of an information element J:

$$Cons(J) = \max\left(\left\{\pi\left(x_m\right) / x_m \in \Omega\right\}\right) \tag{4.17}$$

An information elements $J = (\Theta, X, \Omega)$ is considered consistent or normal if its degree of consistency *Cons*(J) equals 1 [i.e., there exists at least one content $x_0 \in \Omega$ for which $\pi(x_0) = 1$]. Otherwise, the information element J is considered inconsistent or subnormal.

4.6.4.2 Fusion Mechanisms

Within the possibility theory framework, combining two possibility distributions π_1 and π_2 defined over a same set Ω can be performed with various techniques operating according to three main modes (similarly to those used within the framework of fuzzy sets theory):

- *Conjunctive mode:* This mode is used when the fusion goal is to reflect the common information between the individual information elements. It clearly requires the information elements to be reliable, with no contradiction. The implementation of this combination mode is provided by t-norm operators [66] (a generalization of a set intersection operator).
- *Disjunctive mode:* In contrast, when we believe that combining conjunctively may lead to losing some meaningful information because of the lack of reliability or the presence of conflict, the disjunctive mode generalizing the set union operator (t-conorm [66]) is more suited.
- *Compromise mode*: It is an intermediate situation between the conjunctive and the disjunctive modes.

4.6.5 Fuzzy-Evidence Theory (Fuzzy-Belief Structures)

Denoeux proposed in [67] a new framework to cope with imprecise belief masses assigned to imprecise propositions. This idea was first introduced by Zadeh [68], who mixed concepts from evidence and possibility theories. In the new framework, the focal elements (propositions) defined in evidence theory as crisp subsets of the frame of discernment are considered as fuzzy sets (fuzzy propositions) over the frame of discernment. Following Zadeh's work, Ishizuka [69], Yager [70], and

Ogawa [71] extended the D-S belief functions by defining a measure of inclusion. Yen [72] first generalized the compatibility relation in the D-S theory to a joint possibility distribution, then, formulated the linear programming problems that compute the belief measures and the plausibility measures.

4.6.5.1 Definition

A fuzzy belief structure (as called by Yager in [73]) is defined as a mapping from defined as a mapping from 2^{Ω} into [0, 1] such that:

$$\begin{gathered}\tilde{m}(\varnothing) = 0 \\ \sum_{A \in F(\Omega)} \tilde{m}(A) = 1\end{gathered} \tag{4.18}$$

where $F(\Omega)$ is the set of all fuzzy sets defined over the frame of discernment Ω and A is a fuzzy set defined over Ω. The extension of other concepts from the evidence theory, such as the belief, the plausibility, and the commonality functions to the new framework, is based on the interpretation of the simple operators of union, intersection, and inclusion defined for crisp sets in terms of fuzzy sets. Several authors have proposed different approaches to these extensions. Following Yager in [73]:

$$««««_f(\quad) = \sum_{i=1}^{n} \tilde{}(\ _i)\ (\ |\ _i) \tag{4.19}$$

$$Pl_1(A) = \sum_{i=1}^{n} \tilde{m}(F_i)\Pi(A \mid F_i) \tag{4.20}$$

where $N(A|F_i)$ and $\Pi(A|F_i)$ are defined as:

$$N(A|F_i) = \min_{w \in \Omega} \max_{C} \left[\mu_A(\omega), \mu_{F_i^C}(\omega)\right] \tag{4.21}$$

$$\Pi(A \mid F_i) = \max_{w \in \Omega} \min_{C} \left[\mu_A(\omega), \mu_{F_i^C}(\omega)\right] \tag{4.22}$$

Yen [72] proposed different definitions for the belief and plausibility measures of fuzzy sets, based on the decomposition of the fuzzy sets into crisps sets (α-cuts decomposition). Each fuzzy focal element F_i of m is decomposed into a set of crisps sets $F_{i,j}$. The belief and the plausibility measures are then defined by Denoeux [67] as follows:

$$Bel_2(\mathrm{A}) = \sum_{i=1}^{n} \tilde{m}(F_i) \sum_{j} (a_j - a_{j-1}) \min_{w \in F_{i,j}} \mu_A(\omega) \tag{4.23}$$

$$Pl_2(\mathrm{A}) = \sum_{i=1}^{n} \tilde{m}(F_i) \sum_{j} (a_j - a_{j-1}) \max_{w \in F_{i,j}} \mu_A(\omega) \quad (4.24)$$

4.6.5.2 Fusion Mechanisms

The combination of fuzzy belief structures is based on Dempster's rule of combination from the evidence theory. A general formulation for the fuzzy belief structure combination is proposed in [67]:

$$\tilde{m}_1 \nabla \tilde{m}_2 (A) = \sum_{B \nabla C = A} \tilde{m}_1 (B) \lozenge \tilde{m}_2 (C) \quad (4.25)$$

where A, B, and C are fuzzy sets defined over Ω. ∇ and $\lozenge$ are different operators over the fuzzy sets such as any t-norms and/or any t-conorms. The normalization process for fuzzy belief structures was extended from Dempster's normalization by both Yen [72] and Yager [74]:

$$\tilde{m}^* (A) = \frac{\sum_{B^* = A} h_B \, \tilde{m}(B)}{\sum_{B \in F\left(\tilde{m}\right)} h_B \, \tilde{m}(B)} \quad (4.26)$$

where h_B is the height of the fuzzy set B, B^* is the normalized fuzzy set obtained from B, and $F(\tilde{m})$ is the set of focal elements associated to the fuzzy belief structure $\tilde{m}$.

4.6.6 Summary on Information Element Representations

As previously stated, three main approaches are widely used for uncertainty representation, processing and fusion: probabilistic, evidential, and possibilistic approaches. Recall that an information element $J = (\Theta, X, \Omega)$ is said to be certain, if the informative function X is punctual and the unique content outcome x_{True} is certainly and precisely known. Different types of uncertain information elements are encountered:

- J^{Pr} is the probabilistic information element. This information element type arises when J is uncertain and the available knowledge about the true information content x_{True} is given as a probability distribution Pr{.}defined over Ω.
- J^{Im} is the imprecise information element (or partial ignorance information element). This information element type arises when J is uncertain and the sole knowledge existing about x_{True} is that x_{True} belongs to a subset $A \subseteq \Omega$.

- J^{Ev} is the evidential information element. This information element type arises when J is uncertain and the available knowledge about x_{True} can be expressed as a basic probability assignment.
- J^{Pos} is the possibilistic information element. This information element type arises when J is uncertain and the available knowledge about x_{True} can be expressed as a possibility distribution.
- J^{FE} is the fuzzy-evidential information element: This information element type arises when J is uncertain and the available knowledge about x_{True} can be expressed as a fuzzy basic probability assignment.
- J^{Tot} is the total ignorance information element (or missing data information element): This information element type arises when J is uncertain and absolutely no knowledge is available about the true information content x_{True}.

As different uncertain information element types may be encountered in the framework of information fusion systems, cross-transformations can be conducted to allow the representation and the processing of different information elements within the same theoretical framework. Cross-transformations can be evaluated with respect to a gain-loss information and semantic perspectives using multiple criteria of information quality. Table 4.2 sums up the different representations that have been discussed in this chapter. The choice of a particular approach depends mainly on the nature of the information. With the various types of the information elements and their mathematical representations of Table 4.2, all the aspects of the circular typology of uncertainty developed in the previous chapter can be fully represented as illustrated in Figure 4.9.

4.7 Conclusion

This chapter presented an approach to characterize information and its imperfections by defining a basic structure for a computable information element. Different methods of representation of the imperfections as well as a concern to integrate criteria of quality of information within that computational information element structure have been described. Table 4.2 shows potential cross transformations at information element level between the various representation modes and covers most of the uncertainty typologies presented in the previous chapter.

Knowledge and belief analysis and uncertainty management are cornerstones to information fusion. The methods for modeling and processing uncertainty naturally differ from one scientific community to another. For example concerning perception, probability theory is adopted by most of electrical engineers. In the case of reasoning, logical approaches are rather used by the artificial intelligence (AI) community (logical AI), philosophers, and logicians. In the information fusion domain, the need for manipulation of numerous theoretical frameworks of different natures rapidly becomes a real problem. Traditionally, this problem is tackled using two means:

1. The first approach consists in making the different theoretical probabilistic-evidential-possibilistic) frameworks communicate between each other. For

Table 4.2 Uncertain Information Elements: Representations and Cross-Transformations

Types of $\mathcal{I}$	Probabilistic Modeling	Evidential Modeling	Possibilistic Modeling	Fuzzy evidential Modeling
Certain $\mathcal{I}^{Certain}$	$p\{x_{true}\} = 1$ and $p\{x\} = 0, \forall x \neq x_{true}$	$m(x_{true}) = 1, x_{true} \in \Omega$ and $m(A) = 0, \forall A \subseteq \Omega, A \neq x_{true}$	$\pi(x_{true}) = 1$ and $\pi(x) = 0, \forall x \neq x_{true}$	$\tilde{m}(x_{true}) = 1, x_{true} \in \Omega$ and $\tilde{m}(A) = 0, \forall A \in \mathcal{F}(\Omega), A \neq x_{true}$
Probabilistic $\mathcal{I}^{Pr}$	$0 \leq p\{x\} \leq 1$ and $\sum_{x \in \Omega} p\{x\} = 1$	$m(\{x\}) = p(x), \forall x \in \Omega$ and $m(A) = 0, \forall A \subseteq \Omega, A \neq \{x\}$ (A is not a singleton)	Probability $\underset{to}{\rightarrow}$ Possibility of $p\{.\}$ to obtain $\pi(.)$ **Refs: [80-83]**	$\tilde{m}(\{x\}) = p(x), \forall x \in \Omega$ and $\tilde{m}(A) = 0, \forall A \subseteq \mathcal{F}(\Omega), A \neq \{x\}$ (A is not a singleton)
Imprecise $\mathcal{I}^{Im}$	$p\{x\} = 1/\lvert A \rvert$ for $A \subseteq \Omega$ and $\forall x \in A$ $p\{x\} = 0, \forall x \notin A$	$m(A) = 1, A \subseteq \Omega$ and $m(B) = 0, \forall B \subseteq \Omega, B \neq A$	$\pi(x) = 1, \forall x \in A$ and for $A \subseteq \Omega$ and $\pi(x) = 0, \forall x \notin A$	$\tilde{m}(A) \geq 0, \forall A \in \mathcal{F}(\Omega)$ such that $\pi(x) = 1, \forall x \epsilon A$ (A is crisp)
Evidential $\mathcal{I}^{Ev}$	Pignistic Transformation of $m(.)$ to obtain $p\{.\}$ **Refs: [85, 86]**	$m(A) \geq 0, A \subseteq \Omega$ and $\sum_{A \subseteq \Omega} m(A) = 1$	bpa $\underset{to}{\rightarrow}$ possibility of $m(.)$ to obtain $\pi(.)$ **Refs: [75, 76]**	$\tilde{m}(A) \geq 0, \forall A \in \mathcal{F}(\Omega)$ such that $\pi(x) = 1, \forall x \epsilon A$ (A is crisp) $\sum_{A \in \mathcal{F}(\Omega)} \tilde{m}(A) = 1$
Possibilistic $\mathcal{I}^{Pos}$	Possibility $\underset{to}{\rightarrow}$ Probability of $m(.)$ to obtain $p\{.\}$ **Refs: [76, 80-83]**	Possibility $\underset{to}{\rightarrow}$ bpa of $\pi(.)$ to obtain $m(.)$ **Refs: [75, 76]**	$\pi(x) \geq 0, \forall x \in \Omega$	Possibility $\underset{to}{\rightarrow}$ bpa of $\pi(.)$ to obtain $m(.)$ and fuzzy belief modeling of an evidential ($\mathcal{I}$). **Refs: [75-77]**
Fuzzy Belief $\mathcal{I}^{FE}$	$Fuzzy\ Belief \underset{to}{\rightarrow} Probability$ of $\tilde{m}(.)$ to obtain $p\{.\}$ **Refs: [67, 77, 84]**	$Fuzzy\ Belief \underset{to}{\rightarrow} bpa$ of $\tilde{m}(.)$ to obtain $m(.)$ **Refs: [77, 84]**	$Fuzzy\ Belief \underset{to}{\rightarrow} bpa$ and (if consonant) $bpa \underset{to}{\rightarrow} possibility$ **Refs: [75, 84]**	$\tilde{m}(A) \geq 0, \forall A \in \mathcal{F}(\Omega)$ and $\sum_{A \in \mathcal{F}(\Omega)} \tilde{m}(A) = 1$ **Refs: [67, 70, 72, 78, 79]**
Total ignorance $\mathcal{I}^{Tot}$	$p\{x\} = 1/\lvert\Omega\rvert$ for $\forall x \in \Omega$	$m(\Omega) = 1,$ and $m(A) = 0, \forall A \subset \Omega$	$\pi(x) = 1, \forall x \in \Omega$	$\tilde{m}(\Omega) = 1,$ and $\tilde{m}(A) = 0, \forall A \in \mathcal{F}(\Omega), A \neq \Omega$

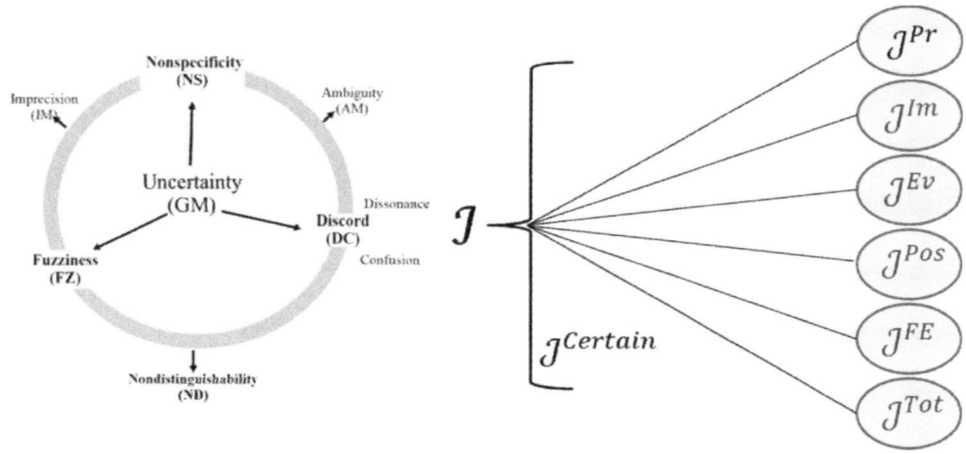

Figure 4.9 Types of *J*s to represent the various aspects of uncertainty.

example, within the same system of analytics and information fusion, one module can process information using probability theory whereas another parallel (or serial) module can use fuzzy sets theory. The difficulty becomes then to define algebraic transformations such that the modules can coherently communicate. The frameworks may not represent the same information nature so one can have a semantic challenge or misalignment.

2. The second approach aims at using general frameworks, and two tendencies appear then: using a single formalism, such as random sets theory for quantitative information or autoepistemic logic for logical information and directly processing two conceptually distinct pieces of information through a single formalism. This latter approach is based on the use of modal logics and possible world semantics to reason on both knowledge and uncertainty. A related approach defines semantics based on the random worlds to integrate statistical data processing and reasoning on knowledge.

Chapter 5 will use the information element structure to define a notion of information fusion cells and discuss how these cells can be assembled into global fusion architectures.

References

[1] Kalman, R. E., and R. S. Bucy, "New Results in Linear Filtering and Prediction Theory," *Journal of Fluids Engineering*, Vol. 83, 1961, pp. 95–108.

[2] Steinberg, A. N., C. L. Bowman, and F. E. White, "Revisions to the JDL Data Fusion Model," *Proceedings of the Joint NATO/IRIS Conference*, Quebec, October 1998, pp. 430–441.

[3] Wald, L., "Some Terms of References in Data Fusion," *IEEE Trans. on Geoscience and Remote Sensing*, Vol. 37, 1999, pp. 1190–1193.

[4] Losee, R. M., *The Science of Information: Measurement and Applications, Vol. 1*, Library and Information Science Series, New York: Academic Press, 1990.

[5] Whitaker, G., "An Overview of Information Fusion," *New Information Processing Techniques for Military Systems*, Vol. 1, 2001.

[6] Losee, R. M., "A Discipline Independent Definition of Information," *Journal of the American Society for Information Science,* Vol. 48, 1997, pp. 254–269.

[7] Solaiman, B., "Information Fusion Concepts: From Information Elements Definition to the Application of Fusion Approaches," *Proceedings SPIE, Vol. 4385, Sensor Fusion: Architectures, Algorithms, and Applications*, 2001, pp. 205–212.

[8] Stonier, T., *Information and the Internal Structure of the Universe*, New York: Springer-Verlag, 1990.

[9] Bossé, É., A. Guitouni, and P. Valin, "An Essay to Characterize Information Fusion Systems," *Proc. of 9th Intl. Conf. on Information Fusion (FUSION2006)*, Firenze, Italy, 2006.

[10] Laudy, C., "Introducing Semantic Knowledge in High Level Information Fusion," Ph.D. Thesis, Université Pierre et Marie Curie, Paris, France, 2010.

[11] Roy, J., "A Knowledge-Centric View of Situation Analysis Support Systems," TR 2005-419) DRDC Valcartier, Technical Report, 2007.

[12] Pravia, M. A., et al., "Generation of a Fundamental Data Set for Hard/Soft Information Fusion," *11th Intl. Conf. on Information Fusion*, 2008, pp. 1–8.

[13] Sambhoos, K., B. Koc, and R. Nagi, "Extracting Assembly Mating Graphs for Assembly Variant Design," *Journal of Computing and Information Science in Engineering,* Vol. 9, 2009, p. 034501.

[14] Sambhoos, K., et al., "Enhancements to High Level Data Fusion Using Graph Matching and State Space Search," *Information Fusion,* Vol. 11, 2010, pp. 351–364.

[15] Laudy, C., J. Dreo, and C. Gouguenheim, "Applying MapReduce Principle to High Level Information Fusion," *2014 17th Intl. Conf. on Information Fusion (FUSION)*, 2014, pp. 1–8.

[16] Najgebauer, A., et al., "The Prediction of Terrorist Threat on the Basis of Semantic Association Acquisition and Complex Network Evolution," *Journal of Telecommunications and Information Technology,* 2008, pp. 14–20.

[17] Rogova, G., and É. Bossé, "Information Quality in Information Fusion," *Proc. of the 13th Intl. Conf. on Information Fusion (FUSION2010)* Edinburg, UK, 2010.

[18] ISO 8402, Quality Management, http://stats.oecd.org/glossary/detail.asp?ID=5150.

[19] O'Brien, J. A., and G. M. Marakas, *Introduction to Information Systems,* Vol. 13, New York: McGraw-Hill/Irwin, 2005.

[20] Wang, R. Y., and D. M. Strong, "Beyond Accuracy: What Data Quality Means to Data Consumers," *Journal of Management Information Systems,* Vol. 12, 1996, pp. 5–34.

[21] Juran, J. M., and J. A. De Feo, *Juran's Quality Handbook: The Complete Guide to Performance Excellence*, New York: McGraw-Hill, 2010.

[22] Bovee, M., R. P. Srivastava, and B. Mak, "A Conceptual Framework and Belief: Function Approach to Assessing Overall Information Quality," *International Journal of Intelligent Systems,* Vol. 18, 2003, pp. 51–74.

[23] P. Eggenhofer-Rehart, et al., "C2 Conceptual Reference Model Version 2.0." http://www.dodccrp.org/files/c2CRM_V2_11.16.04.pdf.

[24] Alberts, D. S., "Agility, Focus, and Convergence: The Future of Command and Control," DTIC Document, *The International C2 Journal*, Vol. 1, No. 1, 2007.

[25] Alberts, D. S., "The Agility Imperative: Precis," Unpublished white paper, CCRP, http://www.dodccrp. org, 2010.

[26] Avizienis, A., J. -C. Laprie, and B. Randell, *Fundamental Concepts of Dependability*, University of Newcastle upon Tyne, Computing Science, Newcastle University Report no. CS-TR-739, 2001.

[27] Avizienis, A., et al., "Basic Concepts and Taxonomy of Dependable and Secure Computing," *IEEE Trans. on Dependable and Secure Computing*, Vol. 1, 2004, pp. 11–33.

[28] Petre, L., K. Sere, and E. Troubitsyna, *Dependability and Computer Engineering: Concepts for Software Intensive Systems*, Hershey, PA: IGI Global, 2012.

[29] Lee, E. A., and S. A. Seshia, *Introduction to Embedded Systems: A Cyber-Physical Systems Approach*, Lee & Seshia, 2011.

[30] Marwedel, P., *Embedded System Design: Embedded Systems Foundations of Cyber-Physical Systems*, New York: Springer Science & Business Media, 2010.

[31] Helfert, M., "Managing and Measuring Data Quality in Data Warehousing," *Proc. of the World Multiconference on Systemics, Cybernetics and Informatics,* Orlando, FL, 2001, pp. 55–65.

[32] Lefebvre, E., M. Hadzagic, and É. Bossé, "On Quality of Information in Multi-Source Fusion Environments," *Advances and Challenges in Multisensor Data and Information Processing,* 2007, p. 69.

[33] Rogova, G., and E. Bossé, "Information Quality Effects on Information Fusion," DRDC Valcartier, Tech Rept., 2008.

[34] Blasch, E., P. Valin, and É. Bossé, "Measures of Effectiveness for High-Level Information Fusion," in *High-Level Information Fusion Management and Systems Design*, E. Blasch, É. Bossé, and D. A. Lambert, (eds.), Norwood, MA: Artech House, 2012, pp. 331–348.

[35] Blasch, E., et al., "The URREF Ontology for Semantic Wide Area Motion Imagery Exploitation," *2012 IEEE National Aerospace and Electronics Conference (NAECON)*, 2012, pp. 228–235.

[36] Blasch, E., et al., "URREF Self-Confidence in Information Fusion Trust," *2014 17th Intl. Conf. on Information Fusion (FUSION)*, 2014, pp. 1–8.

[37] Blasch, E., et al., "URREF Reliability Versus Credibility in Information Fusion (STANAG 2511)," *16th Intl. Conf. on Information Fusion (FUSION)*, 2013, pp. 1600–1607.

[38] Costa, P. C., et al., "Towards Unbiased Evaluation of Uncertainty Reasoning: The URREF Ontology," *15th Intl. Conf. on Information Fusion (FUSION)*, 2012, pp. 2301–2308.

[39] Shafer, G., *A Mathematical Theory of Evidence*, Princeton, NJ: Princeton University Press, 1976.

[40] Jousselme, A. -L., P. Maupin, and É. Bossé, "Uncertainty in a Situation Analysis Perspective," *Proc. of the 6th Intl. Conf. of Information Fusion*, 2003.

[41] Smets, P., "Imperfect Information: Imprecision and Uncertainty," in *Uncertainty Management in Information Systems*, New York: Springer, 1997, pp. 225–254.

[42] Zadeh, L. A., "Fuzzy Sets," *Information and Control,* Vol. 8, 1965, pp. 338–353.

[43] Zadeh, L., "Fuzzy Sets as the Basis for a Theory of Possibility," *Fuzzy Sets and Systems,* Vol. 1, 1978, pp. 3–28.

[44] Duda, R. O., P. E. Hart, and D. G. Stork, *Pattern Classification*, 2nd ed., New York: Wiley, 2001.

[45] Fei, X., et al., "Research of the Reliability Coefficient in Information Fusion," *Intl. Conf. on Signal Acquisition and Processing 2009 (ICSAP 2009)*, 2009, pp. 85–88.

[46] Wang, P., "Confidence as Higher-Order Uncertainty," *ISIPTA*, 2001, pp. 352–361.

[47] Saracevic, T., "Relevance: A Review of the Literature and a Framework for Thinking on the Notion in Information Science," *Advances in Librarianship*, Vol. 6, 1976.

[48] Saracevic, T., "Relevance: A Review of the Literature and a Framework for Thinking on the Notion in Information Science. Part II," *Advances in Librarianship,* Vol. 30, 2006.

[49] Saracevic, T., "Relevance: A Review of the Literature and a Framework for Thinking on the Notion in Information Science. Part III: Behavior and Effects of Relevance," *Journal of the American Society for Information Science and Technology,* Vol. 58, pp. 2126–2144, 2007.

[50] Pichon, F., D. Dubois, and T. Denœux, "Relevance and Truthfulness in Information Correction and Fusion," *International Journal of Approximate Reasoning,* Vol. 53, 2012, pp. 159–175.

[51] Pichon, F., et al., "Truthfulness in Contextual Information Correction," in *Belief Functions: Theory and Applications*, New York: Springer, 2014, pp. 11–20.

[52] Breton, R., et al., "Framework for the Analysis of Information Relevance (FAIR)," *2012 IEEE Intl. Multi-Disciplinary Conf. on Cognitive Methods in Situation Awareness and Decision Support (CogSIMA)*, 2012, pp. 210–213.

[53] Hadzagic, M., et al., "Reliability and Relevance in the Thresholded Dempster-Shafer Algorithm for ESM Data Fusion," *15th Intl. Conf. on Information Fusion (FUSION)*, 2012, pp. 615–620.

[54] White, H. D., "Relevance Theory and Citations," *Journal of Pragmatics*, Vol. 43, 2011, pp. 3345–3361.

[55] Cholvy, L., "Evaluation of Information Reported: A Model in the Theory of Evidence," in *Information Processing and Management of Uncertainty in Knowledge-Based Systems. Theory and Methods*, New York: Springer, 2010, pp. 258–267.

[56] Cholvy, L., and S. Roussel, "Towards Agent-Oriented Relevant Information," in *Artificial Intelligence: Methodology, Systems, and Applications*, New York: Springer, 2008, pp. 22–31.

[57] Cholvy, L., and S. Roussel, "Reasoning with an Incomplete Information Exchange Policy," in *Symbolic and Quantitative Approaches to Reasoning with Uncertainty*, New York: Springer, 2007, pp. 683–694.

[58] Cholvy, L., and S. Roussel, "A Formal Characterization of Relevant Information in Multi-Agent Systems," DTIC Document, 2009.

[59] Kolmogorov, A. N., *Foundations of the Theory of Probability*, 2nd ed., New York: Chelsea Publishing, 1956.

[60] Dezert, J., and A. Tchamova, "On the Validity of Dempster's Fusion Rule and Its Interpretation as a Generalization of Bayesian Fusion Rule," *International Journal of Intelligent Systems*, Vol. 29, 2014, pp. 223–252.

[61] Dezert, J., et al., "Why Dempster's Fusion Rule Is Not a Generalization of Bayes Fusion Rule," *16th Intl. Conf. on Information Fusion (FUSION)*, 2013, pp. 1127–1134.

[62] Smarandache, F., and J. Dezert, *Advances and Applications of DSmT for Information Fusion (Collected Works)*, Vol. 2, Ann Arbor, MI: Infinite Study, 2006.

[63] Smarandache, F., and J. Dezert, *Advances and Applications of DSmT for Information Fusion, Collected Works*, Vol. 3, https://hal.archives-ouvertes.fr/hal-01080187/document, 2009.

[64] Smarandache, F., and J. Dezert, *Applications and Advances of DSmT for Information Fusion*, Rehoboth: AM. Res. Press, 2004.

[65] Dubois, D., and H. Prade, *Possibility Theory: An Approach to Computerized Processing of Uncertainty*, New York: Plenum Press, 1988.

[66] Klement, E. P., R. Mesiar, and E. Pap, *Triangular Norms*, New York: Springer, 2000.

[67] Denœux, T., "Modeling Vague Beliefs Using Fuzzy-Valued Belief Structures," *Fuzzy Sets and Systems*, Vol. 116, 2000, pp. 167–199.

[68] Gupta, M. M., R. K. Ragade, and R. R. Yager, *Advances in Fuzzy Set Theory and Applications*, New York: North-Holland Publishing Company, 1979.

[69] Ishizuka, M., K. S. Fu, and J. T. Yao, "Inference Procedures Under Uncertainty for the Problem-Reduction Method," *Information Sciences*, Vol. 28, 1982, pp. 179–206.

[70] Yager, R. R., "Generalized Probabilities of Fuzzy Events from Fuzzy Belief Structures," *Information Sciences*, Vol. 28, 1982, pp. 45–62.

[71] Ogawa, H., K. Fu, and J. T. P. Yao, "An Inexact Inference for Damage Assessment of Existing Structures," *International Journal of Man-Machine Studies*, Vol. 22, 1985, pp. 295–306.

[72] Yen, J., "Generalizing the Dempster-Schafer Theory to Fuzzy Sets," *IEEE Trans. on Systems, Man and Cybernetics*, Vol. 20, 1990, pp. 559–570.

[73] Yager, R. R., "Arithmetic and Other Operations on Dempster-Shafer Structures," *International Journal of Man-Machine Studies*, Vol. 25, 1986, pp. 357–366.

[74] Yager, R. R., "On the Normalization of Fuzzy Belief Structures," *International Journal of Approximate Reasoning,* Vol. 14, 1996, pp. 127–153.

[75] Dubois, D., and H. Prade, "On Several Representations of an Uncertain Body of Evidence," *Fuzzy Information and Decision Processes,* 1982, pp. 167–181.

[76] Dubois, D., H. Prade, and S. Sandri, "On Possibility/Probability Transformations," in *Fuzzy Logic*, New York: Springer, 1993, pp. 103–112.

[77] Dubois, D., and H. Prade, "A Set-Theoretic View of Belief Functions Logical Operations and Approximations by Fuzzy Sets," *International Journal of General System,* Vol. 12, 1986, pp. 193–226.

[78] Smets, P., "The Degree of Belief in a Fuzzy Event," *Information Sciences,* Vol. 25, 1981, pp. 1–19.

[79] Lucas, C., and B. N. Araabi, "Generalization of the Dempster-Shafer Theory: A Fuzzy-Valued Measure," *IEEE Trans. on Fuzzy Systems,* Vol. 7, 1999, pp. 255–270.

[80] Dubois, D., et al., "Probability-Possibility Transformations, Triangular Fuzzy Sets, and Probabilistic Inequalities," *Reliable Computing,* Vol. 10, 2004, pp. 273–297.

[81] Klir, G. J., and B. Parviz, "Probability-Possibility Transformations: A Comparison," *International Journal of General System,* Vol. 21, 1992, pp. 291–310.

[82] Slimen, Y. B., R. Ayachi, and N. B. Amor, "Probability-Possibility Transformation," in *Fuzzy Logic and Applications*, New York: Springer, 2013, pp. 122–130.

[83] Sudkamp, T., "On Probability-Possibility Transformations," *Fuzzy Sets and Systems,* Vol. 51, 1992, pp. 73–81.

[84] Destercke, S., "Fuzzy Belief Structures Viewed as Classical Belief Structures: A Practical Viewpoint," *IEEE Intl. Conf. on Fuzzy Systems (FUZZ)*, 2010, pp. 1–8.

[85] Smets, P., "Constructing the Pignistic Probability Function in a Context of Uncertainty," *UAI*, 1989, pp. 29–40.

[86] Sudano, J., "Pignistic Probability Transforms for Mixes of Low- and High-Probability Events," *Proc. of Fusion 2001,* 2001.

CHAPTER 5

Information Fusion Cells and Processing Strategies

The purpose of this chapter is to define the concept of information fusion cells as the smallest granular component of information fusion systems. Information fusion cells are then considered an intermediate conceptual level between different individual information elements and the global information processing systems. This behavior will be exhibited in the following sections according to several fusion strategies.

5.1 Introduction

Information fusion approaches mainly focus on the establishment of adequate belief models (BM) characterizing intrinsic information imperfections, as well as on the development of reasoning tools allowing imperfect information processing and fusion, through joint merging of these belief models. The most widely used belief models dealing with single information elements (J) [i.e., having single element (SE) output set] are: probability distributions Pr{.}, evidential mass functions $m(.)$, and possibility distributions $\pi(.)$. As previously shown, these belief models allow representing, reasoning, processing, and merging all forms of intrinsic imperfections that may affect punctual information elements. Membership functions $\mu(.)$ constitute the most adequate belief model allowing to represent and to process multiple and fuzzy information elements [i.e., having multiple element (ME) or fuzzy element (FE) output set].

In this chapter, the concept of information fusion cells (JFC) is introduced as the smallest granular component of information fusion systems. The information fusion cell is then considered an intermediate conceptual level between different individual information elements and the global information processing systems. This behavior will be exhibited in the following sections according to several fusion strategies.

The quality of fusion (QoF), as illustrated in Figure 5.1, measures the quality of the fusion process itself and not the quality of the produced information. It may be constraints or indicators relative to the fusion process (not the information that have their own quality description, QoI). Information fusion and analytics provide services to users. Such issues as efficiency and effectiveness are important in delivering actionable information. To afford interactions between future FIAT-based designs and users' information needs, metrics are required. The metrics chosen include timeliness, accuracy, throughput, confidence, and cost. These metrics [1] are similar to the standard quality of services (QoS) metrics in communications theory and human factors literature.

5.2 Defining an Information Fusion Cell

Gathering all the concepts previously introduced, an information element $J = (\Theta, X, \Omega)$ is considered as fully characterized if the following features are known, besides the input and output sets Θ and Ω:

1. The information input scope (ISc) indicates how the informative relation considers the elementary objects from the definition set (single-arity, plural-arity, fuzzy-arity).
2. The information output scope (OSc) indicates how the informative relation considers the elementary objects from the contents set (single-arity, plural-arity, fuzzy-arity).
3. The imperfection status (St) indicates if the information element is considered to be perfect (precise and certain) or imperfect (uncertain, imprecise, or ambiguous)/
4. Imperfection available knowledge is in terms of belief model (BM).
5. Information reliability weight (R).
6. Information consistency degree (*Cons*).

In this case, a fully characterized information element J is denoted as:

$$J = \left[(\Theta, X, \Omega), \mathrm{IS}_C, \mathrm{OS}_C, St, BM, R, Cons \right] \tag{5.1}$$

Most information processing systems consider information elements as fully reliable (i.e., having unit reliability weights) and fully consistent (having unit

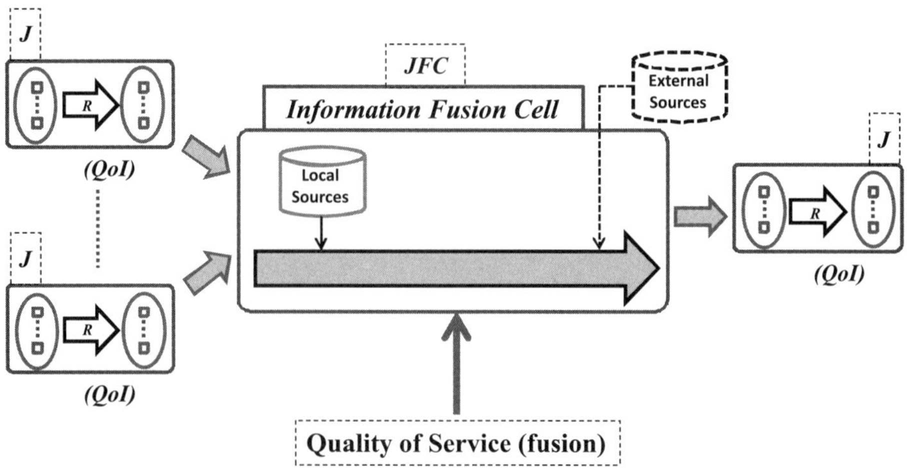

Figure 5.1 Information fusion cell concept.

consistency degrees). Otherwise, these contextual imperfection indicators are expressed and computed using available prior information. An information fusion cell is defined as the basic fusion platform allowing the fusion of information elements:

$$J_1 = \left[(\Theta_1, X_1, \Omega_1), \mathrm{IS}_{C1}, \mathrm{OS}_{C1}, St_1, BM_1, R_1, Cons_1 \right] \tag{5.2}$$

$$J_2 = \left[(\Theta_2, X_2, \Omega_2), \mathrm{IS}_{C2}, \mathrm{OS}_{C2}, St_2, BM_2, R_2, Cons_2 \right] \tag{5.3}$$

$$J_{1\oplus 2} = \left[(\Theta_{1\oplus 2}, X_{1\oplus 2}, \Omega_{1\oplus 2}), \mathrm{IS}_{C1\oplus 2}, \mathrm{OS}_{C1\oplus 2}, St_{1\oplus 2}, BM_{1\oplus 2}, R_{1\oplus 2}, Cons_{1\oplus 2} \right] \tag{5.4}$$

To generate a new information element exhibiting better characteristics, according to the targeted information fusion cell objective. Figure 5.2 illustrates the information fusion cell concept and introduces the following notations:

- $\Theta_{1\oplus 2}$ (resp. $\Omega_{1\oplus 2}$) denotes the resulting information definition (content) set.
- $IS_{C1\oplus 2}$ (resp. $OS_{C1\oplus 2}$) denotes the resulting information input (output) scope.
- $ST_{1\oplus 2}$ is the resulting information imperfection status.
- $BM_{1\oplus 2}$ is the resulting belief model associated with the information element $J_{1\oplus 2}$.
- $R_{1\oplus 2}$ (resp. $Cons_{1\oplus 2}$) denotes the resulting reliability weight (consistency degree).

The qualifier of better characteristics is strongly related to the specific aim of the information fusion cell (information quality, relevancy, and performance).

The proposed information fusion cell model definition has the following advantages:

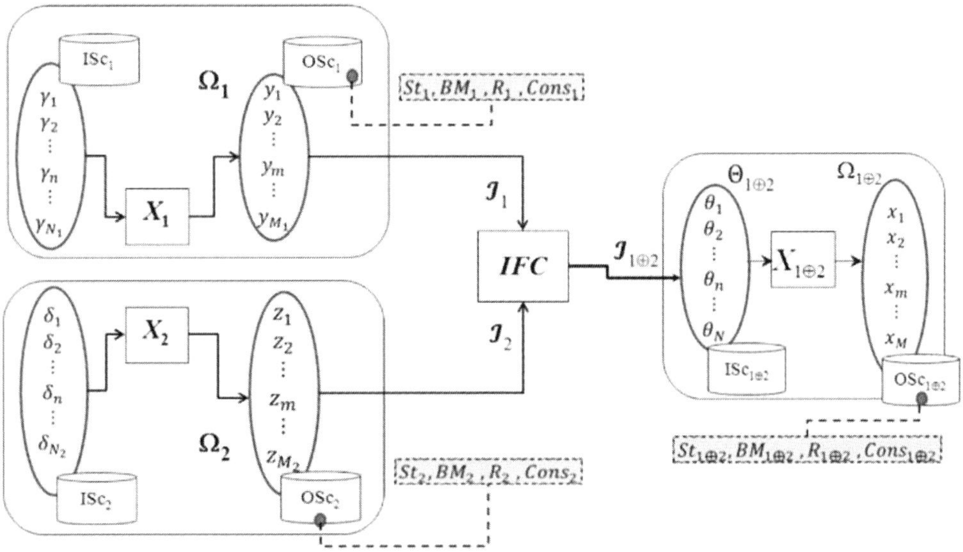

Figure 5.2 Information fusion cell concept and notations.

- *Self-contained:* The resulting product of an information fusion cell is an information element, as its own inputs. Therefore, evaluating the quality of an information fusion cell will use the same criteria as its constituting information elements.
- *Fully characterized*: All features (definition/content sets, input/output scopes, imperfection status, belief models, reliability weights and consistency degrees) characterizing individual input information elements as well as the resulting information element are clearly distinguished and expressed.

Therefore, the proposed information fusion cell model constitutes the most general conceptual framework for the fusion of two information elements. Indeed, the structural and conceptual gaps between this general and simple information fusion cell concept and the existing information fusion systems can be filled within two steps:

1. Defining subcategories of information fusion cells;
2. Considering the global system as a higher structural and conceptual information processing level where elementary information fusion cells are fit together homogeneously in order to fulfill the objectives targeted by the global information fusion system.

Thus, depending on the information fusion cell structural and semantic meaning aspects of both the input and the resulting information elements, four information fusion cell categories, called information fusion cell types, can be identified. These four types widely correspond to semantic objectives encountered for fusion strategies. These processing information fusion strategies will be presented as:

- Processing strategy 1: information fusion cell, Type 1 (data fusion);
- Processing strategy 2: information fusion cell, Type 2 (parallel belief fusion);
- Processing strategy 3: information fusion cell, Type 3 (sequential belief fusion);
- Processing strategy 4: information fusion cell, Type 4 (competitive belief fusion).

5.3 Processing Strategy 1: Information Fusion Cell (*JFC*) Type 1 (Data Fusion)

The first encountered information fusion cell type, called data fusion, considers the situation where both input information elements J_1 and J_2 (see Figure 5.3):

- Share the same definition set Θ (with potentially different input scopes);
- Have punctual informative functions where the corresponding outcomes are assumed to be perfectly known (i.e., precise and certain).

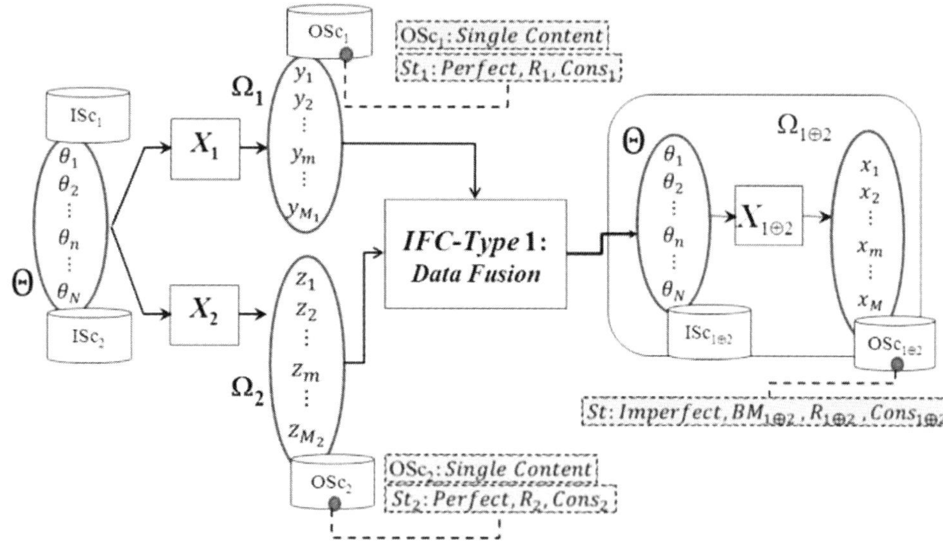

Figure 5.3 Information fusion cell, Type 1 (data fusion).

This information fusion cell type corresponds to the multisensor fusion case where each sensor captures, or measures, a particular feature of an observed object or a situation, related to the physical process of its own informative function. As J_1 and J_2 are considered as perfect information elements, this information fusion cell type exploits mainly the information content outcomes and can generally be expressed using a deterministic procedural approach. In some applications where the informative functions have physical significances, the resulting joint informative function is associated with qualifiers like knowledge-based or intelligent informative relation.

There are multiple potential tasks realized by data fusion information fusion cells:

1. Construction of a new joint information content set $\Omega_{1\oplus2}$;
2. Determination of the joint information output scope that may or may not remain punctual $OS_{1\oplus2}$;
3. Construction of the joint informative function $X_{1\oplus2}$;
4. Construction of a belief model $BM_{1\oplus2}$ associated with the joint information element;
5. Determination of the joint reliability weight $R_{1\oplus2}$ as well as the joint consistency degree $Cons_{1\oplus2}$.

The first three tasks (i.e., constructing the new content set, determining the scope and the joint informative function) are closely related and are generally conducted simultaneously.

5.3.1 Building the Fused Information Element

Based on the available knowledge as well as the targeted objective of the information fusion cell, the content set $\Omega_{1\oplus2}$ may be of three types:

- $\Omega_{1\oplus2}$ is the Cartesian product of the two content sets: $\Omega_{1\oplus2} = \Omega_1 \times \Omega_2$ It involves considering an augmented vector $x_m = [y_m, z_m] \in \Omega_1 \times \Omega_2$ of information contents. In this case, the resulting information content has a higher informational content level than the one characterizing the information input objects. A simple remote sensing example of this information fusion cell type is illustrated on Figure 5.4, where three-band images are merged, using a simple RGB representation space transformation, to produce a color-based informational content image.
- $-\Omega_{1\oplus2}$ can be described through a functional link involving the two input information content sets: $x_m = Function(y_m, z_m)$. It allows easing further information processing or interpretation steps. For instance, in pixel-based remote sensing imaging, where several images (corresponding to different image acquisition modalities) are available, a new image (i.e., new information pixel-based content) can be obtained as a simple color image or a linear combination of the input images (i.e., input information pixel-based content) [2–4]. In this case, the resulting joint informative function $X_{1\oplus2}$ is based on external available knowledge as well as on the physical interpretation of the input informative functions X_1 and X_2. The vegetation index (VI) widely used in remote sensing phonologic monitoring and vegetation classification is a straightforward example of this situation. It is defined as the arithmetic combination of two or more bands (i.e., information elements) related to the spectral characteristics of vegetation [5]. This example is illustrated in Figure 5.5.
- $-\Omega_{1\oplus2}$ is a higher-level decision content set where the resulting contents correspond to semantic categories of information input objects (i.e., decisions and classes). In this case, the joint informative function $X_{1\oplus2}$ is considered as a decisional informative function (based on external available knowledge in terms of models and/or training data and on the input informative func-

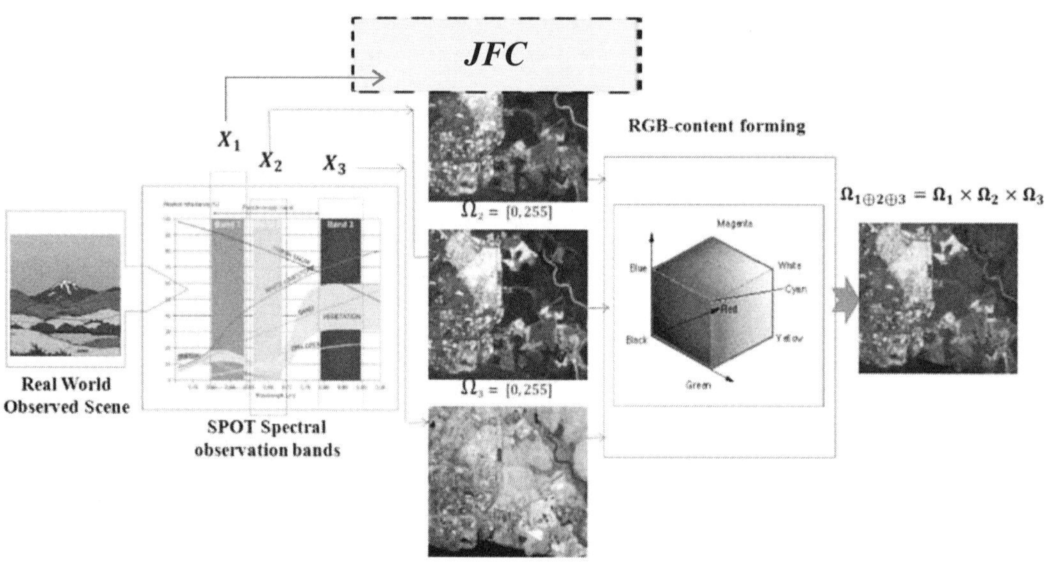

Figure 5.4 RGB information fusion cell.

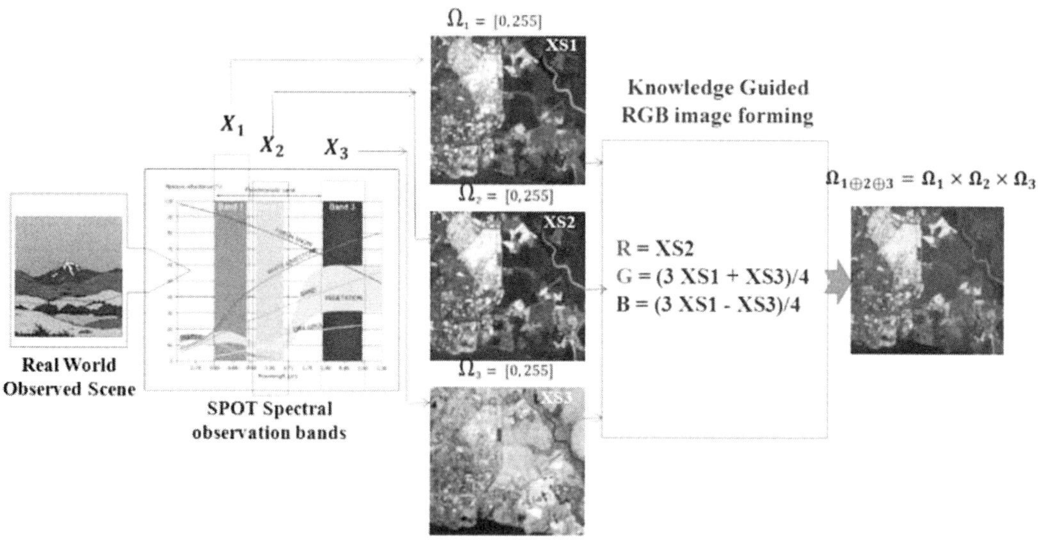

Figure 5.5 Knowledge-guided RGB information fusion cell adapted for vegetation index extraction.

tions) and implements a decision procedure using the two input contents and leading to the targeted decisions.

5.3.2 Remarks on the Construction of the Joint Informative Relation

The construction of the joint informative function $X_{1\oplus 2}$ may be limited to the only input information content outcomes (assumed available and perfect) or extended to the use of information fusion cell available knowledge (related to the informative functions X_1 and X_2) and to available external sources of knowledge. As already mentioned, an information fusion cell exploiting other knowledge sources than the only information content outcomes is generally referred to as a knowledge-based information fusion cell.

It is important to notice that the fact that both input information elements are assumed to be punctual and perfect does not exclude, in some systems, to get a nonpunctual and/or imperfect joint information element. For instance, let us assume in an automatic target recognition (ATR) system that the punctual and perfect outcome of the information element J_1 (resp. J_2) is: detected target uses A-type antenna (detected target speed is 850 km/h). The informative relation $X_{1\oplus 2}$ may then produce detected target is of type Tar_1 or Tar_2, which is not a punctual information, or may produce a probability distribution on the set of potential target classes, which is an uncertain joint information element. Therefore, the determination of the belief model associated with the joint information is a task frequently encountered in information fusion applications.

Finally, if the input information elements J_1 and J_2 are not fully reliable or not fully consistent, then determining the joint reliability weight $R_{1\oplus 2}$ as well as the joint consistency degree $Cons_{1\oplus 2}$ constitutes crucial issues for the characterization of the information fusion cell and is still an open research topic. Resulting from the previous explanations, this type of information fusion cell is obviously strongly application-dependent. This is its major drawback and therefore, from a modeling point of view, it is not realistic to go beyond the conceptual model given in Figure 5.3 and the general spectrum of potential tasks that may be encountered in existing systems. The basic task of symbolic registration fits in a Type 1 information fusion cell as X_1 and X_2 produces symbolic features from one scene observation. The task goal consists in associating features corresponding to the same scene components (objects, location, anchors, and so forth), features possibly coming from several modalities. The information fusion cell resulting information corresponds to an informative relation linking each scene component or subcomponent to a set of multimodal features (coming from both X_1 and X_2 output sets).

5.4 Processing Strategy 2: Information Fusion Cell, Type 2 (Parallel Belief Fusion)

The information fusion cell, Type 2, information fusion cells, called parallel belief fusion, have the same conceptual framework as the data fusion type framework. As shown in Figure 5.6, the fundamental difference is that the two input information elements J_1 and J_2 are considered imperfect information elements (i.e., punctual outcomes are not known), and both are represented through their associated:

- Belief models BM_1 and BM_2;
- Reliability weights R_1 and R_2;
- Consistency degrees $Cons_1$ and $Cons_2$.

Therefore, the focus of this type of information fusion cell is not limited to the constructions of the new joint informative relation $X_{1\oplus 2}$ and joint content set $\Omega_{1\oplus 2}$, but is extended to include the determination of the resulting joint belief

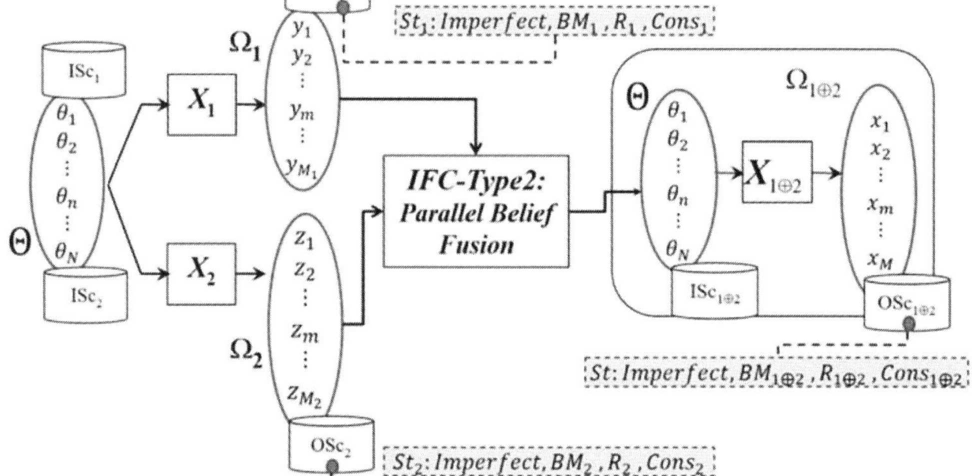

Figure 5.6 Information fusion cell Type 2 (parallel belief fusion).

model $BM_{1\oplus 2}$ defined on $\Omega_{1\oplus 2}$ as well as the computation of the joint reliability weight $R_{1\oplus 2}$ and consistency degree $Cons_{1\oplus 2}$.

A simple example of this information fusion cell is the one where J_1 and J_2 are two sensor-based probabilistic information elements. In this case, $\Omega_{1\oplus 2} = \Omega_1 \times \Omega_2$ is the Cartesian product of Ω_1 and Ω_2, and $J_{1\oplus 2}$ is the joint probabilistic information where the joint probability distribution is the associated joint belief model. More generally, all existing mathematical fusion frameworks (probabilistic, evidential, fuzzy, and possibilistic) include this belief fusion task as a major theoretical cornerstone component under the nomination of joint, multidimensional, or conditional belief model computation.

5.4.1.1 Example: Bayesian Decision-Making

Bayesian decision-making is a fundamental statistical approach to the generic pattern recognition and decision-making problems. It makes the assumption that the solution to the pattern classification problem is purely based on probabilistic belief values and that all the relevant probabilistic belief models are known [6].

In this case, the following notations are used:

- $\Theta = \{\theta_1, \theta_2, \cdots \theta_N\}$ is the set of observed patterns.
- $\Psi = \{y_1, y_2, \cdots y_K\}$ is the features set.
- $\Omega = \{x_1, x_2, \cdots x_K\}$ is the finite set of M classes.
- $\Pi\{x_m\} = m = 1 \cdots M$, $m = 1 \ldots M$, the prior probability distribution, $\Pi\{x_m\}$, is the probability that a randomly selected object $\theta \in \Theta$ belongs to the class x_m.
- $\Pr\{y_k \mid x_m\}$, the likelihood probability distribution, $\Pr\{y_k \mid x_m\}$, is the probability of observing the feature $y_k \in \Psi$ when the observed object θ is randomly selected from the restricted set Θ_m of objects belonging to the class x_m.

The Bayesian decision rule, corresponding to the minimum error rate classifier, consists in choosing the class for which the posterior probability $\Pr\{y_k \mid x_m\}$ is maximum with:

$$\Pr\{x_m | y_k\} = \frac{\Pr\{y_k | x_m\} \bullet \Pi\{x_m\}}{\Pr\{y_k\}} = \frac{\Pr\{y_k | x_m\} \bullet \Pi\{x_m\}}{\sum_{k'=1}^{K} \Pr\{y_k | x_m\} \bullet \Pi\{x_m\}} \tag{5.5}$$

Let J_1 and J_2 denote the prior and the likelihood information elements:

$$J_1 = \begin{bmatrix} (\Theta, X_1, \Omega), \mathrm{IS}_{C1} \equiv \textit{Single element}, \mathrm{OS}_{C1} \equiv \textit{Single element}, St_1 \equiv \textit{Uncertain}, \\ BM_1 \equiv \Pi,\ R_1 = 1, Cons_1 = 1 \end{bmatrix} \tag{5.6}$$

$$J_2 = \begin{bmatrix} (\Theta, X_2, \Omega), \mathrm{IS}_{C2} \equiv \textit{Multiple elements}, \mathrm{OS}_{C1} \equiv \textit{Single element}, St_1 \equiv \textit{Uncertain}, \\ BM_2 \equiv \Pr\{y_k | x_m\},\ R_2 = 1, Cons_2 = 1 \end{bmatrix} \tag{5.7}$$

Using these two information elements, it is straightforward to model the Bayesian decision rule as a parallel belief fusion information fusion cell (see Figure 5.7).

The resulting joint informative function $X_{1\oplus2}$ is based on the use of the two individual belief models $BM_2 \equiv \Pi, BM_2 \equiv \Pr\{y_k|x_m\}$, and computes the posterior probability belief model $BM_2 \equiv \Pr\{y_k|x_m\}$. One can notice that J_2 input scope is a multiple element scope with a partition of Θ corresponding to the output classes (defined by Ω): all the patterns θ belonging to the same class x define a subset of Θ. Thus, the informative function X_2 relies on external statistical knowledge for associating Θ and Ψ.

Symbolic registration also fits in an information fusion cells Type 2 as X_1 and X_2 may produce uncertain symbolic features from one scene observation; for instance, these features may result from classification operators identifying texture regions or specific objects. In this case, the information fusion cell resulting information corresponds to an informative relation linking each scene component or subcomponent to a set of probably heterogeneous symbolic classes (coming from both X_1 and X_2 output sets).

5.5 Processing Strategy 3: Information Fusion Cell Type 3 (Sequential Belief Fusion)

The third information fusion cell Type 3, called sequential belief fusion, represents certainly the major conceptual structure encountered in classical information

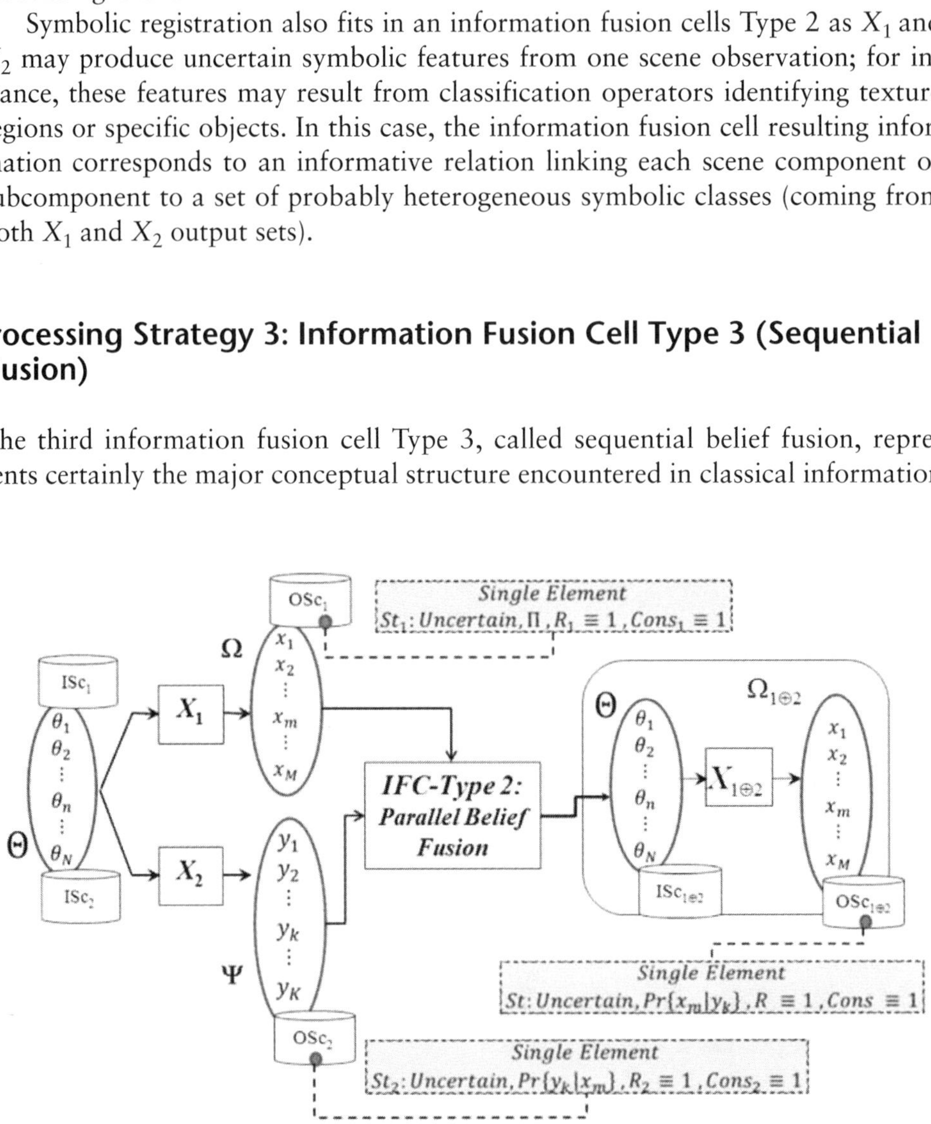

Figure 5.7 Information fusion cell Type 2 (example, Bayesian classification).

processing systems where the outcome of an information element is sequentially used as the input of another information element (see Figure 5.8).

With this type of information fusion cell, the question that naturally rises is: Should we consider this situation as an information fusion one or not? To answer this interesting question, two points of view should be distinguished.

The first point of view considers both information elements as independent and put in a cascade so that $X_{1\oplus 2}$ is structured as the application of X_1 (leading thus to information content outcomes from Ω_1) followed by the application of the informative function X_2. This vision shows that considering this situation as an information fusion one is an artificial vision.

The second point of view considers $X_{1\oplus 2}$ as the informative function linking objects from the definition input set Θ directly to outcomes from the content set Ω, avoiding explicitly going through the set Ω_1. $X_{1\oplus 2}$ certainly depends on X_1 and X_2, but to model a joint behavior of these informative functions, it may require external sources of knowledge. Therefore, this situation is not a simple sequential concatenation of two individual information elements but should be considered as producing a global and synthetic information element. This vision clearly means that this situation should be considered as an information fusion one as shown in the following examples of already existing sequential belief information fusion cells.

5.5.1 Example 1: Fuzzy Relations Composition

The first and straightforward example of application of the sequential belief fusion type concerns the composition of fuzzy relations. A fuzzy relation between two sets is defined as a fuzzy set over the Cartesian product of these two sets. Consider

$$J_1 = (\Theta, X_1, \Omega_1) \tag{5.8}$$

$$J_2 = (\Omega_1, X_2, \Omega) \tag{5.9}$$

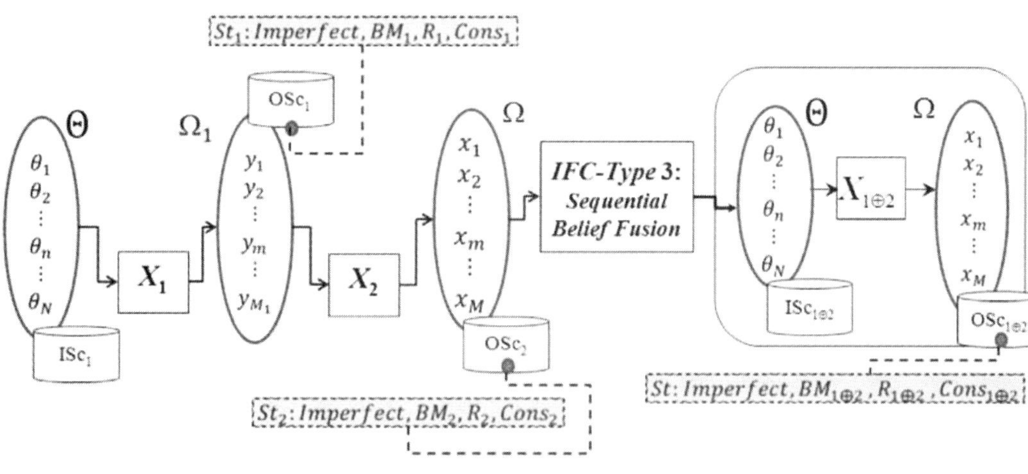

Figure 5.8 Information fusion cell Type 3 (sequential belief fusion).

as two fuzzy information elements for which the informative relations X_1 and X_2 define fuzzy relations associating elements from Θ to elements from Ω_1 and elements from Ω_1 to elements from Ω:

$$\mu_{X_1} : \Theta \times \Omega_1 \rightarrow [0,1] \qquad (\theta, y) \rightarrow \mu_{X_1}(\theta, y) \tag{5.10}$$

$$\mu_{X_2} : \Omega_1 \times \Omega \rightarrow [0,1] \qquad (y,x) \rightarrow \mu_{X_2}(y,x) \tag{5.11}$$

In this case, the fuzzy information element

$$J_{1\oplus 2} = J_2 \circ J_1 = (\Theta, X_{1\oplus 2} = X_2 \circ X_1, \Omega) \tag{5.12}$$

has the fuzzy informative relation $\mu_{X_1 \circ X_2}$ defined as the resulting max-min composition of the two elementary fuzzy relations:

$$\begin{aligned} &\Theta \times \Omega \rightarrow [0,1] \\ &(\theta, x) \rightarrow \mu_{X_1 \circ X_2}(\theta, x) = \max_y \left[\min\left[\mu_{X_1}(\theta, y), \mu_{X_2}(y,x) \right] \right] \end{aligned} \tag{5.13}$$

5.5.2 Example 2: Belief Transformation

Belief transformation denotes the general situation where $J_1 = (\Theta, X_1, \Omega_1)$ is an imperfect information element (for which a belief model BM_1 is available) and $J_2 = (\Omega_1, X_1, \Omega)$ is a deterministic information element. The problem to solve here is to define the global imperfect information $J_{1\oplus 2} = (\Theta, X_{1\oplus 2}, \Omega)$ and the resulting joint belief model $BM_{1\oplus 2}$. Three common actual cases are considered here: Zadeh's extension principle, random variables transformation, and evidential mass refinement and coarsening.

5.5.2.1 Zadeh's Extension Principle

By allowing the combination of fuzzy and deterministic information elements, Zadeh's extension principle [7] is a typical example of sequential belief fusion. Let us consider $J_1 = (\Theta, X_1, \Omega_1)$ a fuzzy information element associated with the membership function $\mu(y)$ defined on Ω_1 and $J_2 = (\Omega_1, X_1, \Omega)$, a mapping from Ω_1 to Ω such that for each $y \in \Omega_1$, $X_2(y) = x \in \Omega$. Zadeh's extension principle allows the determination of the membership function $\mu'(.)$ defined on Ω and associated with the fuzzy information element $J_{1\oplus 2} = J_2 \bullet J_1 = (\Theta, X_{1\oplus 2} = X_2 \bullet X_1, \Omega)$

$$\mu'(\mathrm{x}) = \sup_{y \in X_2^{-1}(x)} \mu(y) \tag{5.14}$$

Notice that Zadeh's extension principle also stands when combining deterministic and possibilistic information elements (the second one is affected by uncertainty and the available knowledge is ambiguous).

5.5.2.2 Random Variables Transformation

In the framework of probability theory, random variables transformation issue refers to the case where a given random variable is transformed, using a known analytical function, into a new random variable. The challenge is then to compute the new probability distribution of the transformed random variable.

Consider a probabilistic information element J_1 (random variable in probability theory) with a known probability distribution function $f_1(.)$ defined on Ω_1 and $J_2 = (\Omega_1, X_2 = g(X_1), \Omega)$, the transformation information element leading, thus, to a new random variable defined on Ω. $g(.)$ is a deterministic function. Probability theory allows the determination of the probability distribution function $g(.)$ defined on Ω and associated with the probabilistic information element $J_{1\oplus2} = J_2 \bullet J_1 = (\Theta, X_{1\oplus2} = X_2 \bullet X_1, \Omega)$ [8]:

$$f(x) = \sum_{y_r} \frac{1}{\left|g'(y_r)\right|} f_1(y_r) \tag{5.15}$$

where y_r denotes all real roots of the equation $g(y) = x$ and g', the derivate of the deterministic function $g(y)$.

5.5.2.3 Evidential Mass Refinement and Coarsening

In evidence theory, it is sometimes useful to consider a mass function with respect to frames of discernment Ω and Ω_1 having different levels of details. Consider the evidential information element $J_1 = (\Theta, X_1, \Omega_1)$ for which we assume known the mass function $m_1 : 2^{\Omega_1} \to [0,1]$ and let $J_2 = (\Omega_1, X_2 = g(X_1), \Omega)$ denotes the deterministic information element allowing the transformation of the mass function m_1 defined on the coarse frame Ω_1 into a mass function m_2 operating on the finer frame Ω. This transformation is called refinement while the opposite transformation from a fine frame Ω_1 to a coarser frame Ω is called coarsening [9], as shown in Figure 5.9.

Refinement requires Ω_1 to be a partition of Ω: that is, every $y \in \Omega_1$ has a corresponding nonempty subset $X_2(y)$ of information contents in the finer frame Ω and these subsets are mutually disjoint. Evidence theory allows such evidential belief transformations by defining over Ω the mass function $m_2 : 2^{\Omega} \to [0,1]$ associated with the evidential information element $J_{1\oplus2} = J_2 \bullet J_1 = (\Theta, X_{1\oplus2} = X_2 \bullet X_1, \Omega)$. For instance, the refinement of $m_1(.)$ is given by the mass function $m_{12}(.)$ over Ω such that:

$$m_2\left(\bigcup_{y \in A} X_2(y)\right) = m_1(A) \text{ for all } A \subseteq \Omega_1 \tag{5.16}$$

Basically, as already shown, typical processing chains from pattern recognition field can be described through information fusion cell Type 3.

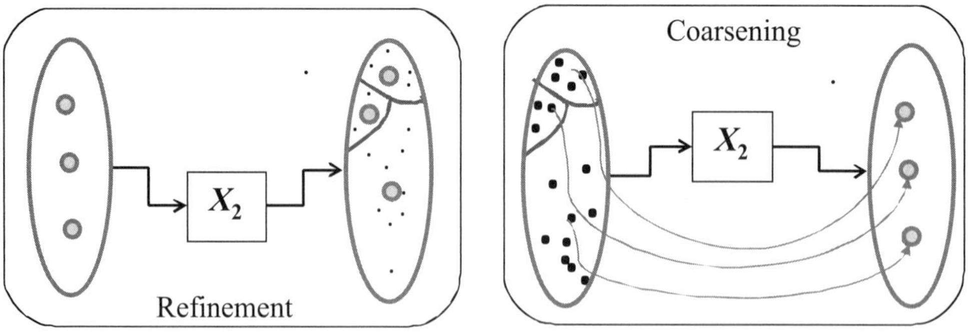

Figure 5.9 Coarsening and refinement in frames of discernment

5.6 Processing Strategy 4: Information Fusion Cell Type 4 (Competitive Belief Fusion)

Type 4 information fusion cells, called competitive belief fusion cells, are certainly the most studied and formalized fusion models by the information fusion community. Indeed, this information fusion cell type targets the conceptual model where both input information elements J_1 and J_2 share the same information input and information content sets as shown in Figure 5.10. Therefore, both individual information elements J_1 and J_2 address the same task but through two complementary or concurrent informative relations X_1 and X_2.

This information fusion cell type is frequently encountered in multiple-expert decision fusion problems (like pattern recognition, multisensor image classification, and scene interpretation). Indeed, each individual information element is assumed to capture a specific aspect of the problem depending on the actual extracted features, on the selected decision strategy or the similarity/dissimilarity measures involved in the process, as illustrated in Figure 5.11. Of course, these information elements have different strong and/or weak characteristics and performances.

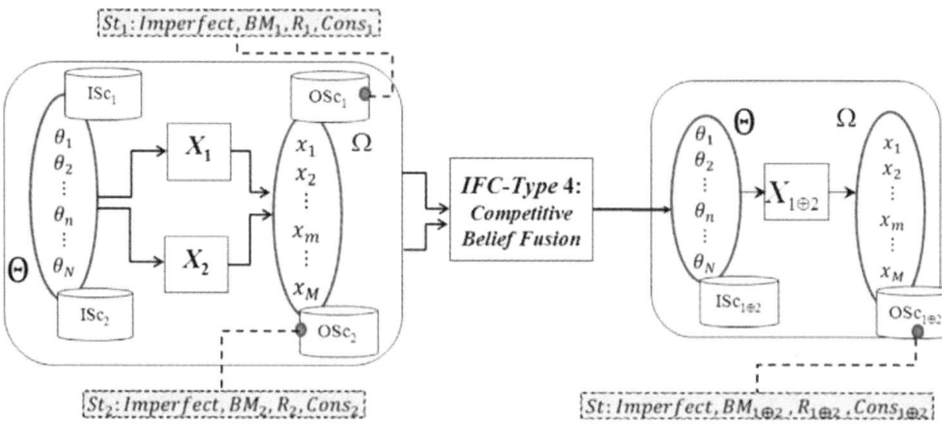

Figure 5.10 Information fusion cell, Type 4 (competitive belief fusion).

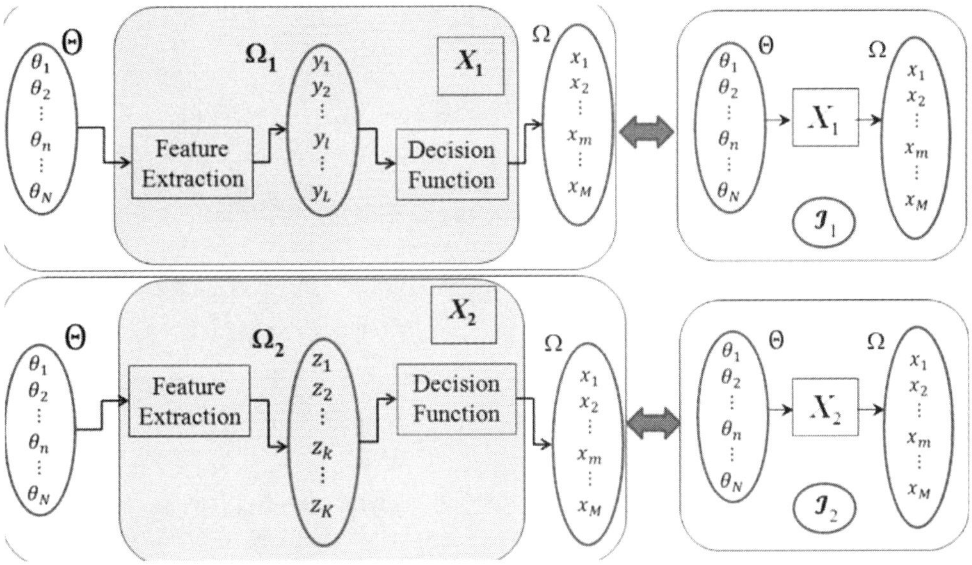

Figure 5.11 Multiple-expert decisional information elements model.

Notice that in some applications, different information elements may use the same feature set ($\Omega_1 = \Omega_2$) but have different decisional strategies or vice versa. All theories and approaches involved in information fusion propose, as a fundamental cornerstone, one or several techniques allowing competitive belief models fusion. Two of them will be described afterwards. Type 4 of the information fusion cell typically concerns all fusion mechanisms consisting on making results fusion to provide an improved information. For instance, two classifications methods may give information elements operating on the same input/output space with different classification operators. These operators probably have a non-uniform capacity of good classification over the whole sets. Indeed, combining their informative functions X_1 and X_2 into a fused one $X_{1\oplus 2}$ may bring a more interesting information.

5.6.1 Dempster Parallel Basic Probability Assignment Combination Rule

As already introduced, in the framework of Dempster-Shafer theory of evidence, the belief model BM_n associated with the information element is defined in terms of basic probability assignments [10]: $J_n = \left[(Q_n, X_n, W_n), IS_{CN}, OS_{CN}, St_n, BM_n, R_n, Cons_n\right]_{n=1,2}$.

The relation allowing the computation of the joint reliability weight $R_{1\oplus 2}$ with the elementary reliability weights R_1 and R_2 remains an open research issue. In fact, most information fusion approaches do not try to compute $R_{1\oplus 2}$ explicitly. In a first step, they take into account all information elements' reliability weights (or weight vectors) to adjust the input belief models. This adjustment operation, called belief discounting, is assumed to lead to what can be considered as fully reliable information elements, for the following processing steps.

Once the integration of the reliability weights has been performed, the second step merges the belief models under the assumption of fully reliable input information elements, thus leading to a fully reliable resulting belief model. Several discounting rules allowing the integratation of reliability weights R_n, $n = 1,2$, are proposed in the literature [10–12]. One of them uses the reliability weight R_n, n

= 1,2, to redistribute the degrees of belief between different content subsets and adjust the occurrence belief of Ω so that the total belief remains equal to the unit:

$$m_n^{discounted}(A) = R_n \bullet m_n(A), \qquad \forall A \subset \Omega, \; n = 1,2 \tag{5.17}$$

$$m_n^{discounted}(\Omega) = (1 - R_n) + R_n \bullet m_n(\Omega), \qquad n = 1,2 \tag{5.18}$$

Once adjusted, both information elements are considered as fully reliable and Dempster rule of combination can thus be used to perform the belief models fusion.

5.6.2 Hybrid Competitive Belief Models Fusion

A crucial and challenging information fusion tasks concerns the situation where the available belief models BM_1 and BM_2 come from different mathematical models. For instance, in the case where the resulting information element is punctual $(OS_{C1\oplus 2} \equiv Single\ element)$, several configurations can be encountered as given by Figure 5.12:

- *Case 1:* Probabilistic and basic probability assignment models fusion;
- *Case 2:* Possibilistic and basic probability assignment models fusion;
- *Case 3:* Probabilistic and possibilistic models fusion.

Concerning Case 1 (probabilistic ↔ evidential), two potential solutions are used:

- The first solution (probabilistic → evidential) considers the probability distribution as a special case of evidential mass functions called Bayesian mass function, where all belief values are distributed over information contents only. No mass is attributed to nonsingleton subsets of Ω. This consideration allows the straightforward fusion of the two evidential mass functions.

BM_2 \ BM_1	Pr_1	m_1	π_1
Pr_2		Case 1	Case 3
m_2	Case 1		Case 2
π_2	Case 3	Case 2	

Figure 5.12 Hybrid competitive belief models fusion.

- The second solution (evidential → probabilistic) is based on transforming the mass function into a probability distribution belief model using, for instance, the pignistic transformation. This automatically leads to consider the fusion of two probabilistic belief models.

When Case 2 is encountered, the first solution (possibilistic ↔ evidential) transforming the belief models can be conducted in both directions. In fact, when the basic probability assignment m is consonant (focal elements are nested), an induced possibility distribution π can be defined from m:

$$\forall x_n \in \Omega . \pi(x_n) = Pl(\{x_n\}) = \sum_{A / A \cap \{x_n\} \neq \emptyset} m(A) \tag{5.19}$$

where $Pl(.)$ is the plausibility function computed from the basic probability assignment and x_n, a singleton from Ω. Therefore, the hybrid concurrent fusion operation is reduced to the fusion of two possibility distributions. Conversely, as any possibility distribution can be transformed into a consonant basic probability assignment [13], the concurrent fusion process becomes a fusion process of two basic probability assignments.

Concerning Case 3 (possibilistic ↔ probabilistic), the first approach consists in transforming both information elements into evidential ones. However, a thorough correspondence also exists between probability and possibility theories [14], considering possibility as an upper probability; indeed, the probability addition operator corresponds to the maximum operator within the possibility framework.

5.7 Remarks on the Design of Distributed Information Fusion Systems (DIFS)

This section introduces the way ahead with respect to the exploitation of the framework presented along this chapter in the design of distributed information fusion system (DIFS). The most general form of a distributed system is a network. It is hence a question of information fusion networks. From a knowledge engineering point of view, DIFS are composed of a smaller set of hopefully autonomic independent submodules, including fusion cells (information fusion cells), usually designed to carry out independent tasks and procedures, which then can be developed, evaluated, and maintained according to their specific requirements.

As illustrated in Figure 5.13, a DIFS has a specific objective (or query) to respond. The DIFS relies on information sources through information elements (J) and processing through information fusion cells to deliver actionable knowledge (i.e., knowledge one can directly use to fulfill objectives). Architecture decomposition of an *information fusion system* (IFS) complex processing systems comes with a number of advantages, most notably with respect to robustness, flexible development, and maintenance of components, as well as the combination of knowledge from different sources, but requires powerful techniques to bring the different contributions together again. A number of quite different reasons can be identified that speak in this favor:

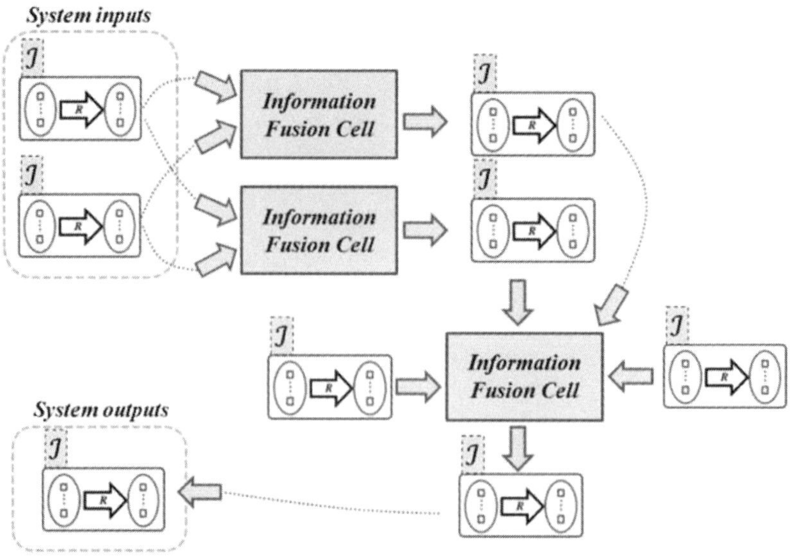

Figure 5.13 Distributed information fusion system.

1. There is a necessity to combine information from different sensory information element channels. However, even in a single-information element modality system, it might become necessary to integrate complementary information.
2. Different processing components have been developed based on completely different paradigms (e.g., the combination of corpus-oriented approaches with others based on more traditional manual modeling techniques). If successful, it would allow a better interface for automatically acquired models with existing knowledge sources like semantic databases. Another application is the integration of direct instruction techniques, which are needed (e.g., in interactive environments).
3. Different components have been designed to separate different knowledge sources to improve on certain aspects of system behavior like portability to other domains or perspicuity with respect to system failure.
4. Information fusion has to be attempted at a proper level of granularity, which allows effectively supporting good local choices.
5. An assessment function is required to evaluate the different system submodules according to their potential contribution to the processing goals. This function can be based on a self-evaluation of the respective processing component (usually given as a kind of confidence score), or an external evaluation based either on purely formal criteria. The availability of such an assessment function is most crucial if alternative (partial) solutions have to be integrated into a single one.
6. The mutual dependence of components should be minimized to avoid system failures due to missing or conflicting information. This requires a flexible integration scheme based on some kind of soft computing. If suitably used, it might prevent a failing component from turning down the whole processing chain.

5.8 Conclusion

This chapter presented a systemic information fusion framework having a functional structure suitable for processing any level of abstraction: data, information, and knowledge. The proposed framework is information-centric as characterization of an element of information *J* is of prime importance. The definition of information fusion cells to process sets of information elements *Js* and their composition into various forms of architecture (DIFS) allow the processing of multiple heterogeneous sources of information: a necessity to understand situations evolving in most of today's complex dynamic environments.

References

[1] Blasch, E., P. Valin, and É. Bossé, "Measures of Effectiveness for High-Level Information Fusion," in High-Level Information Fusion Management and Systems Design, E. Blasch, É. Bossé, and D. A. Lambert, (eds.), Norwood, MA: Artech House, 2012, pp. 331–348.

[2] Gungor, O., and J. Shan, "An Optimal Fusion Approach for Optical and SAR Images," Proc. of the ISPRS Commission VII Mid-Term Symposium: Remote Sensing: From Pixels to Processes, 2006, pp. 111–116.

[3] Yin, Z., and A. A. Malcolm, "Thermal and Visual Image Processing and Fusion," *Singapore Inst. Manuf. Technol., Singapore, Tech. Rep. AT/00/016/MVS,* 2000.

[4] Sun, F., S. Li, and B. Yang, "A New Color Image Fusion Method for Visible and Infrared Images," IEEE Intl. Conf. on Robotics and Biomimetics 2007 (ROBIO 2007), 2007, pp. 2043–2048.

[5] Huete, A., et al., "A Comparison of Vegetation Indices over a Global Set of TM Images for EOS-MODIS," Remote Sensing of Environment, Vol. 59, 1997, pp. 440–451.

[6] Duda, R. O., P. E. Hart, and D. G. Stork, *Pattern Classification*, 2nd ed. New York: Wiley, 2001.

[7] Zadeh, L. A., "Fuzzy Sets," *Information and Control*, Vol. 8, 1965, pp. 338–353.

[8] Casella, G., and R. L. Berger, *Statistical Inference*, Vol. 2, Duxbury Pacific Grove, CA: Duxbury, 2002.

[9] Kohlas, J., and P. -A. Monney, "A Mathematical Theory of Hints (an Approach to the Dempster-Shafer Theory of Evidence)," *Lecture Notes in Economics and Mathematical Systems*, 1995.

[10] Shafer, G., *A Mathematical Theory of Evidence*, Princeton, NJ: Princeton University Press, 1976.

[11] Lefevre, E., O. Colot, and P. Vannoorenberghe, "Belief Function Combination and Conflict Management," Information Fusion, Vol. 3, 2002, pp. 149–162.

[12] Smets, P., "Data Fusion in the Transferable Belief Model," Proc. of 3rd Intl. Conf. on Information Fusion 2000 (FUSION 2000), vol. 1, 2000, pp. PS21–PS33.

[13] Dubois, D., and H. Prade, "On Several Representations of an Uncertain Body of Evidence," Fuzzy Information and Decision Processes, 1982, pp. 167–181.

[14] Dubois, D., et al., "Probability-Possibility Transformations, Triangular Fuzzy Sets, and Probabilistic Inequalities," Reliable Computing, Vol. 10, 2004, pp. 273–297.

CHAPTER 6

A Holonic Processing of Information Fusion Cells

This chapter proposes a holonic processing framework of information fusion cells. Several concepts borrowed from complex systems theory, informational philosophy, and computer sciences have been integrated to conceptualize that framework. The concepts of holon and informon developed by Koestler et al. are exploited here to develop an information fusion processing framework. The proposed functional holonic structure is suitable for processing the four levels of information fusion cells proposed in the U.S. Joint Directors of Laboratories (JDL) model. The framework comprises the notions of information elements and information fusion cells described in previous chapters and proposes a functional holonic processing structure to network the information fusion cells and achieve fusion of information. The framework advocates a goal-driven approach with notions coming from business sciences to take into account quality of information for managing the fusion process. The framework is illustrated with an elaborated case in remote sensing. Finally, a discussion follows on a holonic interpretation of the JDL data and information fusion model.

6.1 Introduction

Let us recall that in the military domain, data and information fusion comprise (more or less) analytics while in civilian domains analytics comprises (more or less) data and information fusion. To meet expectations of adequately supporting decision-makers, information fusion (IF), in a large sense encompassing analytics, needs to go beyond the JDL data fusion model and its revisions [1–5] that have dominated the last 25 years of contributions in that field. Lambert [6] stated: "Data fusion is currently dominated by a machine based sensor fusion paradigm. Machine based higher-level fusion is acknowledged in the dominant JDL model for data fusion through the inclusion of amorphous 'level 2' and 'level 3' modules, but these higher levels are not yet supported by a standardized theoretical framework."

There is a need to extend beyond machine-based data fusion to fusion systems involving both people and machines. To do so, we need a general IF framework to design all fusion levels but especially for high-level information fusion (HLIF) systems where the focus is more the relations between objects and their impact on situations. Lambert [6] stated: "The traditional roots of the data fusion community are in sensor fusion, where the 'data sources' are established sensors and the 'aspects of interest in the environment' are moving objects, each typically represented by a set of state vectors. The broader definitions reflect an increasing emphasis toward generalizing sensor fusion into so called higher-level fusion, in which 'the aspects of interest in the environment' are not restricted to objects.'

Kokar et al. [7] pointed out the lack of theoretical tools in information fusion domain to analyze the performance of fusion systems before building them. Discussions in [8] called for a framework in which knowledge and information can be represented, combined, interpreted and managed; where decision theories are exploited to impact situations or the environment; and multi-agent systems theories to formalize the distributed aspect. In a 2010 survey of information fusion frameworks, Llinas [9] concluded: "While a number of efforts have been made to define a robust, extended-domain Information Fusion Framework, our review of past works indicates that no such Framework has been well-defined."

In this chapter, we use the term "framework" in the same sense as in Llinas [9], that is, a software framework: "a conceptual structure... an abstraction, not populated by domain-specific components... an abstraction in which common code providing generic functionality can be selectively overridden or specialized by user code providing specific functionality."

In addition, the Board of the International Society for Information Fusion (ISIF) [10] created in 2011 the Fusion Process Model and Frameworks Working Group (FPMFWG), recognizing the insufficiency of the JDL model for the design of information fusion systems for complex dynamic environments such as cyber physical and social systems (CPSS). The trend of adding layers to the JDL model to cover missing aspects when confronted to CPSS complexity could help to define such framework. This is even more evidenced by Lambert [6], who had to deconstruct JDL to propose his blueprint for high-level fusion.

In the past decades, government agencies, universities, and industries deployed a lot of efforts to develop several implementations of fusion systems using a diversity of architectures, software platforms, and languages [11–19]. A consequent and legitimate question is why the information fusion community still lacks that general framework after almost two decades and all these substantial efforts. The answer is partially within the great challenges identified in Lambert [20] and revisited in [21]: paradigm, semantic, epistemic, interface, system, design, and evaluation. To adequately address these challenges, the information fusion community needs more than the traditional ways to solve hard sciences problems where reductionism has been the dominant approach [22]: "beginning with the simplest and most easily understood objects, and gradually ascending to the knowledge of the most complex." The design of higher levels of fusion in CPSS environment where people and machines are interconnected cannot be boxed into any single discipline but requires an interdisciplinary understanding (holism instead of reductionism) based on scientific foundations that have not yet been invented. New sciences such as systems biology [23], evolutionary economics [24], network theory [25], and the science of complex systems [26] should take an increasing significance to the information fusion community.

Having that holism perspective in mind, this chapter provides the following contributions towards the definition of a general framework to support the design of information fusion systems for dynamic complex environments:

- Definition and characterization of a basic element of information using the concepts of informon from archetypal dynamics [27] and holon [28] from complex systems theory;
- Definition and characterization of an information fusion processing cell: its core functions, properties, and holonic structure;
- Definition of an information fusion functional processing framework and its illustration through a remote sensing case;
- Discussion on a holonic functional interpretation of the JDL data fusion model.

6.2 Interpretation of Information Elements (J_s) as Holons and Informons

The success of information fusion systems is strongly related to the way its basic components are defined, to the quality of their associated knowledge as well as to the exploitable knowledge produced by the fusion process itself. Unfortunately, few papers in the information fusion community address this topic. Moreover, there is no agreement on what information is and what defines its characteristics. Several studies discussed the differences between data, information, and knowledge for years [29, 30] but the results are still controversial as discussed in Chapter 3. These studies tried to find the differences among concepts properties/characteristics. For instance, words such as data, facts and information are often used interchangeably with knowledge [30]. Knowledge, Information, and Data (KID) can be classified by their degrees of abstraction and by their quantity. The terms are hierarchical in nature, with data being the foundation upon which information builds meaning to a pinnacle of knowledge [15]. Using the notion of informon from archetypal dynamics does not require a clear distinction between levels of the KID hierarchy. Informons can be at any level: data, information, knowledge, or even wisdom [31].

The most basic unit of information in archetypal dynamics is termed "informon." The word *informon* is the fusion of the Latin *informo* and the Greek *on* (being, entity). Sulis [27] stated: "An informon simply refers to any aspect of reality that possesses the capacity to inform. An informon is an aspect of the reality that exists prior to any semantic frame. An informon must be attached to, coupled to, or in resonance with, a semantic frame in order to be given form, meaning, and behavior." Alonso et al. [32] stated: "Informon is the basic element of information that has sense for a holon and that allows it to make the right decisions and to execute the proper actions."

The term *holon* has been introduced by Koestler [28] to characterize organisms and societies that are a "multi-levelled hierarchy of semi-autonomous sub-wholes, branching into sub-wholes of a lower order." The term *holon* is from the Greek *holos* (whole) plus the suffix *on*, as in *informon*. Holons group in temporal hierarchies (networks) of collaboration called holarchies. When a holarchy is formed by companies, each of the possible enterprises corresponding to a different configuration of the network is called a virtual enterprise [33, 34].

Why are holons promising for information fusion in future CPSS environments? Holons possess two main properties [28]. The first property of having a hierarchic organization ensures that holons are stable forms that are able to withstand disturbances. The second property is that holons possess a certain degree of autonomy and can handle contingency plans although subject to control from higher authorities (e.g., hierarchy of the JDL model).

6.2.1 Holon Definition Within the Context of Information Fusion

The human sense-making mechanism consists of three elements: an object (or event), a framework (semantic frame), and an association between them [35]. There is a similarity in archetypal dynamics where meaning is an ongoing process arising from interpretation [27]: "Meaning is a local construct, requiring an interplay between semantic frame, agent/user, system and context." The context plays an essential role in our ability to interpret and to use information that is presented to us. A very often quoted definition of context among all the definitions is by Dey [36] (see Chapter 2): "Context is any information that can be used to characterize the situation of an entity. An entity is a person, place, or object that is considered relevant to the interaction between a user and an application, including the user and applications themselves." Laudy [37] reinforced the necessity to integrate the context in situation understanding by stating: "No understanding is possible without knowing the context in which the process of perception of a situation occurs."

For instance, consider the following statements that are potential system's inputs. First, state a given value of $x = 41$. There is no more to understand from x than that it is a positive integer. Second, consider the statement $x = 41°C$ (105.8°F). It denotes an element from a set of meaningful semantics. The value is related to temperature. Given that 41°C is the temperature of somebody, one can start reasoning about it and concluding a fever and a need to be examined by a medical doctor. Moreover, it can be of great importance to know the sensor (mercury, electronic, plastic strip thermometers), the method of measurement (oral, rectal, gut), the sensor specifications or the values of any specific tuning on the measurement tools. Further interpretation or adequate use of "John's temperature is 41°C," for instance, may be greatly impacted. Thus, all these contextual information define an internal context intrinsically related to the informative relation (here the sensor). Interpretation of "John's temperature is 41°C" also depends on the acquisition conditions. Indeed, it is rather different in terms of subsequent interpretations to know that the temperature is taken while John is in a sauna or outside, for instance. All these contextual information define an external context as it may introduce different interpretations even if John's temperature is actually 41°C. External context extent also depends on the overall goals of the complete system as for instance, it may not be important to know exactly why John's temperature is 41°C but only that it is over 37°C (100.4°F)

According to Losee [38], information can be defined in terms of a process or function: "Information is the value currently attached or instantiated to a characteristic or variable returned by a function, or a process. The value returned by a function is informative about the function's argument, or about the function, or about both." This definition has the merit of positioning information as a relational or functional concept linking data sets [39]. The examples above mentioned highlight that observing data from a given set is not enough to make it an informative entity. Information, hence, requires a content set, how its outcome is obtained and to what it refers. In this way, Stonier [40] highlighted "the full information extent cannot be grabbed when working only on the contents set." This leads to the following definition of information which extends the definition proposed by Losee [38] by including a formal structure of a holon (H).

The definition of a holon, H, is an epistemic entity composed of a definition set and a content set linked by a functional relationship called informative relation, associated with internal and external contexts.

The holon H is denoted as H = information definition set, informative relation, information content set, internal context, and external context. Holons are making use of contexts to give meaning to informons T. This definition of a holon is the same as the information element J detailed in Chapter 4. Therefore, as shown in Figure 6.1, the main components of a holon H are:

1. A definition set giving the potential information input elements (to what the information refers): informons $(T_{n=1,N})$.
2. A content set encoding the possible knowledge produced by the information (e.g., measurements or estimations of physical parameters, decisions, hypothesis): informons $(T_{m=1,M})$.
3. An informative relation (input-output relationship) representing the model (e.g., mathematical, physical, conceptual) that associates the input elements with the produced information contents.
4. An internal context gathering intrinsic characteristics, constraints, or controls about the informative relation itself: semantic frames.
5. An external context containing data, information, or knowledge useful to the elaboration of the meaning or the interpretation of the information element.

The meaning of the three components of Figure 6.1 is rather straightforward. Objects and contents (informons) represent entities linked through the informative relation. However, the nature of these entities may be either hard (quantitatively defined with numbers, individuals) or soft (qualitatively defined with words, opinions, predictions). An informative relation may be purely impersonal (or hard) when it does not depend on external conditions to link objects to contents. However, it may also behave in a softer way by using cognitive factors like subjective judgments, opinions, and perception (e.g., from human experts). To deal with both kinds of entities/relations, holons must use semantic frames to give meaning to informons.

As already mentioned, informons can be at any level of the KID hierarchy. It may have a fractal structure, in the sense that it may be considered formed by many informons, depending on the degree of generalization needed by the holon that uses it: for a certain holon a record of a certain database, D will be an informon, while for another holon, the whole ontology that includes that database D will be its informon. As long as informons are processed by holons, informons of a higher

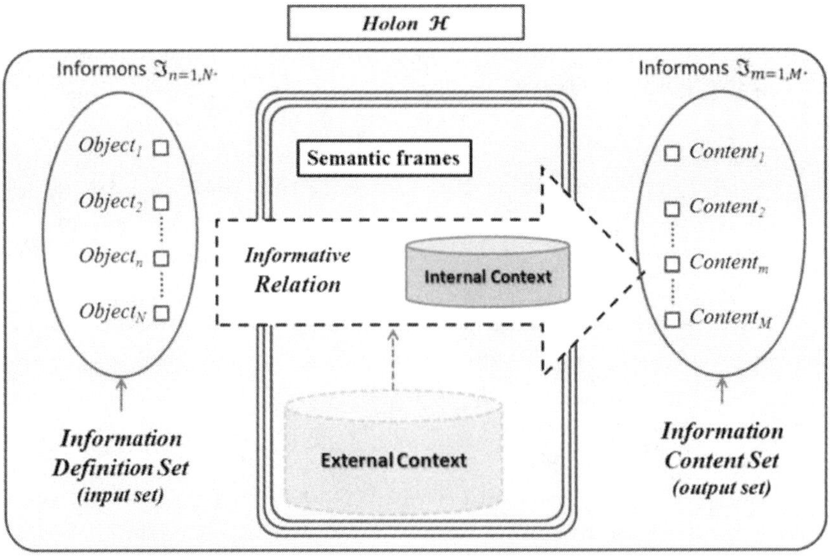

Figure 6.1 Holon *H* structure.

semantic level emerge. A holon always uses at least one informon (the description of the system state in which it operates) and, in turn, an informon must be used by at least one holon to have sense; it is a symbiotic relationship between the concepts of holon and informon as phrased in Paggi [41]: "There exists a relationship of dependent existence between the concepts of holon and informon, meaning that there cannot exist a holon without informons ... neither informons without holons (because of the definition of informon). Given that every holon can be considered as an informon, but not conversely...." This obviously applies to information element, *J*, defined in Chapter 4: *J* is a holon that works on informons from information definition set $(T_{n=1,N})$ and information content set $(T_{m=1,M})$, as pictured in Figure 6.1. These informons are not necessarily holons.

6.3 Holon Characterization

The characterization is done considering the same fundamental axes as in Section 4.4:

1. Information inner sets scope (e.g., resolution and level of abstractions): how the sets of informons are built (level of details, problem representation, image resolution);
2. Informative relation arity: how sets of informons are considered through the informative relation;
3. Informative relation objectivity: on which ground the informative relation links input and output sets;
4. Information imperfections.

We will discuss only the fourth point on information imperfections with respect to the notion of holon.

6.3.1 Types of Holons According to Uncertainty Representations

Holons are entities that decide based on informons. These decisions are coping with two main kinds of uncertainty [42]: randomness and epistemic. Randomness derives from variability of a phenomenon or a system under study and epistemic uncertainty derives from a lack of knowledge about a quantity, a system, or a phenomenon under study. We refer the reader to Paggi and Alonso [43], who presented a survey-type discussion on relating uncertainty, beliefs, holons, and informons.

A holon $H = (\Theta, X, \Omega)$ is said to be certain, if the informative relation X is punctual and the unique content outcome x_{True} is certain and precisely known. Three main approaches are widely used for uncertainty representation: probabilistic [44], evidential [45], and possibilistic [46, 47]. These representations have been used to define the various types of uncertain holons. The list in Figure 6.2 is not exhaustive and other holons can be defined depending upon the way uncertainty is represented. We propose to use the representations of Table 4.2 and Figure 4.9 since information elements and holons are equivalent.

6.4 Information Fusion Cell as a Network of Holons (*H*)

In this section, the concept of information fusion cell introduced in Chapter 5 is translated into the notion of holons to perform all functions required to fuse information coming from more than one source. An information fusion cell is equivalent to the notion of fusion node in the revised JDL [2] (see Figure 2.20) and to some extent to the fusion engine described in Bossé et al. [48] to characterize an information fusion system. An information fusion cell is composed of information fusion core functions. We say the functions are holonic since they can be applied to any level of abstraction of the JDL model and they can be described as holons. As illustrated in the top part of Figure 6.3, the revised JDL [2, 49], presented a fusion node construct upon three functions: common referencing, data association, and state estimation. Another source, Nicholson [50], used five core functions: detection, classification, prediction, correlation, and assimilation (function of combining) for a situation awareness processing engine.

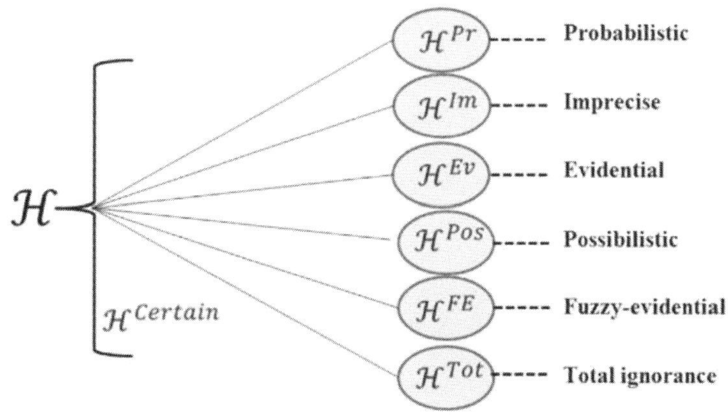

Figure 6.2 Types of holons matched with uncertainty representations.

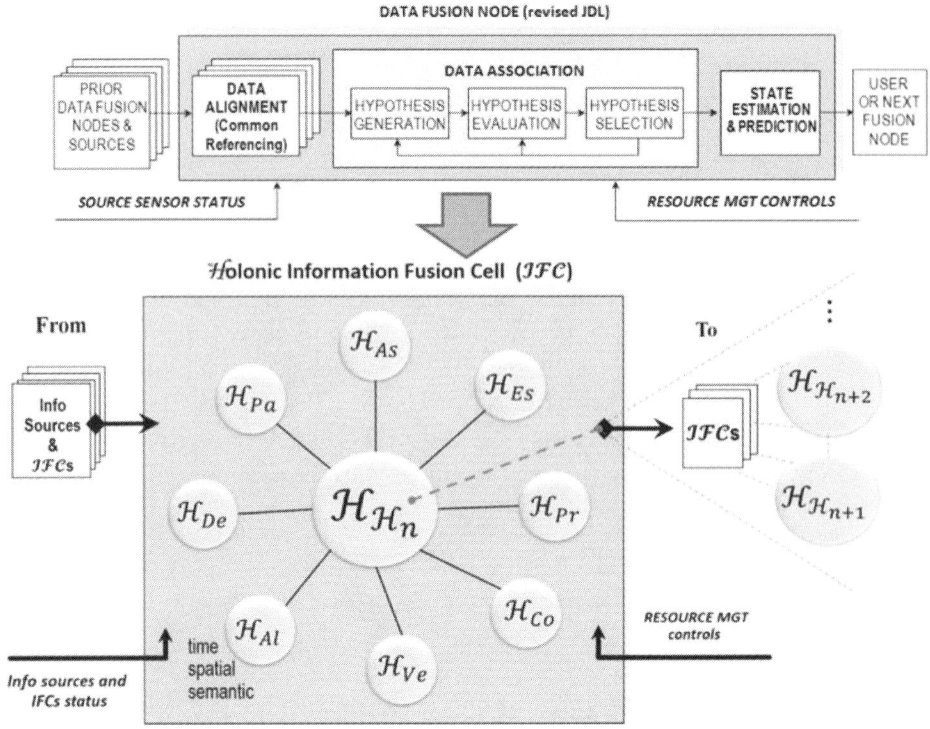

Figure 6.3 From the JDL data fusion node to a holonic network of information fusion cells.

The idea here is to define a more exhaustive set of core functions for an information fusion cell with a better characterization of its system properties so that, for instance, formal approaches such as in Kokar et al. [7] using category theory, ISSA [51], STDF [6], and causal tapestries from Sulis' archetypal dynamics [27] could contribute to adapt and advance current software engineering methodologies for specifying and designing future complex systems.

6.4.1 Holonic Information Fusion Core Functions

The following set of functions, that includes the three functions of the revised JDL, is being proposed to compose an information fusion cell:

1. $H_{Al=alignment}$ (Alignment): time-space-semantic. The world is observed in time and space. To make sense out of observations, one needs to understand the context. Common referencing under these three dimensions (time, space, semantics) is necessary. Registration in time and space is well known in image processing. Context might be represented using ontologies so ontology alignment is an example of semantic alignment.
2. $H_{De=detection}$ (Detection): Informons $T_{n=1,N}$. Applying the term detection to the levels of the KID pyramid, this generic function covers the field of data mining or knowledge discovery, that is, the process of automatically searching large volumes of data for complex patterns down to the bottom of the KID pyramid. Semantic labeling is an example of this function. Granular computing might be considered as a detection of complex patterns.

3. $H_{Pa=partition}$ (Partition): $Part_{k=1,K}\left(Object_{n=1,N} \rightarrow Class_{m=1M}\right)$. Facing the complexity of the world, partitioning eases the KID processing to understand that world. Clustering and classification are examples of partitions.
4. $H_{Co=combination}$ (Combination): Informons $\sum_{n=1,N,T_n}$. After the informons have been well characterized, this generic function combines those elements to gain better awareness. Aggregation and integration are other terms that can perform that function of combination.
5. $H_{Ve=veracity}$ (Veracity): (*true*, ¬*true*, *degree of truth*) →Elements, objects, relations. This is the action of putting forward some statement or proposition as true. The statement, or proposition, can also be put forward not as assertion, but as a supposition or hypothesis, as possibly true, and so on. Assertion using logical approaches is an example of that function. Reliability and relevance are also notions that can be calculated under that function.
6. $H_{Es=estimation}$ (Estimation). To track the dynamics and predict the evolution of a system as well as applying some control, the notion of state is essential. This function means the estimation and prediction (analysis) of an unknown true state by combining observations and system dynamics (model output). Simple object tracking as well as more complex object like event tracking or group tracking can be considered under this generic function.
7. $H_{Pe=prediction}$ (Prediction): $T(t) \rightarrow T(t+1)$. To track the dynamics and predict the evolution of a system as well as applying some control, the notion of state is essential. This function means the estimation and prediction (analysis) of an unknown true state by combining observations and system dynamics (model output). Simple object tracking as well as more complex object like event tracking or group tracking can be considered under this generic function.
8. $H_{As=association}$ (Association): $(H_{n=a} \leftrightarrow H_{n=b})$, $(T_{n=c} \leftrightarrow H_{n=d})$. This is identification and characterization of any link or relation between holons and informons considering all dimensions. Correlation between informons, association between detections and object tracks, application such as link analysis, and relations in social networks are all examples of implementation of that generic function.

The eight functions that compose an information fusion cells are identified as holons. Holons form dynamic hierarchies called holarchies that are a network of information fusion cells where node n is represented by the centered H_{H_n} in Figure 6.3. Holarchies are characterized by making efficient use of the available resources (JDL level 4), being highly resilient to disturbances and adaptable to changes in their environments. Two processes characterize holons: emergence and dissolution. Holons emerge from others and disintegrate later in its components in a continuous process. Holons are supposed to do what gives them more according to a set of criteria (e.g., utility, preferences, goals) and that the conditions for the auto-organization of the system are present. A definition of those holarchies requires a substantial research effort from what is being presented in this chapter.

In our approach, the way to interpret Level 4, Process Refinement, of the JDL model is to define those holarchies. For instance, this definition would include how uncertainty affects the way the holons structure to answer the six basic epistemic questions of Figure 6.4 and how to obtain an optimal design in the sense of optimization criteria (e.g., multiobjective multicriteria). This complex task is left to further researches. Since those holarchies $H_{H(JDL\text{-}level)}$ are not explicitly defined and the principles of emergence and dissolution of holons not demonstrated, we will use the term information fusion cells for the rest of the chapter.

A network of information fusion cells will compose a distributed information fusion system. The set of core holons presented above composes an information fusion cell. A holon $H = (\Theta, F, \Omega)$ is being characterized by the following features, besides the input and output sets Θ and Ω being known:

- Information input scope indicating how the informative relation consider the informons from the definition set (single-arity, plural-arity, fuzzy-arity);
- Information output scope indicating how the informative relation consider the informons from the contents set (single-arity, plural-arity, fuzzy-arity);
- Imperfection holon type indicating if the holon is considered as perfect or imperfect and its representation (see Chapter 5);
- Information quality input $Q(I_j)$, and quality output $Q(O_k)$;
- External and internal contexts translated to goals and constraints (C_{jk}, G_{jk}).

Figure 6.5 is a simplified representation of an information fusion cell using sets **I**, **O**, **F**, **G**, and **C**, defined as follows [48]:

- **I** is the set of all plausible categories of inputs I_j defined as a set of informons available at the entry of an information fusion cell.

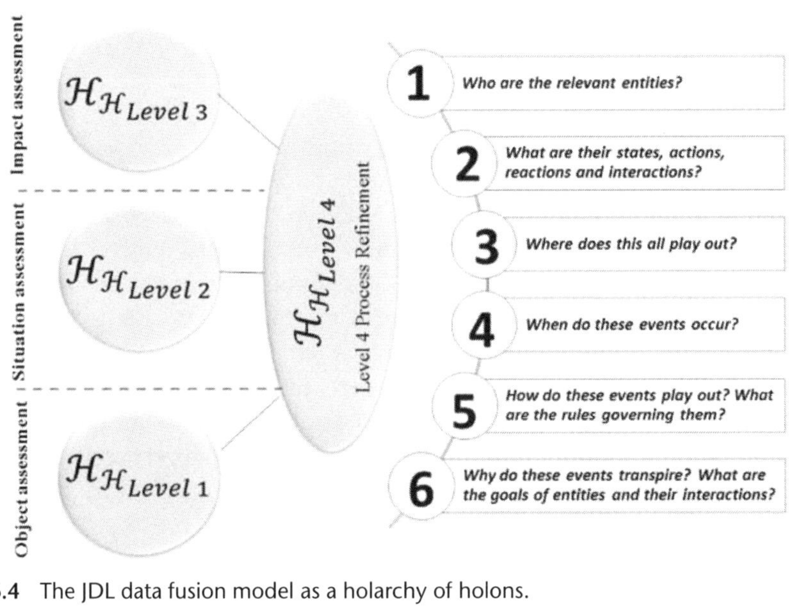

Figure 6.4 The JDL data fusion model as a holarchy of holons.

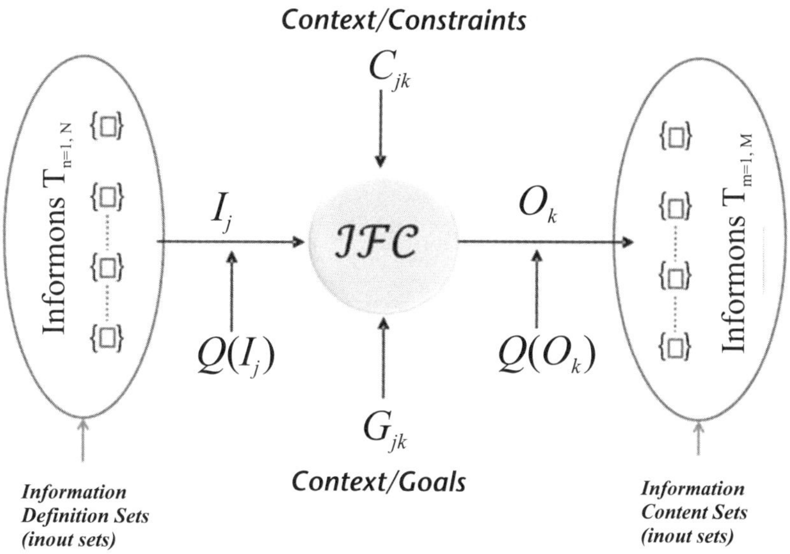

Figure 6.5 A simplified representation of an information fusion cell.

- O is the set of all plausible categories of outputs O_k defined as a set of informons that represent the solution expected from the information fusion cells (holons processing informons).
- F is the set of all informative relations that might implement all holons of an information fusion cells: *JFC*. $\{H_{De}, H_{Al}, H_{Es}, H_{Pr}, H_{As}, H_{Ve}, H_{Pa}, H_{Co}\}$. $f: \mathrm{I} \rightarrow \mathrm{O} / f(I_j)\big|_{(C_{jk}, G_{jk})} = O_k$ is defined on the set of inputs **I** to the set of outputs **O**.
- G is the set of all plausible goals/queries G_{jk}:G_{jk} is a set of goals and queries instantiated to produce the output O_k given that the input is I_j.
- C is the set of all plausible controls including the background knowledge (external and internal contexts) C_{jk}. C_{jk} is in itself a set of control parameters, models, and background knowledge instantiated to produce the output O_k given that the input is I_j.

6.5 Quality and Performance of an Information Fusion Cell

Consider that it is possible to qualify any input I_j and output O_k by a qualification function Q. The function Q: **X**→**Q**/$Q(x) = q_x$ is defined on the set of inputs **I** or **O** to the set of qualities **Q**. For example, if q_x is a confidence level, then **Q** could be a real interval [0, 1]. Wald's definition [52] of information fusion states that "... it aims at obtaining information of greater quality...." Each information fusion cell and its holons should by design direct (holarchy) towards that overall goal. In [53] Talburt related quality of information (QoI) to the value of the application it supports. The quality of an informon depends on the holon that it stimulates. The same informon can have different quality when used for different holons. One of the most cited frameworks (see Chapter 4) to classify QoI dimensions is from Wang-Strong

[54] where QoI has been classified into four major categories: intrinsic, contextual, representational, and accessibility.

The intrinsic level characterizes an information element in terms of its nature, interpretation, and modeling while considering it as an independent entity, out of any global fusion context. Examples of intrinsic QoI criteria are accuracy, believability, objectivity, and reputation. Conversely, the contextual level characterizes an information element in terms of its impact, completeness, relevance, conflict, and redundancy within a global fusion context. Examples of contextual QoI criteria are value-added, relevancy, timeliness, completeness, and amount of data.

The representational (in a human-factor sense as opposed to representations within machines) level addresses the formal representation of the problem (and its context). It encompasses both the inner element representation (e.g., discrete versus continuous output set) and semantic alignment that allows informons to be considered as either input or output for fusion processing (i.e., to be involved in a processing elements flow). Examples of representational QoI criteria are interpretability, ease of understanding, representational consistency, conciseness of representation, and manipulability. The accessibility level characterizes information accessibility. This aspect shall be considered from many perspectives. Accessibility might be physical (e.g., with regard to a sensor's availability) as well as organizational (e.g., holding the right to access and process specific information elements). Examples of accessibility QoI criteria are access and security.

The quality and the performance of an information fusion cell depend not only on the inputs, the controls, and evolving goals (system flexibility and adaptability), but also on the quality of these inputs. For instance, for a given goal set, if the inputs and their qualities are given, then one might control the information fusion cell through a good choice of the control set. This control set might include the formal theories and models to be used by the information fusion cells (Level-4 process refinement).

6.5.1 Desired Information Fusion Cell Properties

In this section, we discuss the desired fundamental properties of an information fusion cell, referred here as a fusion function f. We recognize that these properties are neither universal nor exhaustive. The following properties are proposed based on multiple criteria decision analysis work [48]. These properties are indicative of the kind of formal study required to characterize an information fusion cell. In [48], a preliminary set of desired information fusion cell properties has been defined as well as an initial formalization. The proposed properties are: neutrality (or unbiasedness), consistency (Condorcet or Pareto principle), monotonicity, significance (preserving the data), and risk of fusion errors (real and empirical risk). Many other properties might be developed. An information fusion cell characterization is context dependent.

- *Property 1:* Neutrality (or unbiasedness). Under the same control conditions and goal query, the output O_k of a fusion function f should only depend on the content of the input I_j. In other words, under the same conditions, f should not be dependent on the order in which the sensors inputs are processed.

- *Property 2:* Consistency (Condorcet or Pareto principle). One could define multiple versions and variations of this property. In the case of a single-sensor reporting, the proposition produced by f should be consistent with the sensor report. Let us consider multiple sensors reporting the same proposition on a given target. f is said to be consistent if and only if O_k is consistent with any report of any reporting sensor. A relaxation of this property might be articulated for a coalition of sensors instead of all sensors. For example, another version of consistency could be defined based on the known governing laws (e.g., the laws of physics).
- *Property 3:* Monotonicity. Let us consider that for a given input I_j, the fusion function f produces the output O_k with a quality $Q(O_k)$. Then, if an additional source of information is added and that source of information supports the output O_k, the O_k should be reinforced and the quality $Q(O_k)$ increased. For example, if the output is a proposition a_1 with a likelihood $Q(a_1)$, and if an additional sensor is added and that sensor produces a report in favor of a_1, then the fusion function should reinforce the proposition a_1. A relaxed version of this property might be defined.
- *Property 4:* Significance (preserving the data). The significance is based on measurement theory concepts [55]. Any processing involved by a fusion function f should respect measurement theory principles. This means that any transformation of the information required by any fusion model in f should respect the significance authorized transformations.
- *Property 5:* Risk of fusion errors (real and empirical risk). Based on the information fusion cell transformation (informative relations), the real risk $R(f)$ should represent a boundary on the expected maximum error risk for a given selection of an information fusion cell. The risk of fusion error is inspired from statistical hypothesis test analysis and machine learning; what is the risk that the information fusion cell might produce a wrong solution?

A rigorous and thorough effort should be employed to characterise each information fusion cell. We foresee that as context and application dependent, therefore, the characterization process should be performed for each information fusion cell of the input/output matrix. The result of this process is to present for each couple (I, O) a set of acceptable properties (normative, descriptive, formal, pragmatic) and be able to distinguish between the different information fusion cells. The typology of the information fusion cell design problem, that contribute to define a holarchy, as discussed previously, can then be represented as shown in Figure 6.6 where a multicriteria multi-objective approach (optimization or satisfaction paradigms) can be used to select an information fusion cell.

Many other properties of an information fusion cell might be introduced. It is possible to develop other properties inspired from QoI dimensions of Wang-Strong [54] such as conviviality (information overload, controls, complexity of the processing), transparency (to be able to explain and replicate the result), measurability, computability, tractability, completeness, scalability, and decomposability. A set of QoI criteria $\mathbf{Q}$ can be derived guided by a multicriteria analysis respectively for the set of inputs $\mathbf{I}$ and outputs $\mathbf{O}$. This important analysis effort is left to future work.

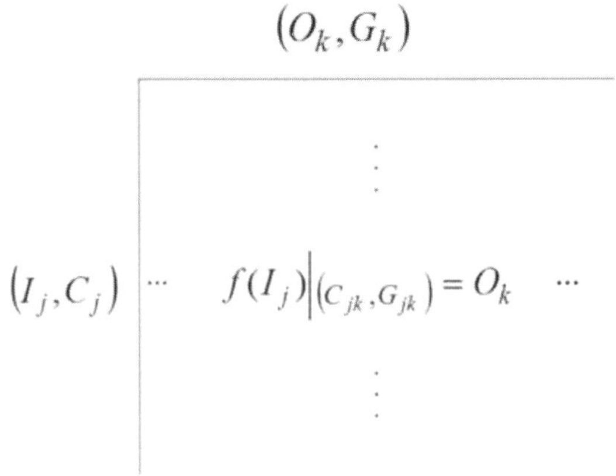

Figure 6.6 Selection of the best information fusion cell in given internal and external contexts.

6.6 Illustration of the Holonic Processing Framework: A Remote-Sensing Case

This section provides an example of how the proposed framework can be used to model an information fusion (IF) solution to a real case of remote sensing. The actual software implementation has not been influenced by the framework. This is a post-analysis using the holonic processing framework. The main objective here is to illustrate the fitness of the proposed holonic architecture with a real-case application. Indeed, one benefit of the approach is to provide a common information fusion cell design that operates generically using its eight core functions, at every level of abstraction of the JDL model.

6.6.1 Context and Generation of Informons

The goal of the presented application is to tell whether a specific shipwreck is lying on a sea-bottom area, using multiple sensors acquisitions from a volume of surveillance of the underwater environment. This is a typical scene interpretation problem that relies on fusion mechanisms to provide a synthetic representation of the scene for decision-making.

Two sensors are used to acquire the sea-bottom reality (generator of informons): a sidescan sonar (SLS) and a multibeam echo sounder (MBES). The SLS provides high-resolution images that inform on the sea-bottom contents and geological nature. The MBES acquires the sea-bottom geometrical shape. Additional sensors characterize the acquisition process: two motion reference units (MRUs) and compasses record towfish and ship attitudes (roll, pitch, yaw, and heading), one GPS collects the geographical ship location, and finally, a sound velocity profiler provides underwater sound velocity according to depth.

Mounted on the GPS unsensored towfish (a towfish is a small underwater vehicle towed by a ship using a variable length cable), no positioning data can be measured with the collected images. On the contrary, the MBES is mounted on the ship, thus allowing a precise global localization; however, georeferencing the individual beams uses sound velocity profile to correctly compute the effective acoustic paths between ship and seabed [56]. SLS images are not affected by acoustic propagation curvatures, as the SLS is meant to fly over the sea bottom with a rather low altitude. Based on the collected data, discovering whether a specific wreck has been observed is not a simple task and clearly operates at several levels of abstraction. Thus, this difficult problem embedding several subtasks may obviously take benefit of the proposed holonic framework. The next section shows how.

6.6.2 Illustrative Example: A Remote-Sensing Case

Based on the proposed framework, Figure 6.7 presents a possible architecture that possesses the capability to search and find an underwater wreck. The system objective is to find the presence of a specific underwater wreck, using multiple sensors acquisitions. Through three levels of abstraction, the figure plots the eight information fusion cells that range from processing raw data to final decision making. The black lines connecting information fusion cells from different levels of abstraction sketches a possible IF system holarchy. For instance, information fusion cell (7), from the impact level of abstraction, relies on information fusion cell (5) (from the situation level) to assess the detection of the specific searched wreck based on a full map of the potential wrecks (objects). To build such a map, information fusion cell (5) embeds information fusion cell (1) to (4) (from objects level) to detect potential objects from raw data sources.

Figure 6.7 shows that each information fusion cell is built upon the same generic architecture defined in previous sections. The eight kind of holons corresponds to the eight functions (holons) of an information fusion cell: prediction

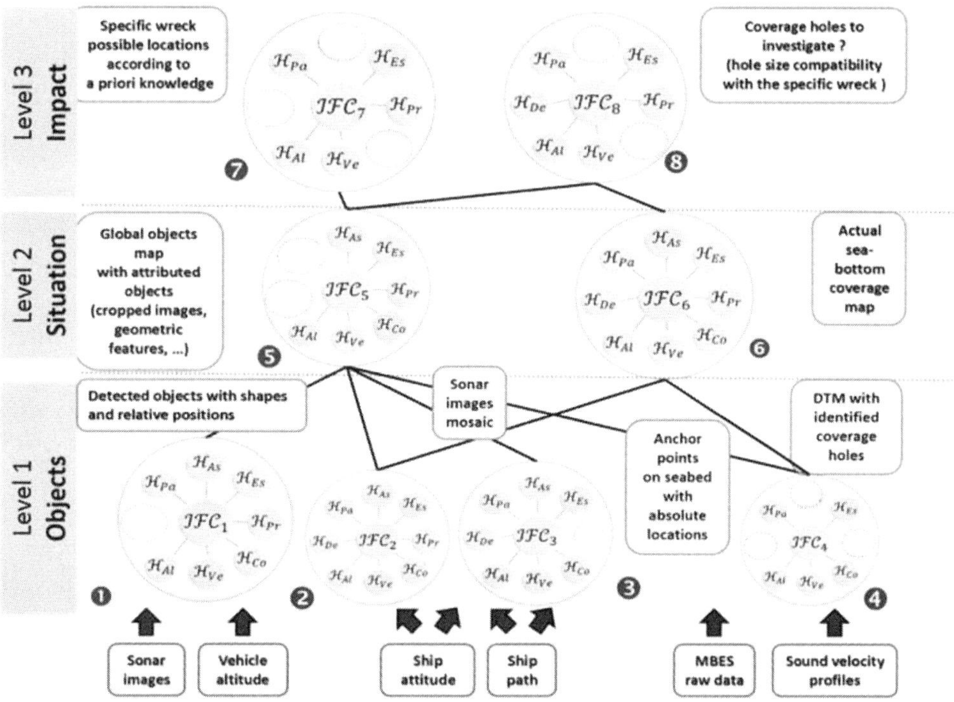

Figure 6.7 Application to a remote-sensing, real-case problem.

(H_{Pr}), detection (H_{De}), association (H_{As}), veracity (H_{Ve}), combination (H_{Co}), estimation (H_{Es}), partition (H_{Pa}), and alignment (H_{Al}). Figure 6.7 also indicates for each information fusion cell, which functions are required to perform the information fusion cell task. This representation clearly illustrates the holonic nature of information fusion cells. Indeed, this list of eight functions {H_{De}, H_{Al}, H_{Es}, H_{Pr}, H_{As}, H_{Ve}, H_{Pa}, H_{Co}} remains the same through the different information fusion cells and levels of abstraction. However, the techniques to implement each holon might be different depending of the level of abstraction of the input informons. To gather a full picture of the solution provided by the holonic framework, Figure 6.7 also represents to some extent the information flow between information fusion cells, explaining what kind of outputs can be expected from each information fusion cell. This guides the choice of informative relations in the holonic structure. Below are details for each level of the JDL.

6.6.3 JDL Level 1: Objects

In this level, the information fusion cell's purpose is to basically extract interesting objects from the raw data and subsequent processed raw data, in order to feed the global situation representation (Level 2).

- *IFC* (1): From raw sonar images, this information fusion cell extracts subimages that are typical candidates for objects lying on the sea-bottom. Usually, this task try to detect on images echo and shadow regions {H_{De}, H_{Ve}, H_{Pa}} that are close enough to be associated in order to create objects (H_{As}) that reflect the possible occurrence of real objects or artifacts (H_{Ve}) on the sea bottom. Combining the vehicle altitude information with the shadow shape (H_{Al}, H_{Co}), an object height profile can be estimated (H_{Es}). Thus, *IFC* (1)produces a list of potential objects with their attributes (height, profile shape, texture, and so forth) from images [57]. The ellipsis elongation in Figure 6.8 is related to the length of the cast shadow. To obtain the effective object profile (and height), the shadow length must be corrected according to the sonar altitude.
- *IFC* (2): Raw sonar images from one track only cover a part of the area to search; however, these images also partly overlap. Thus, building a global view of the area (i.e., a mosaic of images covering the whole area to be searched) requires to align (registration) (H_{Al}) and merge (H_{Co}) these images. Registration techniques {H_{De}, H_{Es}, H_{As}, H_{Ve}, H_{Pa}} either iconic or symbolic try to associate images based on common intensity or features properties between images, along with these anchors for relative spatial distributions (H_{Pr}) [58].
- *IFC* (3): From MBES raw data and sound velocity profiles, elevation can be estimated for each beam with an intrinsic confidence degree {H_{Al}, H_{Es}, H_{Ve}, H_{Pa}, H_{Co}}. Looking at elevation behavior (local gradient) within the beam neighborhood (H_{As}), some sea-bottom locations can be extracted as potential objects location along with elevation features {H_{De}, H_{Es}, H_{Ve}}.
- *IFC* (4): From MBES raw data and sound velocity profiles, elevation can be estimated for each beam with an intrinsic confidence degree {H_{De}, H_{Es}, H_{Ve},

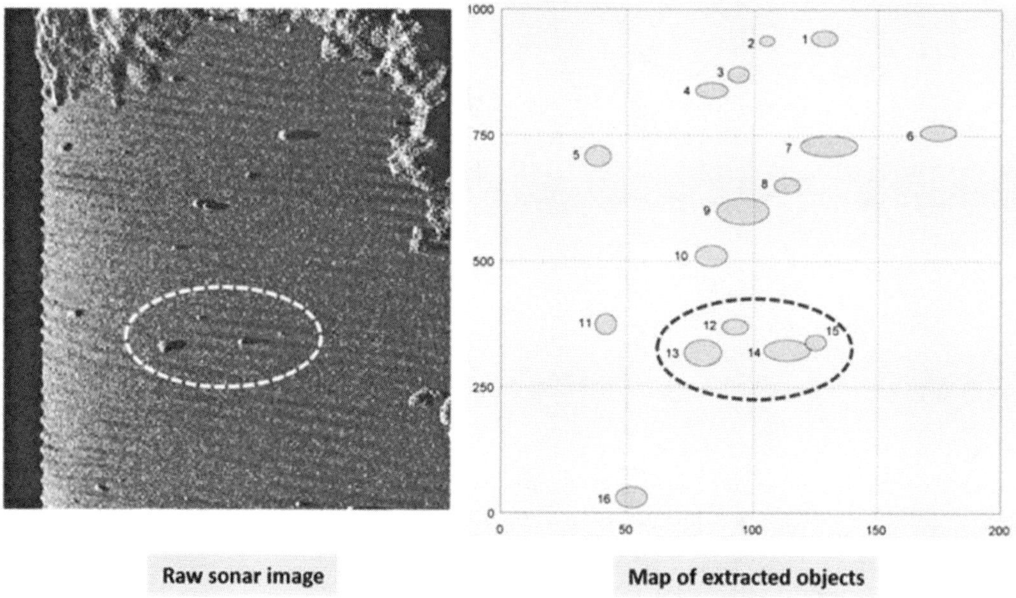

Figure 6.8 Example of extracted objects (or contacts) from a SLS image.

$H_{Pa,} H_{Co}\}$ to build a regular grid or DTM giving an area elevation map. This global elevation map also indicates areas not covered by the sensors but also informs on the amount of beams (possibly coming from several tracks) driving to the estimated elevation $\{H_{Es,} H_{Ve}\}$.

6.6.4 JDL Level 2: Situations

The purpose of this level is to make sense by analyzing all types of relations between objects (and their attributes and topological properties) to build a global view of a situation. Such a view allows further interpretation to act towards achieving the overall goal or the problem to solve.

- *IFC* (5): This information fusion cell collects all the objects resulting from *IFC* (1) and *IFC* (3) to build a global map of the area potential objects. To perform the association between images-based and elevation-based objects, DTM (from *IFC* (4)) and mosaic (from *IFC* (2)) act as global guides $\{H_{Al,} H_{As,} H_{Ve}\}$. Several image-based objects or elevation-based objects may correspond to the same sea-bottom object, providing a more detailed representation of such an object due the multiple observations according to several viewpoints $\{H_{Es,} H_{Pr,} H_{Co}\}$ [59]. Multiview classification or characterization algorithms may be advantageously involved for this task [57]. Thus, in the resulting representation, each potential object is then fully characterized (attributes) and geographically located [60].
- *IFC* (6): *IFC* (2) and *IFC* (4) produce global views of the surveyed area allowing to estimate the actual observed sea-bottom areas (with the amount

of observations) for each sensor family. From these views, *IFC* (6) purpose is to produce a final map (H_{Co}) giving the extent of the known world (H_{Pa}) and to what extent the known world is actually known $\{H_{Es}, H_{Ve}\}$. Prior to estimating the actual coverage, DTM and mosaic are aligned based on the association of elevation features from the DTM and the shadows occurring on the mosaic $\{H_{De}, H_{Al}, H_{Pr}, H_{As}, H_{Ve}\}$. Figure 6.9 illustrates such a coverage map built only from two MBES track coverage maps [61].

6.6.5 JDL Level 3: Impact

This level is strongly application-related as its purpose is to deliver relevant answers to the problem relying on the current situation (level 2). For this application, two questions need to be answered: (1) Does the observed search volume contain the wreck of interest? (2) Is the search complete, that is, is there any unobserved search volume large enough to contain the wreck remaining? Moreover, this level also evaluates how the updated global current situation may impact the delivered answers.

- *IFC* (7): Having a map of potential objects from *IFC* (5), this information fusion cell evaluates the chances $\{H_{Es}, H_{Ve}\}$ for each object to be the searched wreck according to a priori knowledge like the wreck's supposed size, wreckage estimated location, ship initial shape, and $\{H_{Al}, H_{Es}, H_{Ve}, H_{Pa}\}$. This evaluation reflects the current global situation knowledge. When processing new

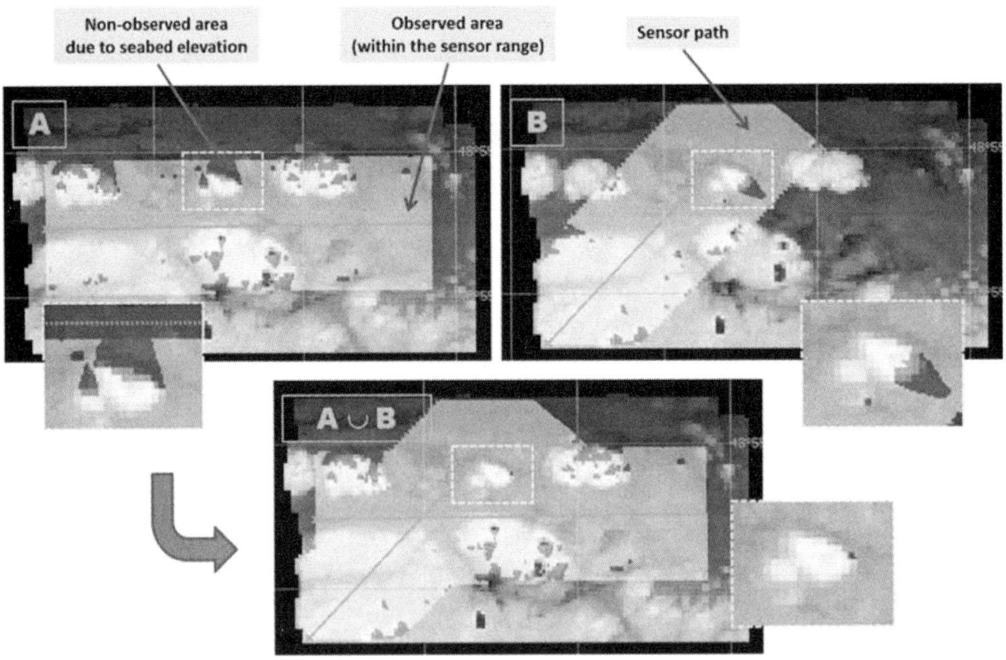

Figure 6.9 Examples of coverage maps manipulated by information fusion cells. (Maps A and B shows the coverage from two individual tracks. Map A ∪ B indicates the survey global coverage.)

images, new objects may be detected or attributes from already detected objects updated (through multiview classification in *IFC* (5), for instance). By updating the global situation, even such local changes may bring every object chances to be reassessed $\{H_{Al}, H_{Es}, H_{Ve}, H_{Pr}\}$.

- *IFC* (8): Knowing the actual sea-bottom coverage from *IFC* (6), the list of potential attributed objects from *IFC* (5) and a priori knowledge about the searched wreck, a set of unexplored coverage zones in which the wreck could be found, can be estimated $\{H_{De}, H_{Al}, H_{Es}, H_{Ve}, H_{Pa}\}$. When performed during the survey, this set can also be used as feedback to the mission planning system to build new paths for the sensors and eliminate relevant coverage holes, depending on the mission objectives and constraints $\{H_{Al}, H_{Pr}\}$.

6.7 Holonic Interpretation of the JDL Model

Three main aspects shall be considered in a holistic fashion for the design of a decision support system (DSS) based on fusion of information and analytics technologies (FIAT) [62] as suggested in Figure 2.22: multi-agent systems (MAS) theories so that the distributed aspect can be adequately formalized, generalized information theory (GIT) [63] for the representation of knowledge and information imperfections, and decision theories (operational research) to explicitly account for the actions and their impact on the environment.

This section stresses the MAS aspect, that is, the potential contributions of holonic multi-agent systems in the integration of FIAT as a DSS. Although holonic system applications are quite common in the manufacturing sector [64–66], there are much less used in services field like information systems businesses for instance health services management [67]. Holonic systems are almost unknown to the FIAT community except some few papers: a work illustrating the holonic nature of the information fusion (IF) processing functions [68] and older papers showing potential applications to sensor management and adaptive IF control [69, 70]. This section presents a holonic interpretation [71] of the JDL data fusion model [72] and proposes a model of data and knowledge representation for an intelligent system in which the knowledge of the fusion agents and the imperfections of information are taken into account.

During the past decade, subsequent revisions were made to the model as well as proposed additional levels [73] to incorporate new understandings of the information fusion issues to end up with the proposed Data Fusion Information Group (DFIG) model [74]. In DFIG, two additional levels were proposed to the original JDL: Level 5 (User Refinement) and Level 6 (Mission Management) (in Chapter 2 of [75]). For the purpose of our discussion here, we will use the four levels (1–4) of the 1998 JDL version [76].

6.7.1 Complex Dynamic Systems

Complex systems generally are also dynamic, that is, the response of the system to the same inputs varies through time. This behavior can be seen as a mode of learning or adaption. One of the more interesting properties of certain complex systems is self-organization, which is understood as the mechanism or process enabling a system to change its organization without explicit command during its execution time [77]. We can generalize this definition for any self-organizing system by rephrasing it as "... during its lifetime" to include living systems, such as human social groups. Self-organization is a desired system's property where changes can cooperate towards achieving system goals. One needs to examine how that property can emerge through the design of an adaptive system for the hierarchical JDL data fusion model. Hierarchical is also with respect to the goals' structure. To that end, three notions from complex systems theory are being used: holons, informons, and holarchies.

6.7.1.1 Holons and Holarchies

Holons form dynamic hierarchies called holarchies, which are characterized by making efficient use of the available resources, being highly resilient to disturbances and adaptable to changes in their environments [78]. Holonic applications in the manufacturing industry are being broadly studied [65, 66]. In the information technology field many applications have been developed: a model of architecture for the management of distributed systems and services [79] and a model for the management of the health services at national level [80]. It has been used also to predict in the short run the coordination and control of multi-agent systems [81], in the management of telecommunications networks [82], in incidents arising in software engineering projects [83]. Examples of natural holons are a cell, a human liver, a workgroup or the solar system; an autonomous multi-agent system or a (holonically organized) factory could be examples of artificial holons.

Holons can be seen as embodied abstract agents. That is, abstract (abstract in the sense of formed by other agents) agents having a physical part interacting with their environment. Such situated abstract agents are called holonic agents. In other words, a holon is an element of a building block scenario for the construction of an autonomous and cooperative information system used to select, acquire, process, analyze, retrieve, and disseminate information based on predetermined goals: a distributed information fusion system, for instance.

The applicability of a holonic paradigm to information fusion field lies in the layered view that the JDL model proposes. A holonic structure would allow modeling more naturally the processes' complexity as discussed in [41]. A holarchy allows the existence of a certain (varying) degree of control of the higher level on the lower level and, at the same time, gives to these levels freedom to reorganize so their participants can reach goals set by higher levels, according to optimality/satisfiability criteria (e.g., cost, time, trustworthiness, dependability, quality of information, ability). This semiautonomy would avoid the necessity of the higher level having to afford all the planning and detailed information management required at lower levels, which happens in the case of a centralized architecture.

The holarchy allows, in turn, the emergence of certain system properties or behaviors such as the famous IBM autonomic principles [84]: self-configuration, self-healing, self-optimizing, and self-immune. These autonomic principles have been advocated and applied in control applications such as IBM's MAPE (monitor, analyze, plan, execute) loop as illustrated in Figure 6.10. MAPE has been created

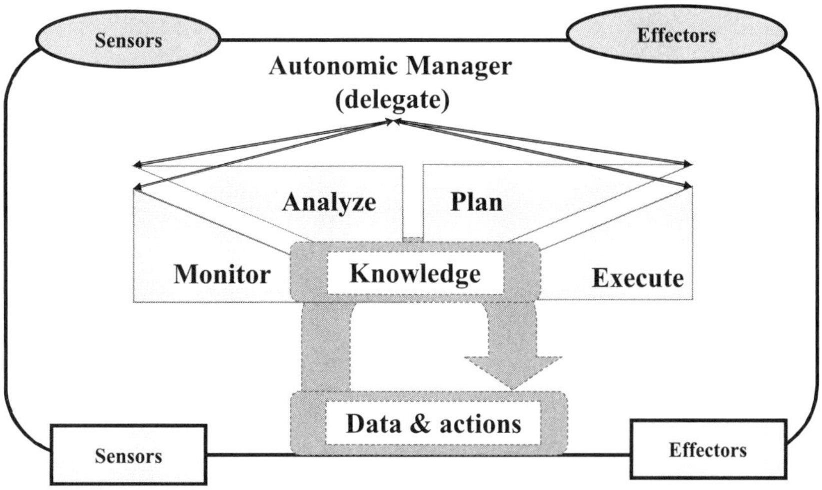

Figure 6.10 The IBM MAPE autonomic manager control loop.

to manage these autonomic principles in real design and applications. We strongly believe that Level 4 of the JDL shall be defined according to autonomic principles and benefit from applications using the MAPE loop. Obvious capabilities should possess holons to reason upon imperfect information and knowledge and to learn according for instance to some schemes (e.g., reinforcement, online).

6.7.1.2 Informons

The holonic paradigm is based on the concept of decision (or reaction): a holon is an entity that decides/reacts based on information (informons). Holons interpret informons obtained from their environment. In the scientific community, there are two main currents in the definition of informon: Zeleznikar's one [85] that relates it to the consciousness where the informon is an emergent and conscious entity, and a second one, proposed by Sulis [27] and followed by [31, 41, 43, 68]. In this work we adopt the second standpoint, more specifically: "an informon is the basic element that has meaning for a holon and that allows it to make the right decisions and to execute the proper actions" [27, 86]. As long as informons are processed by holons, higher semantic level informons emerge: a nice feature that applies to the increasing semantic processing levels of the JDL data fusion model.

6.7.1.3 Information Fusion Cells and Holarchies

An information fusion cell has been proposed above. An information fusion cell, as illustrated in Figure 6.5, computes some functions of an information fusion process on input sets of informons according to a set of goals and constraints to deliver sets of output informons. This information fusion cell is a holon formed by networked holons, each of them having similar or different goals.

A network of information fusion cells, perhaps represented either through probabilistic, evidential or possibilistic frameworks, can be organized in holarchies to constitute a distributed information fusion system. A holon $\{H\}$ is being characterized (in any moment) by the following features: information definition set, informative relation, information content set, internal context, and external context as described in the previous sections.

A temporal series of informons (a stream of informons) related to the same semantic frame forms a thread, for example, the sequence of coordinates of an object and the sequence of reference coordinates. Each holon implements one or more weaving operators by means of which complex patterns of threads are generated (e.g., the informons corresponding to the degrees of truth of the expression "an object is approaching us"). These patterns are, in turn, informons according to Sulis' archetypal dynamics [27]. If we consider that we know a priori all possible sets of informons, one can speak of a set of input informons and in this case a holon implements a relationship ($I \rightarrow 0$) that was called its informative relation. The question is still open in the case that we do not know the complete sets of informons. Intelligent learning holons should be capable to complete the set of outputs but the impact of the input sets still requires more definition work.

6.8 A Holonic Approach for Information Fusion

Using a holonic approach one can model a distributed information fusion system and benefit from the physical execution capability of a holon. That is, if we do not employ holons, the system would prescribe the steps to follow and, in the best case, would have to execute the ones that can be triggered automatically by the system, such as a program execution, the online backup of data, and so forth, in concurrence to many other actions (typically those that involve activities in the physical world, for example, any managing actions on sensors or on effectors).

6.8.1 A Holonic View of the JDL Model

A way to interpret the JDL model is to define each level to be processed by a holon or a holarchy that, in fact, is also a holon. The four levels of JDL show a holarchy to which Level 4 is the meta-level or the manager. This is justified by the fact that to achieve each goal in the different levels, a variable number of information fusion cells working collaboratively is required. For example, in Level 1, goal might be "estimate and predict entity states on the basis of observations-to-track association, continuous state estimation (e.g., kinematics) and discrete state estimation (e.g., target type and identification)" [76]. This can be achieved by an elementary holon, for instance, a radar operator or by a whole team of radar operators. This can be schematized as depicted in Figure 6.11.

Holons corresponding to each level of the JDL model (bold hexagons in Figure 6.11) are formed by information fusion cells, while a special type of holon constitutes the meta-level (planning and control holon, in a dashed line). The main functions of such planning holons could be designed as:

1. A goal *g* coming from higher levels decomposes *g* in subgoals to be achieved by holons in its respective JDL level.
2. Coordinate assignment of subgoals.
3. Replan in case of failure of any assigned holon.

The plan is formed according to optimality/trustworthiness criteria. Recall that the planning holon does not create an assignment since it emerges from interaction (negotiation) among holons at same level: it only coordinates assignments. The bold lines in Figure 6.11 represent control and coordination. Segmented ones mean required information flow and dotted ones represent optional information flow. The oval shapes mean collections. The use of optional information flows is conditioned by the holon information processing capability of higher level indicated as dotted lines. For example, suppose that Level 1 is performed by using two sales databases with their respective administrators and data mining systems. Assume also trying to detect speculation or fraud in sales so a holon performing Level 4 could contact the database administrators (who, in turn, would query the databases) and ask/receive additional information. However, instead of two databases, let us have hundreds of radars detecting unidentified ships. The flow of information to Level 4 should be managed.

The holarchy structure will be influenced by quality of information, behavior and performance of holons (e.g., effectiveness), and system trustworthiness optimality/satisfiability criteria preestablished at a holarchy metasystem level (e.g., at Level 4 of JDL). Research on general conditions under which a holarchy might emerge is left to future work. Defining holarchies and principles of emergence and

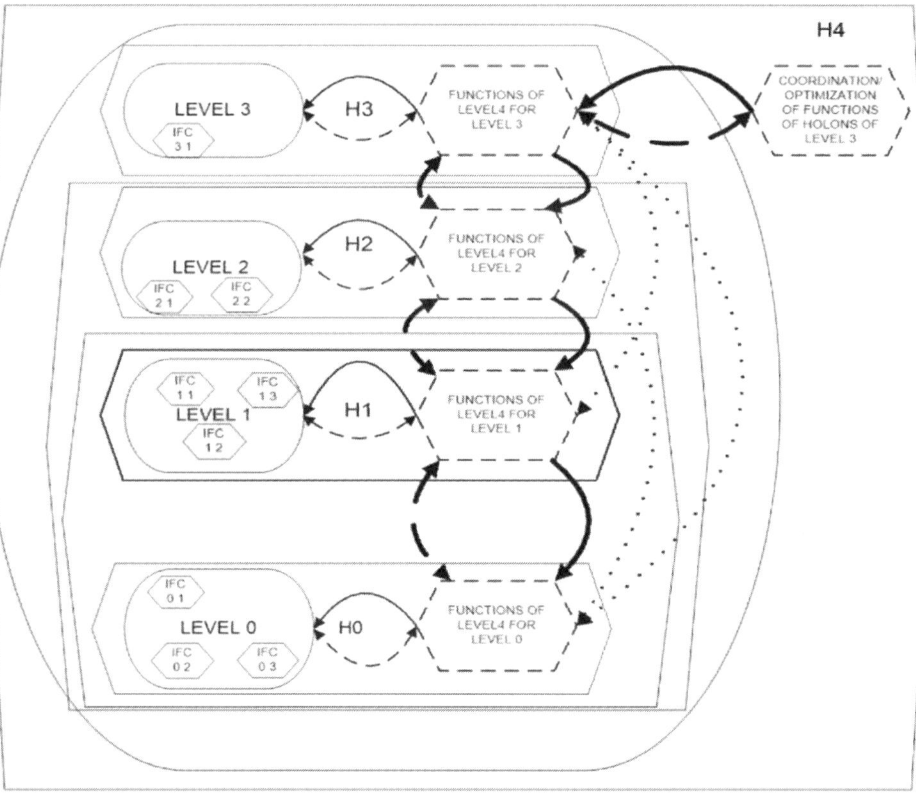

Figure 6.11 A holonic interpretation of the JDL data fusion model.

dissolution of holons is a complex and challenging task requiring multi-objective, multicriteria optimization is beyond the scope of this work.

6.8.1.1 Goals and Models

In the model presented in [48], goals are "the set of all plausible goals/queries instantiated to produce the output O_k given the input I_j." Goals can be handled by using ontologies. They are grouped in classes (for example, goals of same class being achieved in similar way) and these classes are set through presumed relationships among goals. By defining and using these ontologies, one can explicit some type of tacit knowledge about the organization or context to which the ontology has been created.

6.8.1.2 Representation of Controls, Models, and Goals

The proposed model is a continuation of the discussions undertaken in [83]. The idea is to start generating an ontology of goals, models, and actions as follows. Goal referred to a model is formed by a series of actions. A goal belongs to a class of goals. The classes are formed with the criterion: a goal class A is a particular case of B, for a given model, "if *a* is a goal of class A and *b* is a goal of class B, then by achieving *a* one achieves *b*, or one is closer to *b* than one would be if *a* were not achieved." We will say then that A is a "son" of B. More formally, A is the "son" of B if B subsumes A or if the probability that a goal is in class A given that is in B is known or it is known that every object in class A belongs to B. Using a description logic ALC [87], we would represent also the solutions, the good and better practices of the domain and of the organization, and using a reasoner (such as Pellet [88] and DeLorean [89]) we could infer the solution's validity (if the solution can be applied in that context). Additionally, part of the models could be represented using adequate modal logics, such as the deontic [90] (expressing permissions and obligations).

6.8.2 Examples of Holons and Informons in the JDL's Four Levels

In Table 6.1, we include an example of the informons, holons, and the levels of the JDL model interacting. Basically, the functions of the Level 4 performed in each level are related with the learning and optimizations of the tasks in the level. Also, examples of these functions instantiated in each level are: for Level 1, adaptive data alignment; for Level 2, multicriteria optimization of the clustering of objects; and for Level 3, algorithms or heuristics that optimize the threats and events detection.

Several holons performing Level 3 functions could be coordinated by the Level 4 holon. Note that given that Level 4 is optimization and resources assignment and that the self-organization is desired, there cannot be holons having all the functions of Level 4; instead, there are holons that help to optimize and to improve the achievement of the goals in each level, among other tasks. The functions of Level 4 are then mostly distributed/embedded in the heads of the holons of the other levels, and only a little part of them is left to the head of $H_{level\,4}$.

Table 6.1 Examples of Informons and Holons

Fusion Levels	*Informons (as Inputs)*	*Holons (as Functions Performed)*	*Informons (as Outputs)*
Level 1 Object Assessment	Physical sensors and sources Temporal references Spatial coordinates Units, system models and mathematical equations, identity databases, objects, entities, object characteristics (shape, color) Observation equations, dynamic maneuver models, state vector definition	Data alignment, temporal adjustment Data/object correlation, hypothesis generation, object tracking, identity estimation, position estimation, a Kalman filter, a Dempster-Shaffer combination rule, maximum likelihood processing, least squares...	Data aligned, data in context, object speed, object course, object identified
Level 2 Situation assessment	Time relationships, social relationships, geometric proximity, functional dependence, context environmental, sociopolitical, business rules, events	Clustering objects aggregation, contextual interpretation, event activity aggregation, associations, assessment of organizations, structures networks, time-event relationship	Type of events, situations Identified networks, structures, activities identified
Level 3 Impact assessment	Time relationships, social relationships, geometric proximity, functional dependence, context environmental, sociopolitical, business rules, events	Intent predictions, Identify threat opportunities Identify own vulnerabilities Timing of critical events	Threat, crises, events, predictions

6.8.3 Future Work on the Holonic JDL Model

Work reported in this section mainly comes from [71] where the concepts of how to use a holonic approach in the design of information fusion systems have been presented but without any validation or demonstration. Future work on this topic will specifically address concepts validation for one of the main domains of application such as smart energy management, distributed health systems, intelligent transportation, or defense and security. That validation will be achieved through the design of a holonic IF prototype system based on the principles elicited in the previous subsections. It will show:

1. How to structure a holarchy of JDL IF levels to obtain optimal results (the description of an algorithm that can be used by each holon so it can select its collaborative holons to obtain that optimal holarchy);
2. How the holonic structure affects the properties of the IF system;
3. To verify quantitatively some of the desired information fusion cell properties.

6.9 Conclusion

This chapter presented a holonic information fusion framework having a holonic functional structure suitable for processing any level of abstraction of the JDL model. The proposed framework can be considered as an evolution of the JDL process model exploitable to design complex IF systems. Concepts from business sciences, complex systems theory, informational philosophy, and computer sciences have been successfully incorporated in the framework. Notions from archetypal dynamics namely the Sulis' framework itself still unknown to the IF community have been demonstrated to be very useful to the development of a universal IF processing framework. The concepts of holon and informon developed by Koestler et al. to model evolving complex dynamic systems have been proved to be key concepts for the proposed IF framework. The framework has been conceptually defined and illustrated but an important effort remains to be done in formalization and compositional testing methods to evolve that framework to a powerful suite of tools for the design, validation, and verification of not only IF distributed systems but also as a computational model for analytics and IF that is foreseen as a significant contributor to deal with the heterogeneous nature and complexity of CPSS.

References

[1] White, F. E., "Data Fusion Lexicon," Joint Directors of Laboratories, Technical Panel for C3, Data Fusion Sub-Panel, Naval Ocean Systems Center, San Diego, CA, 1987.

[2] Waltz, E., and J. Llinas, *Multisesnor Data Fusion*, Norwood, MA: Artech House, 1990.

[3] Llinas, J., et al., "Revisiting the JDL Data Fusion Model II," *Proc. of 7th International Conference on Information Fusion (FUSION 2004)*, Stockholm, Sweden, 2004.

[4] Steinberg, A. N., and C. L. Bowman, "Rethinking the JDL Data Fusion Levels," *Proc. of the Natl. Symp. on Sensor and Data Fusion, John Hopkins Applied Physics Laboratory*, 2004.

[5] Blasch, E., and S. Plano, "Level 5: User Refinement to Aid the Fusion Process," *SPIE Proceedings, Vol. 5099, Multisensor, Multisource Information Fusion: Architectures, Algorithms, and Applications*, 2003.

[6] Lambert, D. A., "A Blueprint for Higher-Level Fusion Systems," *Information Fusion,* Vol. 10, 2009, pp. 6–24.

[7] Kokar, M. M., J. A. Tomasik, and J. Weyman, "Formalizing Classes of Information Fusion Systems," *Information Fusion,* Vol. 5, 2004, pp. 189–202.

[8] Bossé, É., A. -L. Jousselme, and P. Maupin, "Situation Analysis for Decision Support: A Formal Approach," *Proc. of 10th Intl. Conf. on Information Fusion (FUSION 2007)*, Quebec City, 2007.

[9] Llinas, J., "A Survey and Analysis of Frameworks and Framework Issues for Information Fusion Applications," *Proc. of 5th Intl. Conf. on Hybrid Artificial Intelligence Systems, Hybrid Artificial Intelligence Systems, Lecture Notes in Computer Science, Vol. 6076*, 2010, pp. 14–23.

[10] *Fusion Process Model and Frameworks Working Group (FPMFWG)*, http://www.isif.org/node/154, 2011

[11] Das, S., *Computational Business Analytics*, Boca Raton, FL: Taylor and Francis, 2013.

[12] Hall, D. L., and J. M. Jordan, *Human-Centered Information Fusion*, Norwood, MA: Artech House, 2010.

[13] Das, S., *High-Level Data Fusion*, Norwood, MA: Artech House, 2008.

[14] Bossé, É., J. Roy, and S. Wark, *Concepts, Models and Tools for Information Fusion*: Norwood, MA: Artech House, 2007.

[15] Waltz, E., *Knowledge Management in the Intelligence Enterprise*, Norwood, MA: Artech House, 2003.

[16] Liggins, M. E., D. L. Hall, and J. Llinas, (eds.), *Handbook of Multisensor Data Fusion*, Boca Raton, FL: CRC Press, 2009.

[17] Hall, D. L., *Mathematical Techniques in Multisensor Data Fusion*, Norwood, MA: Artech House, 1992.

[18] Klein, L. A., *Sensor and Data Fusion Concepts and Applications*, New York: SPIE Optical Engineering Press, 1993.

[19] Luo, R. C., and M. G. Kay, (eds.), *Multisensor Integration and Fusion for Intelligent Machines and Systems*, New York: Ablex Publishing Corporation, 1995.

[20] Lambert, D. A., "Grand Challenges of Information Fusion," *Proc. of 6th Intl. Conf. on Information Fusion (FUSION 2003)*, Cairns, Australia, 2003.

[21] Blasch, E., "Situation Assessment and Situation Awareness," in *High-Level Information Fusion Management and Systems Design*, E. Blasch, É. Bossé, and D. A. Lambert, (eds.), Norwood, MA: Artech House, 2012, pp. 13–31.

[22] Descartes, R., *A Discourse on the Method*, Oxford, U.K.: Oxford University Press, 2006.

[23] Wikipedia, "Systems Biology," http://en.wikipedia.org/wiki/Systems_biology, 2013.

[24] Wikipedia, "Evolutionary Economics," http://en.wikipedia.org/wiki/Evolutionary_economics, 2013.

[25] Wikipedia, "Network Theory," http://en.wikipedia.org/wiki/Network_theory, 2013.

[26] Mitchell, M., *Complexity: A Guided Tour*, Oxford, U.K.: Oxford University Press, 2009.

[27] Sulis, W. H., "Archetypal Dynamics: An Approach to the Study of Emergence," *Formal Descriptions of Developing Systems, NATO Science Series*, Vol. 121, 2003, pp. 185–228.

[28] Koestler, A., *The Ghost in the Machine*, New York: Macmillan, 1967.

[29] Stenmark, D., "The Relationship Between Information and Knowledge," *Proc. of IRIS 24*, Ulvik, Norway, 2001.

[30] Giarratano, J., and G. Riley, *Expert Systems: Principles and Programming*, New York: PWS Publishing Company, 1998.

[31] Paggi, H., and F. Alonso, "A Holonic Model of System for the Resolution of Incidents in the Software Engineering Projects," *Proc. of 2009 Intl. IEEE Conf. on Computer and Automation Engineering*, Bangkok, Thailand, 2009.

[32] Alonso, F., et al., "Fundamental Elements of a Software Design and Construction Theory: Informons and Holons," *Proc. of the Intl. Symp. of Santa Caterina on Challenges in the Internet and Interdisciplinary Research (SSCCII)*, 2004.

[33] McHugh, P., G. W. Merli, and W. A. Wheeler, *Beyond Business Process Reengineering: Towards the Holonic Enterprise*, New York: Wiley, 1995.

[34] Ulieru, M., S. S. Walker, and R. W. Brennan, "The Holonic Enterprises a Collaborative Information Ecosystem," *Proc. of the Workshop on Holons: Autonomous and Cooperating Agents for Industry*, Montreal, Canada, 2001.

[35] Weick, K., *Sensemaking in Organizations*, Thousand Oaks, CA: Sage Publications, 1995.

[36] Dey, A., "Understanding and Using Context," *Personal and Ubiquitous Computing*, Vol. 5, 2001, pp. 4–7.

[37] Laudy, C., "Introducing Semantic Knowledge in High Level Information Fusion," Ph.D. Thesis, Université Pierre et Marie Curie, Paris, France, 2010.

[38] Losee, R. M., "A Discipline Independent Definition of Information," *Journal of the American Society for Information Science*, Vol. 48, 1997, pp. 254–269.

[39] Solaiman, B., "Information Fusion Concepts: From Information Elements Definition to the Application of Fusion Approaches," *Proceedings SPIE, Vol. 4385, Sensor Fusion: Architectures, Algorithms, and Applications*, 2001, pp. 205–212.

[40] Stonier, T., *Information and the Internal Structure of the Universe*, New York: Springer-Verlag, 1990.

[41] Paggi, H., and F. Alonso, "Uncertainty and Randomness: A Holonic Approach," *2nd Intl. Conf. on Computer Engineering and Applications*, 2010.

[42] Dubois, D., "Uncertainty Theories: A Unified View," *SIPTA School 08, UEE 08*, Montpellier, France, 2008.

[43] Paggi, H., and F. Alonso, "Beliefs, Certainty and Complex Systems Structure," *Proc. of 2010 2nd Intl. Conf. on Computer Engineering and Applications (ICCEA)*, 2010.

[44] Pearl, J., *Probabilistic Reasoning in Intelligent Systems Networks of Plausible Inference*, Boston, MA: Morgan Kaufmann Publishers, 1988.

[45] Shafer, G., *A Mathematical Theory of Evidence*, Princeton, NJ: Princeton University Press, 1976.

[46] Zadeh, L., "Fuzzy Sets as the Basis for a Theory of Possibility," *Fuzzy Sets and Systems*, Vol. 1, 1978, pp. 3–28.

[47] Dubois, D., and H. Prade, *Possibility Theory: An Approach to Computerized Processing of Uncertainty*, New York: Plenum Press, 1988.

[48] Bossé, É., A. Guitouni, and P. Valin, "An Essay to Characterize Information Fusion Systems," *Proc. of 9th Intl. Conf. on Information Fusion (FUSION2006)*, Firenze, Italy, 2006.

[49] Bowman, C. L., and A. N. Steinberg, "Systems Engineering Approach for Implementing Data Fusion Systems," in *Handbook of Multisensor Data Fusion*, M. E. Liggins, D. L. Hall, and J. Llinas, (eds.), Boca Raton, FL: CRC Press, 2009, pp. 561–596.

[50] Nicholson, D., "Defence Applications of Agent-Based Information Fusion," *The Computer Journal*, Vol. 54, 2011, pp. 263–273.

[51] Maupin, P., and A. -L. Jousselme, "Interpreted Systems for Situation Analysis," *Proc. of 10th Intl. Conf. on Information Fusion (FUSION2007)*, Quebec City, Canada, 2007.

[52] Wald, L., "Some Terms of References in Data Fusion," *IEEE Trans. on Geoscience and Remote Sensing*, Vol. 37, pp. 1190–1193, 1999.

[53] Talburt, J. R., *Entity Resolution and Information Quality*, Boston, MA: Morgan Kaufmann, 2011.

[54] Wang, R. Y., and D. M. Strong, "Beyond Accuracy: What Data Quality Means to Data Consumers," *Journal of Management Information Systems*, Vol. 12, 1996, pp. 5–34.

[55] Suppes, P., *Foundations of Measurement, Vol. 2*: New York: Elsevier, 2014.

[56] Guériot, D., et al., "The Patch Test: A Comprehensive Calibration Tool for Multibeam Echosounders," *IEEE Conference and Exhibition Oceans (2000 MTS)*, 2000, pp. 1655–1661.

[57] Daniel, S., and B. Solaiman, "Object Recognition on the Sea-Bottom Using Possibility Theory and Sonar Data Fusion," *Proc. of the 1st Intl. Conf. on Information Fusion (FUSION1998)* Las Vegas, NV, 1998.

[58] Chailloux, C., et al., "Intensity-Based Block Matching Algorithm for Mosaicing Sonar Images," *IEEE Journal of Oceanic Engineering*, Vol. 36, 2011, pp. 627–645.

[59] Daniel, S., et al., "Side-Scan Sonar Image Matching," *IEEE Journal of Oceanic Engineering*, Vol. 23, 1998, pp. 245–259.

[60] Guériot, D., "Bathymetric and side-scan data fusion for sea-bottom 3D mosaicing," *IEEE Conference and Exhibition OCEANS 2000 MTS*, 2000.

[61] Guériot, D., C. Sintes, and B. Solaiman, "Sonar Data Simulation & Performance Assessment Through Tube Ray Tracing," *Underwater Acoustic Measurements: Technologies & Results (UAM'09)*, Nafplion, Greece, 2009.

[62] Bosse, E., A. L. Jousselme, and P. Maupin, "Situation Analysis for Decision Support: A Formal Approach," *10th Intl. Conf. on Information Fusion*, 2007, pp. 1–3.

[63] Klir, G. J., and M. J. Wierman, *Uncertainty-Based Information: Elements of Generalized Information Theory*, Berlin, Germany: Physica-Verlag HD, 1999.

[64] Botti, V., and A. Giret, "ANEMONA: A Multi-Agent Methodology for Holonic Manufacturing Systems," *Advanced Manufacturing*, 2008.

[65] Marík, V., J. L. M. Lastra, and P. Skobelev, "Industrial Applications of Holonic and Multi-Agent Systems," *Proc. of DEXA International Workshop,* Greenwich, UK, 2000, pp. 212–213.

[66] Marik, V., V. Vyatkin, and A. W. Colombo, *Holonic and Multi-Agent Systems for Manufacturing*, New York: Springer, 2007.

[67] Ulieru, M., S. Walker, and R. Brennan, "Holonic Enterprise as a Collaborative Information Ecosystem," *Workshop on Holons: Autonomous and Cooperative Agents for the Industry*, Montreal, Canada, 2001, pp. 1–13.

[68] Solaiman, B., et al., "A Conceptual Definition of a Holonic Processing Framework to Support the Design of Information Fusion Systems," *Information Fusion,* 2013.

[69] Benaskeur, A. R., F. Rhéaume, and É. Bossé, "Adaptation Hierarchy for Data Fusion and Sensor Management Applications," *Proc. Cognitive Systems with Interactive Sensors,* Paris, France, 2006.

[70] McGuire, P., et al., "The Application of Holonic Control to Tactical Sensor Management," *IEEE Workshop on Distributed Intelligent Systems: Collective Intelligence and Its Applications 2006 (DIS 2006)*, 2006, pp. 225–230.

[71] Paggi, H., et al., "On the Use of Holonic Agents in the Design of Information Fusion Systems," *17th Intl. Conf. on Information Fusion (FUSION)*, 2014, pp. 1–8.

[72] Blasch, E., et al., "Revisiting the JDL Model for Information Exploitation," *16th Intl. Conf. on Information Fusion (FUSION)*, 2013, pp. 129–136.

[73] Blasch, E., and S. Plano, "DFIG Level 5 (User Refinement) Issues Supporting Situational Assessment Reasoning," *8th Intl. Conf. on Information Fusion*, 2005.

[74] Blasch, E. P., et al., "High Level Information Fusion (HLIF): Survey of Models, Issues, and Grand Challenges," *IEEE Aerospace and Electronic Systems Magazine*, Vol. 27, pp. 4–20, 2012.

[75] Blasch, E., É. Bossé, and D. A. Lambert, *High-Level Information Fusion Management and Systems Design*, Norwood, MA: Artech House, 2012.

[76] Steinberg, A. N., C. L. Bowman, and F. E. White, "Revisions to the JDL data fusion model," *The Joint NATO/IRIS Conference*, Quebec City, 1999, pp. 430–441.

[77] C. Bourjot, et al., "Self-Organization in Multi-Agent Systems," Institut de Recherche en Informatique de Toulouse, Toulouse IRIT/2005-18-R, 2005.

[78] Botti, V., and A. Giret, *ANEMONA: A Multi-Agent Methodology for Holonic Manufacturing Systems*, New York: Springer-Verlag, 2008.

[79] Soriano, F. J., "Modelo de Arquitectura para Gestión Cooperativa de Sistemas y Servicios Distribuidos Basado en Agentes Autónomos," Doctor en Infromática, Facultad de Informática., Universidad Politécnica de Madrid, Madrid, Spain, 2003.

[80] Urra, P., "Reflexiones sobre la autonomía e interoperabilidad de los sistemas en la era de las redes," *Foro Internacional sobre Tecnologías e Información*, Cuba, 2004.

[81] Valckenaers, P., et al., "Emergent Short-Term Forecasting Through Ant Colony Engineering in Coordination and Control Systems," *Design of Complex Adaptive Systems*, Vol. 20, July 2006, pp. 261–278.

[82] Chou, L. -D., et al., "Implementation of Mobile-Agent-Based Network Management Systems for National Broadband Experimental Networks in Taiwan," in *Holonic and Multi-Agent Systems for Manufacturing*, New York: Springer, 2003, pp. 280–289.

[83] Paggi Straneo, H., and F. Alonso Amo, "A Holonic Model of System for the Resolution of Incidents in the Software Engineering Projects," *2009 Intl. Conf. on Computer and Automation Engineering*, Bangkok, Thailand, 2009.

[84] Agoulmine, N., *Autonomic Network Management Principles: From Concepts to Applications*, New York: Elsevier Science, 2010.

[85] Zeleznikar, A., "Informon: An Emergent Conscious Component," *Informatica,* 2002.

[86] Alonso, F., et al., "Fundamental Elements of a Software Design and Construction Theory Informons and Holons," *Intl. Symp. of Santa Caterina on Challenges in the Internet and Interdisciplinary Research*, Amalfi, Italy, 2004.

[87] Baader, F., et al., *The Description Logics Handbook*, Cambridge, U.K.: Cambridge University Press, 2007.

[88] Parsia, B., and E. Sirin, "Pellet: An OWL DL Reasoner," *3rd Intl. Semantic Web Conf.*, 2004.

[89] Bobillo, F., M. Delgado, and J. Gómez-Romero, "DeLorean: A Reasoner for Fuzzy OWL 2," *Expert Systems with Applications,* Vol. 39, 2012, pp. 258–272.

[90] Chellas, B. F., *Modal Logic: An Introduction,* Cambridge, U.K.: Cambridge University Press, 1980.

CHAPTER 7

FIAT-Based Decision Support for Complex Environments

This chapter mainly contributes to provide solutions (partial) for the Variety and Veracity dimensions of big data. First, we present two critical notions for systems interoperability: cooperability and openness. Second, we present an application of an Abstract State Machine framework to support the development of integrated FIAT-based decision support systems. The contexts of application are defense coalitions and a simple scenario for surveillance.

7.1 Introduction

Applications of fusion of information and analytics technologies (FIAT) to support decision-making are numerous. However, very few of them, if none (to the extent of our knowledge), result from the integration of FIAT concepts through an explicit framework: a framework reusable from one domain to another (where standardization is highly exploited), as well as to seize and tailor applications for large systems such as CPSS. Unfortunately, the art is not at that state yet despite laudable efforts, discussed in previous chapters.

All throughout this book, we have advocated the necessity of better integration at all CPSS levels (system of systems, techniques, information, knowledge, meaning, sensing, and actuation). We point out integration because, for instance, it is believed that the fundamental problem of networked military operations (e.g., network-enabled operations, network-centric warfare) is that of information and system integration as NATO studies on systems interoperability have revealed [1–3]. The military domain [4] is still certainly the most constrained environment, but future civilian networked environments are pressing also on demand for integration such as in Smart Grids, intelligent transport, and health systems. We are using future here since none of these systems, including the military, are currently connected or integrated enough to be called representative of CPSS.

The design of FIAT-based decision support systems (DSS) depends entirely on contexts and the domain of application. In addition to what has been detailed in the previous chapters about decision support, Figure 7.1 summarizes our line of thought. Figure 7.1(a) proposes to assemble FIAT into three main classes (also done in Figure 2.22). These classes should be considered in a holistic manner (integration) for the design of FIAT-based support systems to either situation analysis (the first concentric layer) or decision actuation (the next two layers). FIAT classes are: (1) multi-agent systems (MAS) theories so that the distributed aspect can be adequately formalized; (2) generalized information theory (GIT) for the representation of knowledge and information imperfections; and (3) decision theories or operational research (OR) to explicitly account for the actions and their impact on the environment. The interaction with the environment (real or virtual) is through sensing and actuation (last layer).

The black dots in Figure 7.1 represent potential points of integration. Big data and Internet of Things (IoT) are contextual to CPSS complex dynamic environments. For instance, addressing the big data Vs (Value, Volume, Veracity, Variety, and Velocity) requires that FIAT be integrated. This multifaceted integration is extremely complex, especially if CPSS comprises a large number of cyber physical systems (CPS) or IoT. As an example, stability and real-time issues still present quite a challenge to CPS designers; imagine the difficulty for a network of CPS. Povendram [5] describes those CPS challenges and, in 2012, Sztipanovits et al. [6] stated the need to develop a science of system integration:

> The need for a science of system integration was first expressed over a decade ago in the aerospace community driven by the pressures of rapidly growing complexity of space systems.... The unique challenges of a CPS integration science emerge from the heterogeneity of components and interactions inside CPS systems. This heterogeneity drives the need for modeling and analyzing cross-domain interactions among physical and computational/networking domains and demands deep understanding the effects of heterogeneous abstraction layers in the design flow.... As CPS-based solutions become ubiquitous, the need for theories, methods, and tools to ensure predictability of system behavior has significantly increased and expanded to most engineering systems.

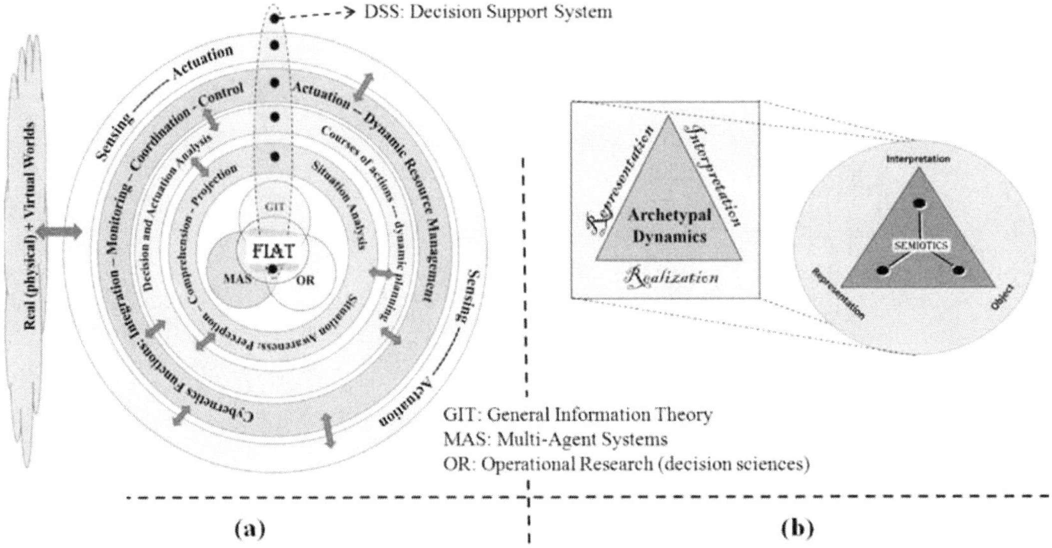

Figure 7.1 FIAT-based decision-support systems.

A framework for integration has been proposed in Chapters 4 to 6 via a FIAT holonic computational model. However, consequent to the previous discussion, only few facets of integration have been addressed. The model presented follows Sulis' archetypal dynamics triad framework. Sulis' triad is in accordance with the famous semiotics triad, depicted in Figure 7.1(b), and, in that model, the meaning of information is gained from what one does with it (actions). The objectives of FIAT are to transform data to information into actionable knowledge to support decisions and actions in complex dynamic environments. These complex environments have been described in Chapter 1, as well as in several dedicated books [7–9], using a terminology that comprises CPSS, CPS, IoT, and big data. The problem is tremendously complex to be the solution. Several FIAT-based contributions already exist to explore the large solution space and that for the four critical domains of applications: energy, transport, health and defense. Several books [10–15] would be required for a full coverage of the solution space but a nonexhaustive list of surveys [16–29] shows the level of efforts already dedicated to that problem. Note that contributions are either from information fusion or from analytics but very few, if none, present an integrated FIAT approach.

This chapter mainly contributes to provide solutions (partial) for the Variety and Veracity dimensions of big data. First, we present two critical notions for systems interoperability: cooperability and openness. These notions are being formally defined and applied to an organizational paradigm such as a military coalition. Second, we propose a strength-weakness analysis of some current frameworks compared to an approach based on Abstract State Machines (ASM) for the design of FIAT-based situation analysis and decision support systems. We illustrate the application of the ASM framework using a simple scenario and demonstrate how the framework facilitates a seamless mapping of the requirements into an abstract executable model that can be validated through simulation. The two contributions are considered essential to support the development of integrated FIAT-based solutions. They are, respectively, system interoperability in heterogeneous environments and integrated frameworks to process imperfect information.

7.2 System Interoperability in Heterogeneous Environments

The distributed nature of most contexts of application, especially CPSS, requires relating heterogeneous ontological specifications and to integrate information from multiple sources. The data/information delivered by different sources must be converted into a certain language (e.g., mathematical representation) or with other means (e.g., textual, visual, graphical, symbols, rules) so that they can be processed to make semantical sense of them. To have cooperative work between applications, multiple data sources must be linked and integrated. The integration of formally represented knowledge from different sources would be easy if all ontologies would be similar, using the same vocabulary for the same intended semantic meaning, the same background assumptions, and the same logical formalism. In large CPSS, like Smart Grids, different networks (generation, transmission, distribution, consumers) may develop their own ontologies. To maximize the benefit, these heterogeneous ontologies need to interwork, but are likely to have different vocabularies and even different underlying formalisms. The capacity to access data or information from multiple information systems is known as information interoperability.

In future CPSS, the capacity of different information systems, applications, and services to communicate, share, and interchange data, information, and knowledge in an effective and precise way, and be integrated with other systems, applications, and services to deliver services is simply mandatory. In CPSS, all types of systems (discrete, discrete event, hybrid, human) as well as all sources of data (e.g., sensors, effectors) are part of a complex network. This emphasizes the challenge of heterogeneity (the Variety dimension in big data) and consequently the problem of interoperability.

7.2.1 Barès' Approach to Formalize Cooperability in Military Coalitions

This section is based on Barès' approach to formalize interoperability in defense and security coalitions. Barès [30–34] proposed a formal modeling approach to the problem of interoperability in large information systems. His approach is relying on three main concepts: openness structure of a federation of systems, a definition of distinct domains for interoperability and cooperability, and mathematical tools in which parameters can be used to assess interoperability.

A great deal of important transformations generated by the evolution of geopolitical contexts implies the responsibility of the international community as soon as a crisis or a conflict is emerging. All nations in a position to share this responsibility are more and more often involved in an international coalition. The different technical systems, networks, and C4ISR (Command, Control, Computers, Communications, Intelligence, Surveillance, and Reconnaissance) that are taking part of a coalition must cooperate for executing a common mission fixed by a coalition authority under particular conditions and temporal constraints. The term "cooperate" is intentionally used to highlight the type of communication required in a coalition. This term completely exceeds the simple exchange of messages, as will be shown later. For now, let us illustrate it by an agent who realizes, either by insufficient know-how or a lack of knowledge, that the agent cannot alone achieve an objective. This agent will request assistance from other agents. In explaining the considered case, the agent specifies a context of openness in a potential cooperative scheme for other agents to bring the requested assistance. When doing so, agents must position themselves in a cooperative scheme. Moreover, that hypothetical coalition cannot be successful if the members are not willing to:

- Exchange their knowledge about situations that necessitate their intervention. This knowledge must be enriched as long as the process is going on (validity of an information may depend on time).
- Exchange the know-how about operating processes and methods of application.
- Share timely, under appropriate conditions, actionable knowledge with all agents of the cooperation (support to the evolution of actuation).
- Add intelligence at the different levels of the interoperability mechanisms (that may be of a great help to other members).

Barès [30, 35] limited the discussion to an organizational paradigm of a coalition. However, his concepts could be adapted to other organizational paradigms such as holarchies, teams, and federations with the support of a substantial legacy of analyses from the multi-agent community [29, 36–42]. In a coalition, the first step to be considered is to make systems cooperate in achieving a mission. To that end, a major problem to face is the fact that these systems are heterogeneous: a heterogeneity, inherent to national design and employment concepts. As a result, they present enormous deficiencies at the interoperability level. One could argue that it is always possible to solve this question by making gateways, but that represents a temporary solution. Moreover, gateways solution cannot be easily and reasonably generalized. Such generalization is getting more difficult as the number of systems increases (combinatory). Therefore, what seems more reasonable is to design new concepts of interoperability embedded in each system.

7.2.1.1 The Concept of Cooperability

The cooperation between a system A and a system C presupposes that both systems are interoperable and the interoperability between A and C presupposes that they are connectable within each other. This is illustrated in Figure 7.2. Cooperability between systems is only possible under certain conditions. In fact, systems must be:

- *Connectable:* Systems are capable to exchange data according to known protocols.
- *Interoperable:* Systems are designed to progressively open their structures to make possible richer exchanges to pursue and to meet common goals.

Interoperability requires the following:

- A common representation and understanding (common semantics) of the actuation world;
- A shared knowledge about the required materials and tools for achieving the task;
- A convergence with respect to viewpoints and processes (convergence space);
- A sharing knowledge environment;
- Teaming up to assemble all means necessary to establish a cooperation.

As illustrated in Figure 7.2, Barès [34] defined a space in which distinction is made between three domains relevant to interoperability. These domains are structured; they have semantic links and their own different techniques and approaches. Interoperability is making the essential link between connectivity and cooperability. Interoperability must be understood more than just being an exchange of simple messages. Modeling and symbolic representations are being involved to carry and share knowledge. For instance, in a coalition, C4IRS systems bring each other a mutual assistance in their cooperative action to reach a common objective. C4IRS systems must possess a joint comprehension of what they are doing. That makes it necessary to introduce semantic notions in the interoperability domain and also

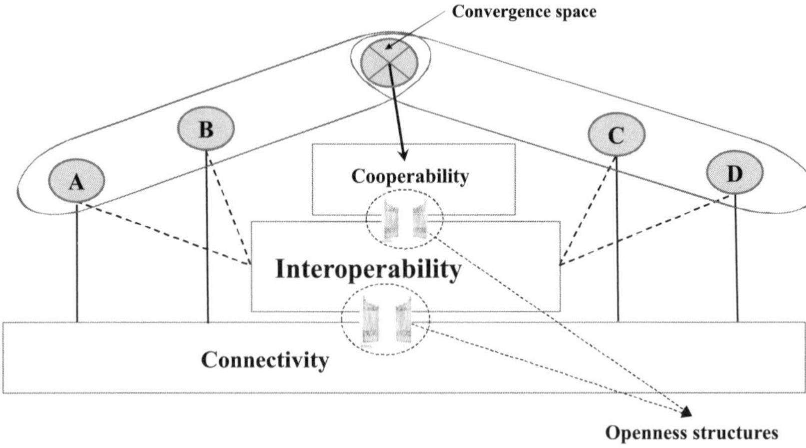

Figure 7.2 Three distinct domains for interoperability. (Adapted from [35].)

to determine modalities that enable to add intelligence and to interpret that intelligence in the exchange mechanisms. At last, the cooperability domain represents the final level to reach, through the definition of a world, in which all (cooperative) systems are able to share all elements constituting their common activity in the coalition.

Cooperability implies other system notions such as concurrency, parallelism, communication, and synchronization. For instance, communication supports synchronization to get timely and useful information for proper system functioning and dependability. These system notions are mainly concerned with two of four cybernetics functions discussed in Chapter 1: coordination and control. A cooperative system has certain abilities when it is placed in a coalition framework. These abilities are:

1. *Openness ability:* Quality of a system, previously connected with others, to share a common understanding with them, relative to some themes of a coalition, for instance, ground evacuation, medical assistance. The openness of a system appears to be a subset of the openness structure of the coalition (illustrated by the gates on Figure 7.2).
2. *Interoperability ability:* Capability of a system to (inter)operate with (interoperable) actions, relevant to the cooperation, more precisely, orders and missions fixed within the coalition.
3. *Cooperability ability:* We will consider that a system is (inter)cooperable when this system is capable to share knowledge and also know-how with its neighboring systems. A system is understood as a network of people and machines (see CPSS definition in Chapter 1). Beyond the interoperability, systems must cooperate by sharing professional experiences, skills, and knowledge (e.g., skills, rules, knowledge of Rasmussen's model [43, 44]) to best perform in doing a common task.
4. *Ability to conduct actions:* Let us assume a system owns all the abilities mentioned in 1, 2, and 3 to perform a mission in a coalition. The mission can fail if the conditions of temporal intervals are not strictly respected. An

action is only valid within a precise temporal interval, sometime referred as opportunity.

7.2.1.2 The Concept of System Openness

The first and necessary step of interoperability is the concept of openness. An openness context is, to specify beyond connectivity, ways and limits of opening systems for interoperability. First, let us define some pertinent notions to a coalition.

- *System in a coalition*: A system i will be designated by S^i where $i \in [1, n]$, n = number of systems placed in the coalition. They are supposed to be able to share a minimum common knowledge and to have common comprehension of fundamental orders.
- *Theme in a coalition*: A theme is a set of knowledge required for the coalition and describing an ability, for instance, medical assistance or civil rescue. A theme t will be designated by T_t, with $t \in [1, q]$, q is the number of themes of the coalition. T_t encompasses a variable number of interoperable actions. An action j will be designated by A_j.

The context of openness is the network of existing relationships between systems and the coalition-themes described in a formal way, the formalization of which is based on some mathematical notions such as relations and Galois connections. The context is formally defined by a triplet (S, T, R):

$$S :: S^1, \ldots, S^n \tag{7.1}$$

$$T :: T_1, \ldots, T_q \tag{7.2}$$

$$R :: R \subset S \times T \tag{7.3}$$

where R is a binary relation.

The context may be given a priori when the coalition is defining the mission of every system. It can be also defined a posteriori when the coalition is running and situations evolving.

Example 1:

Three different nations are asked to interoperate within a coalition framework for rescuing civil people in an African state. The coalition is defined as follows: three nation systems S^1, S^2, S^3 are concerned, and three coalition themes are defined: ground evacuation operations (T_1), airborne transportation (T_2), and logistical medical aid (T_3). This supposes first that systems can interoperate on different actions relevant to the coalition themes and second that they exchange knowledge required to achieve their respective missions as:

$$R(S^1,T_1),R(S^1,T_2),R(S^1,T_3),$$
$$R(S^2,T_1),R(S^2,T_2),R(S^3,T_3),$$
$$R(S^3,T_1),R(S^3,T_2),R(S^3,T_3)\subset S\times T \tag{7.4}$$

Example 1, in Figure 7.3, is a particular case where relation R on $\{S^1, S^2, S^3\} \times \{T_1, T_2, T_3\}$ is total. Considering strictly the semantic point of view, systems are totally open to the themes involved in this coalition. This example describes an ideal situation that rarely takes place in reality. Consequently, we get a unique totally open couple in the sense of a Galois connection [45, 46]. The subset $\{S^1, S^2, S^3\}$, must be considered as totally open on the subset $\{T_1, T_2, T_3\}$. In the second example of Figure 7.3 (Example 2), two couples are absent. The absence of certain couples, (S^1, T_3), (S^3, T_2) makes $R \subset S^1, S^2, S^3\} \times \{T_1, T_2, T_3\}$ not total. The coalition must reduce its openness context.

Example 1 describes an ideal case, since all systems of S are totally open to all themes of T. A condition of openness is defined as:

$$\exists i,t \mid S^i \in S\, and\, \mathrm{T}_t \in \mathrm{T}, and \mid \exists(S^i,T_t)\subset \mathrm{R} \subset \mathrm{S}\times\mathrm{T}$$
$$i\in[1,2,\ldots,n], t\in[1,2,\ldots,q] \tag{7.5}$$

A totally interoperable group (IG) is defined as:

$$< \mathrm{IG}-\#(<\mathrm{S}>\rho<\mathrm{T}>)> \tag{7.6}$$

ρ means that R is a total relation on $S \times T$, in other words, there exists only one dependency between the subset S and the subset T. The subset S is totally open on the subset T. All IG must be numbered (IG-#) to construct a lattice for the openness structure of a coalition.

Relation R	T_1	T_2	T_3
S_1	■	■	■
S_2	■	■	■
S_3	■	■	■

Example 1

Relation R	T_1	T_2	T_3
S_1	■	■	
S_2	■	■	■
S_3	■		■

Example 2

Figure 7.3 Illustration of a context of openness.

7.2.1.3 Openness Structure of a Coalition

The notion of IG supports a formal structure to represent a coalition openness: a lattice with its interesting properties. Let us illustrate that with an example. Figure 7.4 presents a coalition C with an openness context composed of eight subsets of systems. This context is translated through an IG structure. From those IG, a graph is built where its nodes correspond to numbered IG (IG-#).

The right part of Figure 7.4 represents the only possible openness structure of the coalition C that, obviously, depends upon the way to fix the openness context. The graph on Figure 7.4 allows the following remarks about the openness of the coalition C:

- Every IG-# inherits all coalition-themes linked up to it in the graph.
- Every node number is constituted of all systems that are linked down to it.
- The graph can be used as a tool to visualize consequences of actions from coalition C decision-makers upon basic interoperability, for example, elimination of a link, assignment of a system to a coalition theme, suppression of themes, or restriction of a node for security reason.

7.2.1.4 Barès' Definition of an Interoperability Space

Space interoperability relies upon different notions; some of them appear in Figure 7.5. First, let us consider an action to be not interoperable with itself but with only systems that are capable to handle it. This designates an interoperable action to be represented by a couple: (S^i, A_j) where $S^i \in S$ and $A_j \in A$ (the set of all allowed actions in a coalition). This couple (S^i, A_j) must encompass temporal dimension

Openness context

Relation R	T_1	T_2	T_3	T_4	T_5	T_6
S^1	■	■			■	■
S^2		■	■	■	■	
S^3	■		■		■	

Interoperable Groups

IG-1	$\{S^1, S^2, S^3\}\ \rho\ \{T_5\}$
IG-2	$\{S^1, S^2\}\ \rho\ \{T_2, T_5\}$
IG-3	$\{S^1, S^3\}\ \rho\ \{T_1, T_5\}$
IG-4	$\{S^2, S^3\}\ \rho\ \{T_3, T_5\}$
IG-5	$\{S^1\}\ \rho\ \{T_1, T_2, T_5, T_6\}$
IG-6	$\{S^2\}\ \rho\ \{T_2, T_3, T_4, T_5\}$
IG-7	$\{S^3\}\ \rho\ \{T_1, T_3, T_5\}$
IG-8	$\{\emptyset\}\ \rho\ \{T_1, T_2, T_3, T_4, T_5, T_6\}$

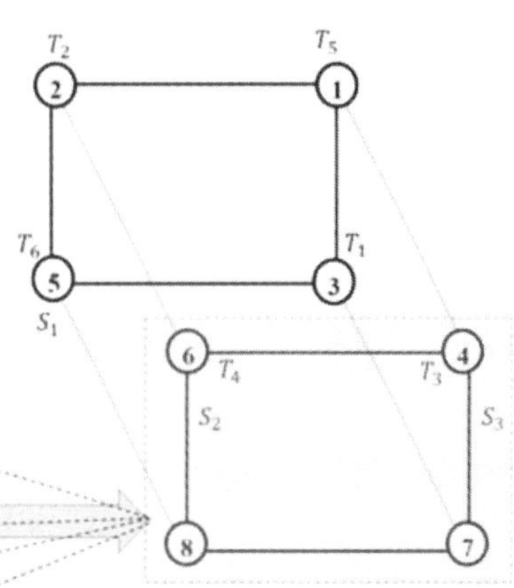

Figure 7.4 Openness context and structure of a coalition C.

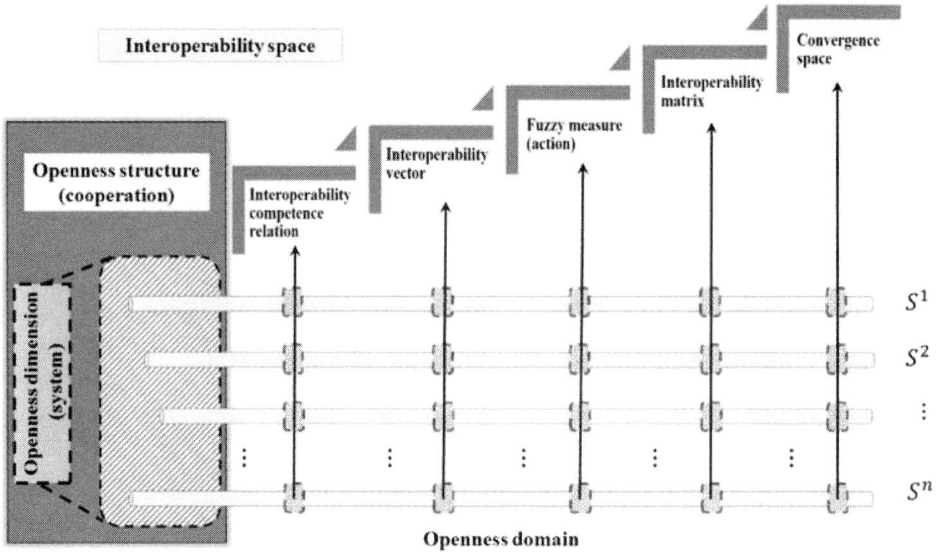

Figure 7.5 Interoperability space.

since situations and systems are dynamics. The validity of (S^i, A_j) will depend on a temporal window, θ_m, or opportunity window denoted as (S^i, A_j, θ_m). System S^i interoperates on action A_j in a temporal interval θ_m assigned to mission M. Time parameters are determined by coalition decision-makers.

7.2.1.5 Interoperability Competence Relation

Let us define a predicative interoperability relation, $\mathcal{R}$, in a propositional calculus view. The arity of that relation is 3 and, by that relation, any system can evaluate its ability to operate on an action of a coalition. This relation must be applied by every system and to every action of the coalition. We form the following proposition:

$$\mathcal{R}\left(S^i,\left\{A_j\right\},\theta_M\right),\forall i\in[1,n],\forall j\in[1,p] \tag{7.7}$$

System S^i considers itself competent to interoperate on actions $\{A_j\}$ where all A_j can be described in a formal way. Each system is bound to determine a first condition, necessary but not sufficient for its interoperability. According to its self-awareness of its neighboring world, a system is able to say if it has the required competence to interoperate on an action. In fact, the relation $\mathcal{R}$ allows to define an effective interoperability in the following fashion: a system S^i can evaluate its competence to operate on any action, in a window time θ, for a specific mission M, under normal and usual conditions. As $\mathcal{R}$ $(S^i, \{A_j\}, \theta_m)$ is considered as a proposition, we can assign a truth value to it:

$$\text{Val}\left[\mathcal{R}\left(S^i,\left\{A_j\right\},\theta_M\right)\right]::\text{True}(T/1) \quad or \quad \text{False}(F/0) \tag{7.8}$$

Meaning, if true, S^i can interoperate on A_j, in time window θ, fixed by mission M. $\forall i \in [1, n]$, $\forall j \in [1, p]$; if false, S^i is not capable to interoperate on A_j. In practice, decision-makers who are responsible for S^i are entitled to apply this relation and thus to decide about the interoperability status of S^i with respect to their mission context. In a formal way, S^i is interpreting its own possible world.

7.2.1.6 Fuzzy Representation of an Interoperable Action

A fuzzy measure refers to a means of expressing uncertainty in the presence of incomplete information. Since we do not dispose of complete information, where probabilities could have been used, the suggestion here is to determine, in a subjective way, numerical coefficients or certainty degrees. These coefficients indicate how it is necessary for a system to interoperate on an action that has been beforehand declared possible. Let us assume that a system only executes one interoperable action at a time. We can therefore define a universe, W, from the following singletons:

$$\begin{gathered} \mathrm{W} = \left\{ \left(S^i, A_1\right), \ldots, \left(S^q, A_p\right) \right\} \\ \text{with } \mathrm{d}\left(S^i, A_n\right) :: \text{degree of possibility}, \; \textit{and} \; \mathrm{d}\left(S^i, A_n\right) \in [0,1] \end{gathered} \tag{7.9}$$

Here the value, $\mathrm{d}(S^i, A_n)$, assesses the possibility that S^i executes the action A_n.

A fuzzy measure is completely defined as soon as a coefficient of possibility has been attached to every subset of a universal set U. If the cardinality of U is n, we must state 2^n coefficients, to specify a measure of possibility. Here, we will proceed by simply observing that each subset of U may be regarded as a union of singletons that it encompasses. So the determination of a measure of possibility can be done from only n elements. Figure 7.6 presents an illustration of an interoperable action using the concept of a fuzzy cube where the dimensions are:

1. A feasibility measure comparable to a possibility;
2. An imperativity measure comparable to a necessity that will be dual to (1);
3. A credibility measure to assess trust on systems in fulfillment of an action.

Dimensions 1 and 2 will be defined by distributions of possibility.

7.2.1.7 Fuzzy Matrices of Interoperability

For a given system S^i, if we successively apply the predicative relation of interoperability, $\mathcal{R}$, to couples (S^i, A_j), j varying from 1 to p, as in (7.10), we obtain a binary vector $V(S^i)$. There are as many vectors as systems in the coalition.

$$\begin{gathered} \mathrm{Val}\left[\mathcal{R}\left(S^i, A_1\right)\right] :: \mathrm{T} \\ \mathrm{Val}\left[\mathcal{R}\left(S^i, A_3\right)\right] :: \mathrm{F} \\ \cdots\cdots\cdots\cdots \\ \mathrm{Val}\left[\mathcal{R}\left(S^i, A_p\right)\right] :: \mathrm{T} \end{gathered} \tag{7.10}$$

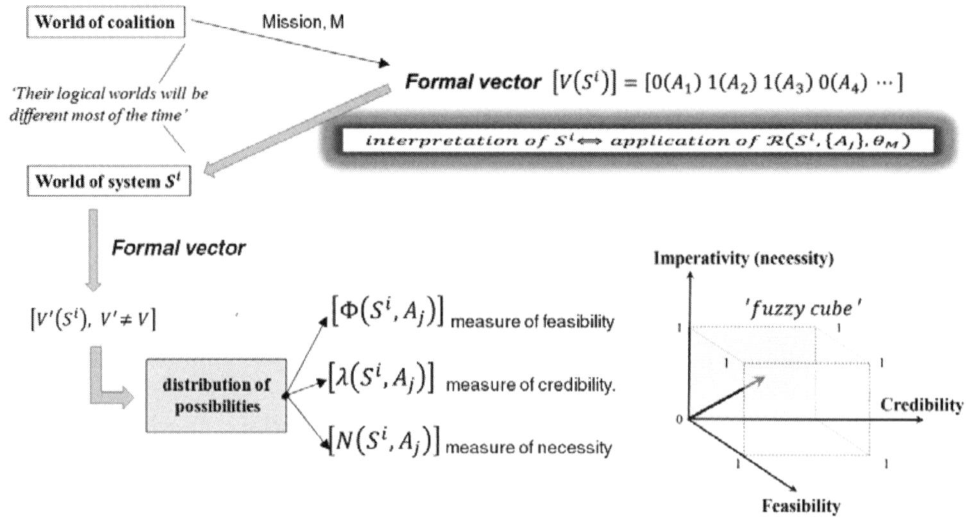

Figure 7.6 Fuzzy representation of an interoperable action

Let us evaluate a component of vector $V(S^i)$ (row j), for instance:

$$\text{Val}\left[V\left(S^i\right)_j\right] :: \text{F} \Rightarrow \neg\exists\ \mathcal{R}\left(S^i, A_j\right) \tag{7.11}$$

False (F), in (7.11), means that there is no openness structure for the coalition, and therefore, S^i has no semantics to evaluate. $V(S^i)_j$ is not supposed to exist. From the binary vector or from the (logic) world resulting of the application of $\mathcal{R}$, it becomes possible to assign fuzzy measures to each vector components for values different of False (F). These fuzzy vectors are computed by taking either couples of the world $\mathcal{R}$ such as:

$$V\left(S^i\right)_{j=1,2,\ldots,p} :: 1 \tag{7.12}$$

or vector's elements $V(S^i)$ such as

$$\left[V\left(S^i\right)_{j=1,2,\ldots,p}\right](\mathcal{R}) :: 1 \tag{7.13}$$

The evaluation of the semantics, $\text{Val}\left[V\left(S^i\right)_{j=1,2,\cdots,p}\right] :: T$, has been made necessary because: either unpredictable events happen in own system's world or changes in the mission could modify the world of S^i. This means a potential shift in the meaning of A_j for system S^i and possibly also for the coalition. In gathering all vectors of interoperability $V(S^i)$, we get what we call an interoperability matrix.

$$\left[\text{I}\left(S^i\right)_{i=1,2,\ldots,n}\right] = \left[V\left(S^1\right)V\left(S^2\right)\ldots V\left(S^n\right)\right] \tag{7.14}$$

This matrix represents only an apparent interoperability and can be used in different ways such as to indicate what is theoretically the most interoperable system, S^i, for action, A_j or to identify the most adequate system, S^i, to operate under special conditions: a mission that imposes a temporal constraint to execute an action. Three kinds of fuzzy interoperability matrices can be constructed:

- *Matrix of feasible interoperability:* This matrix gives a dimension of feasibility of the interoperability of $\{S^i\}$. The matrix is denoted by:

$$\left[\mathrm{I}-\Phi\left(S^i\right)_{i=1,2,\ldots,n}\right] \tag{7.15}$$

- *Matrix of imperative interoperability:* The matrix of necessary interoperability is built with fuzzy vectors of necessity as described above and on Figure 7.6. It informs about the necessary interoperating conditions that are imposed to some systems. This matrix is denoted by:

$$\left[\mathrm{I}-N\left(S^i\right)_{i=1,2,\ldots,n}\right] \tag{7.16}$$

- *Matrix of credible interoperability:* This matrix brings visibility on systems that are in the best position to interoperate successfully. It will be denoted by:

$$\left[\mathrm{I}-\lambda\left(S^i\right)_{i=1,2,\ldots,n}\right] \tag{7.17}$$

Unfortunately, space restrictions prevent us from presenting detailed examples of the usage of these three matrices, but the reader can find more details in Barès et al. [30, 35].

7.2.1.8 Barès' Definition of a Cooperability Space

This section defines concepts of a cooperability space. The same methodology is being used as for the interoperability space above. Space restrictions again prevent us from being as detailed as earlier, but details can be found in [30–35]. However, it is worth mentioning that for cooperability, in addition to system self-ability to execute any action, a system must interpret other systems ability for interoperating on actions. This means that in addition to the predicative relation of interoperability of S^i (see Section 7.2.2.5), a predicative relation of cooperability of system S^i is also required.

Assume that a system S^i has a competence in cooperability when S^i can judge the ability of adjoining (cooperative) systems to interoperate on a set of actions $\{A_j\}$, in a time window θ_m. This competence is denoted by a quadruplet:

$$\left[S^i \because S^k, \left\{A_j\right\}, \theta_M\right] \forall i,k \in [1,n], j \in [1,p] \tag{7.18}$$

(the symbol $\because$ indicates the way of interpretation)

A predicative relation of cooperability, $\mathcal{R}'$, is defined according to the following conditions: $\mathcal{R}'$:: ≪is able to cooperate≫, and $\mathcal{R}'$:: ≪interpret the other systems' aptitude to interoperate on $\{A_j\}$≫. Then the following predicate relation can be written:

$$\mathcal{R}'\left[S^i \because S^k,\left\{A_j\right\},\theta_M\right]\forall i,k \in [1,n], j \in [1,p] \tag{7.19}$$

This means that S^i judges that S^k is able to interoperate on $\{A_j\}$ in the time window θ_m. This evaluation is made with a fuzzy measure of credibility, λ. In a predicate calculus view, the relation $\mathcal{R}'$ defined under these conditions, is equivalent to a propositional function: representing the variables (S^i, S^k, A_j), (θ_m may be considered here as a constant). So, for a given S^i, the truth value of the predicate: S^k is interoperable on each $A_{j=1,2,\ldots,p}$ can be evaluated.

$$\text{Val}\left[\mathcal{R}'\left[S^i \because S^k,\left\{A_j\right\},\theta_M\right]\right]::True \tag{7.20}$$

if true means S^i interprets that S^k is able to interoperate on the actions $\{A_j\}_{j=1,p}$.

$$\text{Val}\left[\mathcal{R}'\left[S^i \because S^k,\left\{A_j\right\},\theta_M\right]\right]::False \tag{7.21}$$

if false means S^i considers that S^k is unable to interoperate on the actions $\{A_j\}_{j=1,p}$.

A cooperabilility matrix is obtained by assembling the vectors of cooperability. Although the interoperability matrix is unique, it is necessary to establish two categories of matrices in the domain of cooperability. The first category, called the cooperability system, is going to indicate how the set of systems interpret their respective interoperability. The second one, called cooperability action, regards actions (i.e., a matrix to comprehend the interoperability of the coalition from its elementary actions).

7.2.1.9 Matrix of Cooperability System

The cooperability matrix of a system S^i is denoted by $[C(S^k)]$. Special computations can be made on rows and columns that provide useful insights to characterize the cooperability capacity of a coalition, that is, the visibility about the quality (e.g., easiness) of system interoperation.

- *Properties of a column:* Let $[C(S^k)]$ be the matrix of cooperability system of S^k, and consider the mth column of that matrix. If we sum up all components of the vector column $\boldsymbol{m}$ of the matrix $[C(S^k)]$, then we get a scalar $a_c(S^k)$ defined as:

$$\alpha_C\left(S^k\right)=\sum_{j=1}^{p}\left[C\left(S^k\right)\right]_{j,m} \tag{7.22}$$

The scalar $\alpha_c(S^k)$ indicates how S^m assesses the interoperability strength of S^k regarding actions. The following are two cases to illustrate the interpretation of that cooperability-system matrix.

- *Case 1:* $\alpha_c(S^k) = 0$. According to its evaluation, S^m considers S^k being not interoperable. S^k must not play any role in the coalition since S^k is unable to execute any action. This does not mean that S^k has no interoperable capacity. At this point, the other systems' assessment is not known and several issues can be raised such as if there is a misjudgment between S^m and S^k, and if there is anything shareable between S^m and S^k.
- *Case 2:* $\alpha_c(S^k) \neq 0$. This means that S^m trusts more or less the ability of S^k to interoperate. If the value of α_c is close to 0, S^m trusts weakly S^k and thinks that S^k is likely to fail. If $\alpha_c(S^k)$ is close to 1, S^m trusts strongly S^k and thinks that S^k will be successful in its interoperations.

Figure 7.7 presents an example of a matrix of cooperability system with system S^2 and four actions. From the columns, one can get a given system point of view, for example, on which action the system C is the most interoperable? From the rows, the points of view of all systems (consensus notion, property of a line l) are shown, for example, which system is in the best position to interoperate on action l.

7.2.1.10 Matrix of Cooperability Action

We now define another kind of fuzzy matrix allowing the required visibility of the cooperability of all systems of the coalition. This special matrix is assumed to indicate the systems that are in the best conditions to interoperate on actions. For this reason, these matrices will be called matrices of cooperability action. If the jth row in each of the previous matrices (Section 7.2.1.9) is considered, we form a new matrix that reports about the systems cooperability capacity relative to action A_j. This matrix is denoted by $[C(A_j)]$ and the process is illustrated in Figure 7.8.

$$[C(S^2)] = \begin{bmatrix} \mathcal{R}'(S^1 \because (S^2,A_1)) & \mathcal{R}'(S^2 \because (S^2,A_1)) & \mathcal{R}'(S^3 \because (S^2,A_1)) \\ \mathcal{R}'(S^1 \because (S^2,A_2)) & \mathcal{R}'(S^2 \because (S^2,A_2)) & \mathcal{R}'(S^3 \because (S^2,A_2)) \\ \mathcal{R}'(S^1 \because (S^2,A_3)) & \mathcal{R}'(S^2 \because (S^2,A_3)) & \mathcal{R}'(S^3 \because (S^2,A_3)) \\ \mathcal{R}'(S^1 \because (S^2,A_4)) & \mathcal{R}'(S^2 \because (S^2,A_4)) & \mathcal{R}'(S^3 \because (S^2,A_4)) \end{bmatrix}$$

Application of the relation $\mathcal{R}'$

$$[C(S^2)] = \begin{bmatrix} 1 & 1 & 0 \\ 0 & 0 & 1 \\ 0 & 0 & 0 \\ 1 & 1 & 0 \end{bmatrix}$$

Measures of credibility

$$c[S^2] = \begin{bmatrix} 0.3 & 0.6 & 0.0 \\ 0.0 & 0.0 & 0.7 \\ 0.0 & 0.0 & 0.0 \\ 0.3 & 0.2 & 0.0 \end{bmatrix}$$

Figure 7.7 Matrix of cooperability system.

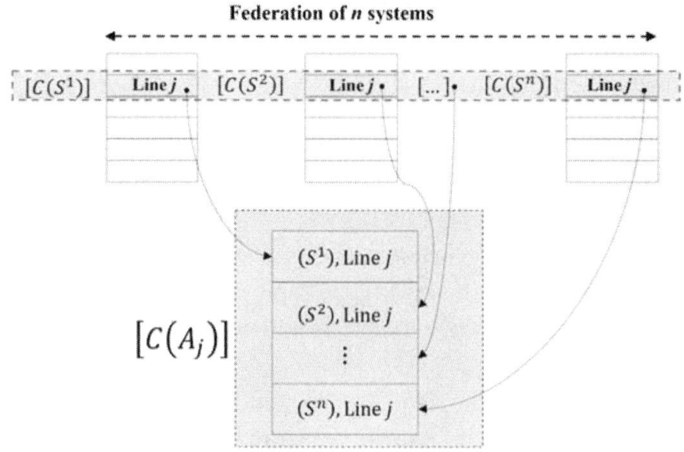

Figure 7.8 Matrix of cooperability action.

The matrix of cooperability action, $[C(A_j)]$, presents interesting features: its structure is square; it allows understanding of what the systems of the coalition are that are in the best position to interoperate on an action A_j; and, more importantly, it gives an assessment of the level of difficulty to interoperate on specific actions. For instance, the computation of the **p** matrices of all the actions $\{Aj\}_{j=1,p}$ of the coalition gives visibility on which actions are difficult to carry out and which actions are likely either to get the coalition into trouble or to force the coalition to face difficult issues.

7.2.1.11 Concluding Remarks on Systems Interoperability

Section 7.2.1 presented very innovative notions of system openness, interoperable groups, and their application through mathematical tools (fuzzy matrices of interoperability and cooperability) to assess and analyze interoperability in an organizational paradigm such as a military coalition. It was possible to formalize the structure of coalition's openness and to develop analyzing tools to get interesting properties. Tools based on FIAT (fuzzy measures, possibility distributions, logic-based predicative relations) can provide quantitative evaluation of interoperability pertaining to large systems such as CPSS. The development of such intelligent FIAT tools enlarges our comprehension of interoperability that is critical to address the Variety dimension of the big data problem.

7.3 Integrated Frameworks to Process Imperfect Information

The previous section exemplifies the dimension of Variety of systems and sources and the question of interoperability. This section emphasizes the Veracity dimension. FIAT provides techniques and tools to analyze data, enables situation awareness, and supports decision-making. Characterizing and interpreting information imperfections are probably the most important and difficult task of the FIAT-based process.

FIAT supports a process to transform data into actionable knowledge for sense-making, to reduce uncertainty and thus to improve dependability in systems. Meaning and making sense are intimately related to actions of any agent, human or machine. A goal-driven, or activity-based, approach sounds appropriate for that process. The development of a framework in which knowledge, information, and uncertainty can be organized, structured, represented, combined, managed, reduced, and updated is a core requirement. Such integration FIAT framework should: (1) provide means to represent knowledge through well-defined notions of situation and awareness; (2) support modeling uncertainty, belief, and belief change; (3) provide the key integration model for FIAT and linkage to actions (users and machines); (4) provide practical support for system design through modularization, refinement, validation, and verification; (5) offer a good compromise between operational and functional modeling in capturing systems behavior; (6) enable rapid prototyping and experimental validation of fairly abstract models; and finally, (7) support modeling of multi-agent systems.

A very limited number of information fusion frameworks have been proposed in the literature that offer partial fulfillment of the above mentioned requirements. A number of significant and powerful analytics tools and techniques have been offered to scientific and engineering communities for organizing, integrating, and visualizing large volumes of data. One can think of *MapReduce* [47], *HADOOP* [48, 49], and *EXALEAD* [50], to name a few. In the domain of information fusion alone, noticeable efforts have recently been dedicated to define an integration fusion framework: the State Transition Data Fusion (STDF) model introduced in [51], the Interpreted Systems for Situation Analysis (ISSA) [52], Abstract State Machine (ASM) methods [53, 54] for high-level design and analysis of distributed systems, OODA-based agents [55], the recombinant design approach [56], fuzzy cognitive maps used for situation assessment [57–59], and recently, the holonic processing framework of [60, 61] based on Sulis' archetypal dynamics and causal tapestries [62].

7.3.1 Integrated Frameworks to Process Imperfect Information

A cognitive map (CM), also known as a mental map or a mental model, is, according to [63]: "a type of mental representation which serves an individual to acquire, code, store, recall, and decode information about the relative locations and attributes of phenomena in their everyday or metaphorical spatial environment." A cognitive map can be described as a qualitative model of how a given system operates. The map is based on defined variables and the causal relationships between these variables. These variables can be physical quantities that can be measured, such as level of nuclear radiation or percent ice cover, or complex aggregate or event, or abstract ideas such as cyber threats or intents.

The concept, introduced formally by Tolman [64] in 1948 and Kosko [65] in 1986, has brought the contemporary concept of fuzzy cognitive maps (FCM) that has been used in several applications [66]. FCM [65] is a collection of nodes and connecting arcs used to represent the system states and interacting causality: a graphical archetype that encodes and processes vague causal reasoning numerically. The fuzzy notion of the states provides a means to deal with uncertainty involved in the process. Leung et al. [57–59, 67] used FCM for FIAT-based applications to support situation assessment, disaster management, coastal surveillance, and distributed sensing.

As with the other potential FIAT frameworks [51, 53, 60, 68] discussed in previous chapters, FCM also possess interesting features but still requires further efforts to become a FIAT integration framework. For instance, analytical research efforts are required to address two main issues with FCM: the lack of design tools to support automated synthesis process in the development and refinement of the maps, and the development of appropriate reasoning schemes in situations of multistimulus. The combination of several inputs or initial states may result in new patterns with unexpected behaviors. Current techniques for constructing and analyzing FCMs are inadequate and infeasible in practice especially for large complex systems. Finally, the formulation of a concrete FIAT framework will require not only improvements to current FCM approaches but also an integration of the interesting and relevant characteristics of other FIAT frameworks such as in [51, 53, 60, 68].

7.3.2 Sulis' Causal Tapestries

Archetypal dynamics [62] is a formal approach to the modeling of information flow in complex systems. It is grounded in the triad of realization (system), interpretation (archetype, mental model), and representation (formal model) as described in Chapter 2. Sulis [69] proposed the use of tapestries as a formal representation model. Tapestries represent information flow by means of multilayered, recursive, interlinked graphical structures that express both form and semantics. Causal tapestry is a specific tapestry model in describing behaving systems from the standpoint of the process theory [70] by means of a reality game. Observables are represented by tapestry informons (see Chapter 4) while subjective components are incorporated into a reality game determining the tapestry dynamics. This results in a random graphical dynamical system with causal tapestries. Nevertheless, this approach still requires substantial scientific development to become an integrating framework for FIAT. For instance, the fuzzy cognitive maps notions of the previous section can be related to the semantic frames of Sulis. Finally, the same remark as mentioned above, the formulation of a concrete FIAT framework will require not only improvements to Sulis and current FCM approaches but also an integration of the interesting and relevant characteristics of other FIAT frameworks such as in [51, 53, 60, 68].

7.3.3 Abstract State Machines (ASM) to Engineer FIAT-Based Systems

This section briefly outlines the basic concepts of the ASM method for high-level design and analysis of distributed systems. This approach can be used to design FIAT-based intelligent support systems as demonstrated in [53]. For further details on ASM, the reader can consult [53, 54, 71–73]. ASMs offer a versatile mathematical method of modeling discrete dynamic systems with the goal of bridging the gap between computation models and specification methods [74]. They combine two well-known and fundamental concepts of transition systems, to model the dynamic aspects of a system, and abstract states, to model the static aspects at any desired level of abstraction. In the last decade, ASMs were further developed into a systems engineering method [54] that guides the development of software and embedded hardware-software systems from requirements capture to their implementation. Today, ASMs are well known for their versatility in computational and mathematical modeling of architectures, languages, protocols, and virtually all kinds of sequential, parallel, and distributed systems with an orientation towards practical applications [53, 54, 71–77].

The ASM method for systems design and analysis builds on the concept of abstract state machines and brings together two tasks of requirement capture and system design. The goal is to improve industrial systems development by integrating precise high-level problem-domain-oriented modeling into the development cycle and systematically linking abstract models down to executable code. The method consists of three essential elements: (1) capturing the requirements into a precise yet abstract operational model, called a ground model ASM; (2) systematic and incremental refinement of the ground model down to the implementation; and (3) experimental validation of the models at each level of abstraction through simulation or testing. The freedom of abstraction offered by ASMs supports expressing the original ideas behind the design of a system at the same level of complexity and enables system designers to stress on the essential aspects of the design rather than encoding the insignificant details.

The operational characteristics of ground models combined with the freedom of abstraction supports design space exploration and experimental validation of design ideas at early stages of development before the expensive task of coding begins. Starting from a ground model ASM and applying the process of step-wise refinement, a hierarchy of intermediate models can be created that are systematically linked down to the implementation (see Figure 7.9). At each step of refinement, the refined model can be validated and the abstract model can be verified. The resulting hierarchy serves as a design documentation and allows one to trace requirements down to the implementation. Validation of the abstract models at every step can be used to correct, or further improve, the design decisions made in the more abstract models.

7.3.3.1 Distributed ASMs

The original notion of ASMs, or basic ASMs, was defined to formalize simultaneous parallel actions of a single computing agent. A basic ASM, M_A, is defined as a tuple of the form (F, J_S, T_R, P_{M_A}) where F is a finite set of function names and symbols, J_S is a set of initial states for signature, T_R is a set of transition rule declarations, and $P_{M_A} \in T_R$ is a distinguished rule, called the main rule or the program of machine M_A.

A state G for F is a non-empty set X together with an interpretation f^G: $X^n \rightarrow X$ for each function name f in F. Functions can be static or dynamic. Values of dynamic functions can change from state to state. Evaluation of a transition rule in a given state produces a set of updates of the form $\langle (f, \langle a_1, \cdots, a_n \rangle), v \rangle$ where f is an n-ary function name in F and $((a_1, \ldots, a_n), \text{v}) \in X$. An update (f, args, v) prescribes a change to the content of location (f, args) in the next state.

A distributed ASM (DASM), $M_{A_{dist}}$ is then defined by a dynamic set agent of computational agents each executing its ASM. This set may change dynamically over runs of $M_{A_{dist}}$, as required to model a varying number of computational resources. Agents of $M_{A_{dist}}$ normally interact with one another, and typically also with

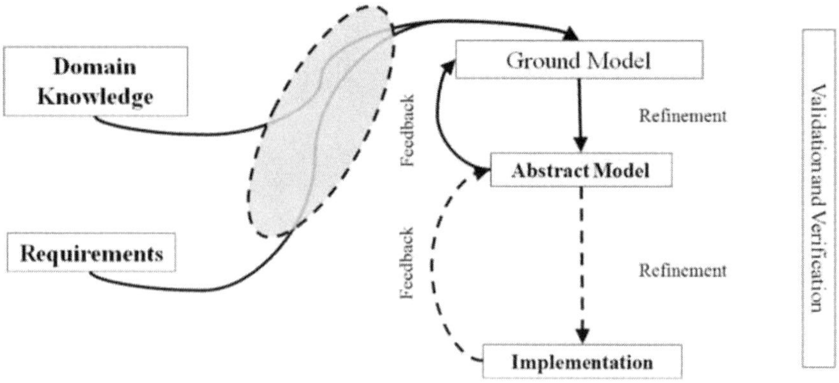

Figure 7.9 The ASM method of systems engineering.

the operational environment of $M_{A_{dist}}$, by reading and writing shared locations of a global machine state.

Intuitively, every computation step of $M_{A_{dist}}$ involves one or more agents, each performing a single computation step according to their local view of a globally shared machine state. The underlying semantic model regulates interactions between agents so that potential conflicts are resolved according to the definition of partially ordered runs [78].

$M_{A_{dist}}$ interacts with its operational environment (part of the external world visible to $M_{A_{dist}}$) through actions/events observable at external interfaces, formally represented by controlled and monitored functions. Of particular interest are monitored functions that are read-only functions controlled by the environment. A typical example is the abstract representation of global system time in terms of a monitored function "now" taking values in a linearly ordered domain time. Values of "now" increase monotonically over runs of $M_{A_{dist}}$.

7.3.3.2 Strengths/Weaknesses Analysis of Three Integrating Frameworks

This section presents a strengths and weaknesses analysis of three potential integrating frameworks for the design of FIAT-based systems: STDF, ISSA, and ASM. Further details on STDF, ISSA, and ASM can be found in [10, 51–53, 68, 71–73, 79].

The STDF model captures the information fusion process in a broader sense. It offers well-defined notions of object, situation, and scenario and characterizes the information fusion process in an abstract functional model (see [51]). However, STDF does not offer a computational model to represent how that functionality can be achieved. The STDF model views the underlying system in terms of a set of abstract states that are transformed in time. This appears to be the right abstraction one would need to focus on modeling the information fusion process but does not provide a comprehensive formal framework to practically model and design the system (in fact, STDF explicitly leaves the choice of such a framework open to the practitioner). To the best of our knowledge, there is no generic tool support provided for STDF.

However, the ISSA approach provides a formal framework for reasoning about knowledge and uncertainty and for dealing with belief change concepts in a distributed systems context. Unlike STDF, ISSA comes with a computational underlying formal framework for modeling a distributed system. Like ASMs, ISSA views distributed computations of concurrent and reactive systems as evolution of states, assuming multiple computational agents interacting with one another as well as with their operational environment representing the external world. The underlying computation model, in both ASMs and ISSA, defines the behavior of a distributed system as the set of all admissible runs originating in a distinguished set of initial system states. However, the underlying view of system behavior in ISSA is geared towards theoretical aspects of systems modeling, building on abstract mathematical concepts that appear to be further away from practical system design and development [53].

Practical design and development of reliable decision support systems for situation analysis require a formal framework that not only offers a formal model of situation analysis concepts and processes, but also supports established system design concepts such as modularization, refinement, validation, and verification of high-level models. Controlled experiments to systematically explore the practicability of situation analysis concepts in meaningful application scenarios call for abstract executable models as a basis for rapid prototyping and experimental validation.

ASMs offer a universal model of computation paired with a systematic approach to high-level design and analysis of systems. ASM models of multiagent distributed systems are executable, in principle, and when combined with CoreASM (an Open Source project available at www.coreasm.org) tool support, enable experimental validation of highly abstract design specifications and facilitate interoperability with other tools and analytical techniques in an open tool environment. Although the ASM modeling approach has also been used as a basis for building knowledge based systems, ASMs do not provide specific support for reasoning about knowledge and uncertainty, and nor do they come with a built-in support for information fusion and situation analysis.

Table 7.1 summarizes the strengths and weaknesses of each of these methods in the context of design and analysis of situation analysis and decision support systems (SADS). In light of this comparison, we propose to combine ASMs, STDF, and ISSA to provide a comprehensive formal framework for design, analysis, and development of SADS. Figure 7.10 illustrates the main components of this framework in which ASM, STDF, and ISSA are positioned with their capabilities.

At the basis, we have the ASM universal computation model of distributed systems. The system view of STDF and ISSA (in terms of transitioning states and especially the freedom of abstraction STDF offers in defining the structure of the states) fits very well with the notion of abstract state machines. The set of states in STDF can be mapped to the global state of a distributed ASM and the notion of state transition would be captured by the notion of computation steps in ASMs. On the top of this foundation, we define the information fusion concepts of object, situation, and scenario based on which the STDF unified model of information fusion process can thus be defined. Sensor signals will be modeled by monitored functions that receive input from the environment. The view of ISSA toward knowledge representation, uncertainty, and belief change is brought in, which enables us to model situation awareness of agents as part of their local states.

Table 7.1 Strengths and Challenges of STDF, ISSA, and ASM

STDF	Strengths	Designed for information fusion
		A unified model of the information fusion process
		Well-defined notions of object, situation, and scenario
		Freedom in choosing an underlying state-based system modeling framework
	Weaknesses	No operational semantics
		No underlying formal framework to model the system (loosely defined notion of state)
		Practical challenges in implementation
		No tool support
ISSA	Strengths	A formal framework for modeling distributed systems (notions of state, run, and state transition)
		A formal framework for reasoning about knowledge and uncertainty
		Well-defined notions of situation, situation awareness, and situation analysis
	Weaknesses	Not a practical framework for system design
		Situation awareness is not computational
		No model for impact or intent
		Lack of strong tool support
ASM	Strengths	A universal model of computation with well-defined notions of state and state transition
		Operational in nature but also supports functional and declarative system modeling
		Practical for system design: abstraction, refinement, modularization, and executability
		Many practical applications
		Supports multi-agent and distributed systems design
		Existing tool support
	Weaknesses	No explicit support for reasoning about knowledge and uncertainty or information fusion and situation analysis

The executable nature of ASMs paired with the Core-ASM modeling environment allows experimental validation of the abstract model in addition to situation analysis through model checking or simulation. The fusion process of STDF is modeled in an operational form by the control state ASM of Figure 7.11, which then can be mapped directly to an ASM rule run by DASM agent programs (to update their situation awareness). Once we have a proper model of the system and its scenarios, various types of queries about truth, knowledge, and time can be used for situation analysis. These queries can be examined either by explicitly encoding the queries as derived functions and running the executable model, or by applying the available model checking techniques for ASM [53].

Semantic values in STDF (i.e., the understanding of an object, a situation, or a scenario) are defined based on the whole history of transitioning states of the corresponding entities. In a DASM at any given state, agents have access to the values of functions and terms only in the current state of the machine. However,

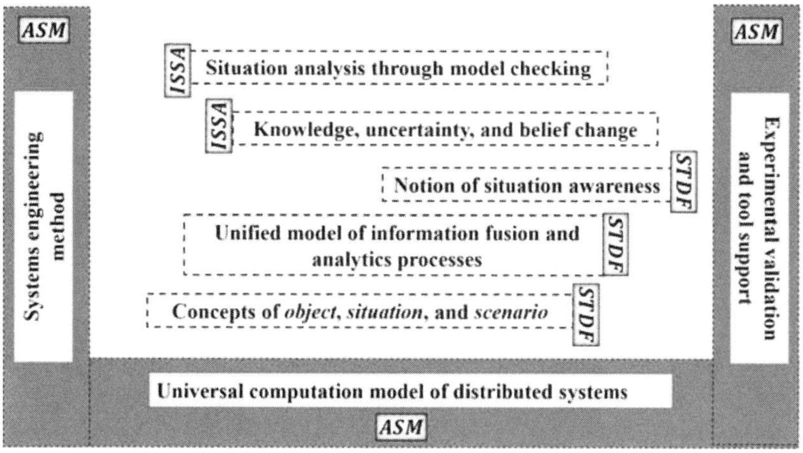

Figure 7.10 Integrating ASM, STDF, and ISSA.

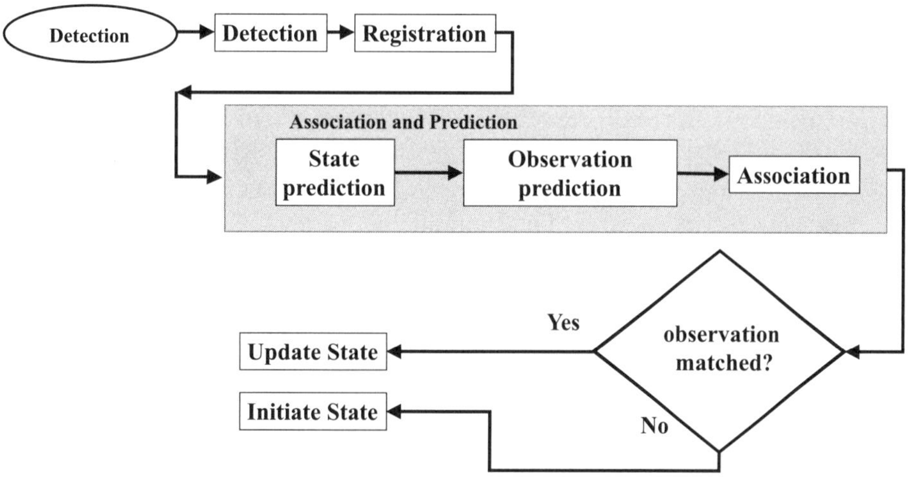

Figure 7.11 Operational model of the STDF fusion process as a control state ASM.

with the freedom of abstraction in ASMs, one can introduce a notion of history of values that would expose, in the current state of the machine, the historical values of functions in the previous states of the machine. A similar notion was introduced earlier in [77] where previous values of a location in ASM states were assumed to be accessible. For this framework, we propose to extend ASMs with the following notation, $f[k](a_1, ..., a_n)$, evaluation of which will result in the value of function f with arguments $a_1, ..., a_n$ in the kth previous state of the machine $(k \in \mathbb{N})$.

In conclusion of this qualitative analysis, a formal computational framework for design and development of FIAT-based systems for SADS is unavoidable if one is interested in representing and reasoning about dynamic situations and producing solid system designs that can be experimentally validated and systematically verified. Such a framework should not only provide a basic model for FIAT and means to represents knowledge and uncertainty, but it should also offer practical support for systems engineering and experimental validation of models. To satisfy the set of requirements for such a formal framework, we propose the integration of two formal approaches, STDF and ISSA, with the systems design and analysis approach of ASMs. This integration has been started in [53] and needs to be pursued considering the notions of archetypal dynamics illustrated in Chapters 5 and 6.

7.3.3.3 Application Scenario: Tracking of Moving Objects

This section illustrates the application of the proposed framework in design and modeling of FIAT-based systems. Let us consider the scenario depicted in Figure 7.12. Here we will be focused on two functions of situation analysis: association and tracking. The scenario involves four nodes positioned on a Cartesian space, representing two observer resources, O_1 and O_2, and two objects of interest, A and B. A and B are moving objects characterized at time t by $\langle x_A^t, y_A^t, v_{x_A}^t, v_{y_A}^t \rangle$ and $\langle x_B^t, y_B^t, v_{x_B}^t, v_{y_B}^t \rangle$. O_1 and O_2 are both equipped with sensors s_{O_1} and s_{O_2} capable of observing locations and velocity vectors of A and B in time intervals Δt_{O_1} and Δt_{O_2}. Each sensor s has an observation error vector of $\langle e_x^s, e_y^s, e_{v_x}^s, e_{v_y}^s \rangle$ defining the range of error, and a variable error coefficient e_d^s, which monotonically increases with the distance d of the observed object from the observer. We define the reliability of sensor s in observing an object T at distance d to be an inverse function of $d : R_T^s \sim d^{-1}$.

Using ASM, we would like to model a DSS for the Figure 7.12 scenario that satisfies the following requirements: (1) the system should provide awareness of the last observed locations and velocities of A and B; (2) the system should provide continuous projection of locations of A and B within observation intervals; (3) observers O_1 and O_2 should maintain the joint reliability of their observations under a certain threshold R_M, if possible; (4) observers O_1 and O_2 should always keep a minimum distance of d_M from all objects of interest; (5) there may be a rendezvous happening between A and B if they stay close for a period of time. A probability of the rendezvous can be defined as:

$$Pr(A,B) = \frac{1}{2}\left(e^{\left(1-\tilde{t}_r(A,B)^{-1}\right)} + e^{\left(1-\tilde{d}_r(A,B)^{-1}\right)} \right) \tag{7.23}$$

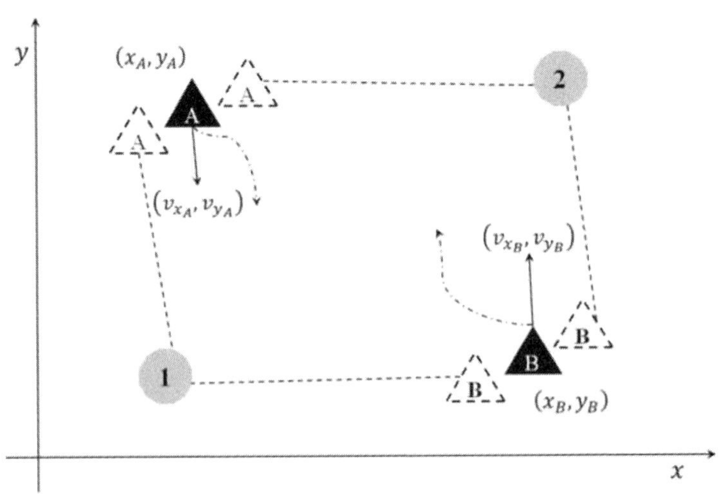

Figure 7.12 Rendezvous scenario settings.

where $\tilde{d}_r(A,B)$ is the normalized distance of A and B in [0,1] for the range of $\left[d_r^{\min}, d_r^{\max}\right]$ and $\tilde{t}_r(A,B)$ is the normalized duration in which the distance of A and B is less than $d_r^{\min}$.

7.3.3.4 The Executable Abstract Model for the Scenario Rendezvous

We model the system as a multi-agent DASM system with four agents representing the objects of interest (A and B) and the observer resources (O_1 and O_2), and a fifth agent M providing the projection of locations of A and B into the future as well as computing the possibility of a rendezvous. The observers and the objects of interest have location and velocity attributes that may change over time, and they also have a common behavior by which they update their location according to their current velocity. In that respect, we can classify the four agents under a domain node. Agents (O_1 and O_2) require sensors to observe the locations and velocities of A and B and a means of communicating their observations to agent M. In this model, we abstract from the communication layer and assume that M is always aware of the observations of O_1 and O_2 by reading shared locations of the global state. Due to space limitations, here we only present the highlights to detect a rendezvous.

The aggregation of observations for every object of interest T can simply be defined as a weighted average of the observed information with respect to the reliability of the observations.

$$\hat{O}_M(T) = \frac{\sum_w \mathcal{R}_T^w \hat{O}_M(T)}{\sum_w \mathcal{R}_T^w}, w \in Observer \tag{7.24}$$

Projection of locations can be modeled using a typical Kalman filter model of the objects of interest. To detect rendezvous, the observed locations of the objects of interest in the last k_b states of the machine will be continuously monitored to compute the probability of rendezvous as given by (7.23). If the probability is above a configurable threshold p_{thres}, the system will report the location of rendezvous together with its probability.

We produced an executable model of the system through refinement of the abstract model and used the CoreASM modeling framework to simulate the model. Such a refinement, with the goal of producing an executable model, not only facilitates experimental validation of design, but it also helps in finding ambiguities, missing pieces, and loose ends of the model and forces the system analyst/modeler to think clearly about the main concepts and their definitions. In our CoreASM model sensors are modeled as derived functions that basically add noise to the actual location and velocity of objects of interests.

The STDF model suggests the use of the possible world distribution approach for observation prediction; however, this approach is practically challenging in implementation. In our abstract model, we use simulation to estimate the possible world distribution. The projection of locations of A and B in the future is modeled using a number of predicting agents that try to imitate the behavior of objects of interested based on the last fused information obtained by the monitoring agent M. When new observation information is obtained for an object of interest, the monitoring agent creates a number of predicting agents positioned on the last observed location with the last observed velocities of A and B. At any moment, the distribution of predicting agents provides an estimation on the current possible locations of the objects of interest.

To observe the state of the system during the simulation, we extended the CoreASM environment with a situation observation plug-in that illustrates the state of the machine at run time, that is, current locations and velocity vectors of the nodes (ground truth), observed locations of the nodes, the fused information, and projection of locations into the future. Figure 7.13 shows a snapshot of the output of this plug-in. Observers O_1 and O_2 are represented by a solid rectangle and a solid triangle, respectively. Objects of interest A and B are represented by solid crosses. Observed locations of A and B by O_1 and O_2 are marked with two rectangles (for O_1 and two empty triangles (for O_2). The system's understanding of locations of A and B (fused position observations of O_1 and O_2) are marked with two solid black circles. Predicting agents that try to estimate the current locations of A and B are illustrated by empty circles.

7.4 Conclusion

This chapter brought two main contributions to support integration in complex systems: system interoperability in heterogeneous environments and integrated frameworks to process imperfect information. The development of integrated FIAT-based solutions in confronted to that complexity. A formal computational framework for the design and development of situation analysis decision support systems is unavoidable if one is interested in representing and reasoning about dynamic situations and producing solid system designs that can be experimentally validated and

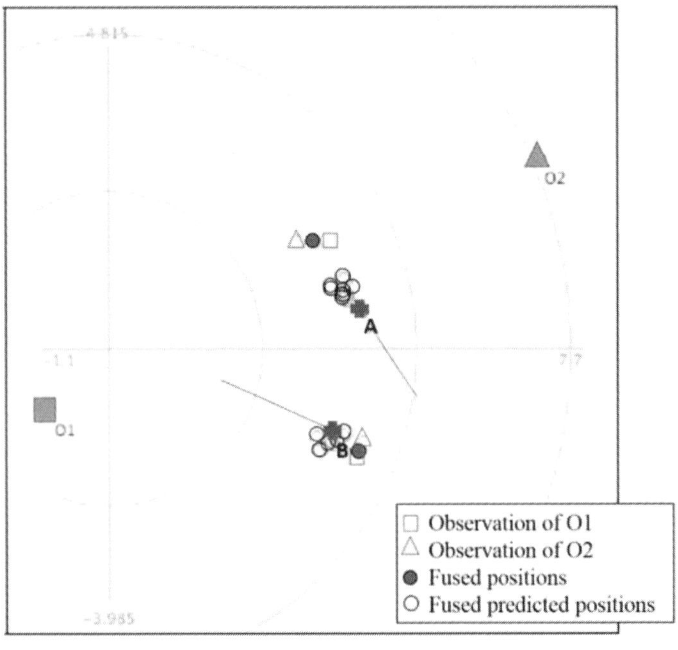

Figure 7.13 Snapshot of a system run in CoreASM

systematically verified. Such a framework should not only provide a basic model for analytics, information fusion, and means to represents knowledge and framework as it will uncertainty, but it should also offer practical support for systems engineering and experimental validation of models. Most of the work presented in previous chapters and discussed in this chapter constitute early steps towards building such a framework. Satisfying the set of requirements for such a formal require the integration of several frameworks with the systems design and analysis approach of ASMs.

References

[1] Alberts, D. S., *The Agility Advantage: A Survival Guide for Complex Enterprises and Endeavors*, Washington DC, CCRP Publication Series.

[2] Alberts, D. S., "Agility, Focus, and Convergence: The Future of Command and Control," DTIC Document, *The International C2 Journal*, Vol. 1, No. 1, 2007.

[3] Alberts, D. S., R. K. Huber, and J. Moffat, "NATO NEC C2 Maturity Model," NATO SAS-065, www.dodccrp.org, 2010.

[4] Liu, Z., et al., "Cyber-Physical-Social Systems for Command and Control," *IEEE Intelligent Systems*, Vol. 26, 2011, pp. 92–96.

[5] Poovendran, R., "Cyber-Physical Systems: Close Encounters Between Two Parallel Worlds [Point of View]," *Proc. of the IEEE*, Vol. 98, 2010, pp. 1363–1366.

[6] Sztipanovits, J., et al., "Toward a Science of Cyber-Physical System Integration," *Proc. of the IEEE*, Vol. 100, 2012, pp. 29–44.

[7] Lee, E. A., and S. A. Seshia, *Introduction to Embedded Systems: A Cyber-Physical Systems Approach*, Berkeley, CA: Lee & Seshia, 2011.

[8] Marwedel, P., *Embedded System Design: Embedded Systems Foundations of Cyber-Physical Systems*: New York: Springer Science & Business Media, 2010.

[9] Minelli, M., M. Chambers, and A. Dhiraj, *Big Data, Big Analytics: Emerging Business Intelligence and Analytic Trends for Today's Businesses*, New York: John Wiley & Sons, 2012.

[10] Blasch, E., E. Bosse, and D. A. Lambert, *High-Level Information Fusion Management and Systems Design*: Norwood, MA: Artech House, 2012.

[11] Bossé, É., J. Roy, and S. Wark, *Concepts, Models, and Tools for Information Fusion*, Norwood, MA: Artech House, 2007.

[12] Das, S., *Foundations of Decision-Making Agents*, New York: World Scientific Publishing, 2008.

[13] Das, S., *Computational Business Analytics*: New York: Taylor & Francis, 2013.

[14] Liggins, M., D. Hall, and J. Llinas, *Handbook of Multisensor Data Fusion: Theory and Practice*, 2nd ed., New York: Taylor & Francis, 2008.

[15] Hall, D. L., and J. M. Jordan, *Human-Centered Information Fusion*, Norwood, MA: Artech House, 2010.

[16] Abbasi, A. A., and M. Younis, "A Survey on Clustering Algorithms for Wireless Sensor Networks," *Computer Communications*, Vol. 30, 2007, pp. 2826–2841.

[17] Al Hasan, M., and M. J. Zaki, "A Survey of Link Prediction in Social Networks," in *Social Network Data Analytics*, New York: Springer, 2011, pp. 243–275.

[18] Saecker, M., and V. Markl, "Big Data Analytics on Modern Hardware Architectures: A Technology Survey, " *Business Intelligence*, Berlin Heidelberg: Springer, 2013, pp. 125–149.

[19] Atrey, P. K., et al., "Multimodal Fusion for Multimedia Analysis: A Survey," *Multimedia Systems*, Vol. 16, 2010, pp. 345–379.

[20] Atzori, L., A. Iera, and G. Morabito, "The Internet of Things: A Survey," *Computer Networks*, Vol. 54, 2010, pp. 2787–2805.

[21] Baldauf, M., S. Dustdar, and F. Rosenberg, "A Survey on Context-Aware Systems," *International Journal of Ad Hoc and Ubiquitous Computing,* Vol. 2, 2007, pp. 263–277.

[22] Berkhin, P., "A Survey of Clustering Data Mining Techniques," in *Grouping Multidimensional Data*, New York: Springer, 2006, pp. 25–71.

[23] Bettini, C., et al., "A Survey of Context Modelling and Reasoning Techniques," *Pervasive and Mobile Computing,* Vol. 6, 2010, pp. 161–180.

[24] Blasch, E. P., et al., "High Level Information Fusion (HLIF): Survey of Models, Issues, and Grand Challenges," *IEEE Aerospace and Electronic Systems Magazine,* Vol. 27, 2012, pp. 4–20.

[25] Bordawekar, R., et al., "Analyzing Analytics, Part 1: A Survey of Business Analytics Models And Algorithms," Technical Report RC25186, IBM TJ Watson Research Center, 2011.

[26] Bravo, C. E., et al., "State of the Art of Artificial Intelligence and Predictive Analytics in the E&P Industry: A Technology Survey," *SPE Journal,* Vol. 19, 2014, pp. 547–563.

[27] Chen, M., S. Mao, and Y. Liu, "Big Data: A Survey," *Mobile Networks and Applications,* Vol. 19, 2014, pp. 171–209.

[28] Fang, Z., and H. Liyan, "A Survey of Multi-Sensor Information Fusion Technology [J]," *Journal of Telemetry, Tracking and Command,* Vol. 3, 2006.

[29] Horling, B., and V. Lesser, "A Survey of Multi-Agent Organizational Paradigms," *The Knowledge Engineering Review,* Vol. 19, 2004, pp. 281–316.

[30] Bares, M., "Formal Approach of the Interoperability of C4IRS Operating Within a Coalition," *Proc. of 5th Intl. Command and Control Research and Technology Symp.*, 2001.

[31] Barès, M., *Maîtrise du savoir et efficience de l'action*, Paris, France: Editions L'Harmattan, 2007.

[32] Barès, M., "Proposal for Modeling a Coalition Interoperability," *6th Intl. Command and Control Research and Technology Symp.*, 2001.

[33] Handley, H. A., A. H. Levis, and M. Bares, "Levels of Interoperability in Coalition Systems," DTIC Document, 2001. Technical Report ADA467953, Fairfax, VA: George Mason University.

[34] Barès, M., *Pour une prospective des systèmes de commandement*, Paris, France: Polytechnica, 1996.

[35] Barès, M., and É. Bossé, "Besoin de coopérabilité pour les C4ISR opérant dans le cadre d'une coalition multinationale," *Symposium Transformation: vers des capacités européennes en réseau*, Paris, France, 2008.

[36] Elmogy, A. M., F. Karray, and A. M. Khamis, "Auction-Based Consensus Mechanism for Cooperative Tracking in Multi-Sensor Surveillance Systems," *JACIII,* Vol. 14, 2010, pp. 13–20.

[37] Khamis, A., "Swarm Intelligence in Robotics," presentation at University of Waterloo, https://www.researchgate.net/profile/Alaa_Khamis/publication/275342512_Swarm_Intelligence_in_Robotics/links/553901e90ef2239f4e7b2210.pdf.

[38] Khamis, A. M., M. S. Kamel, and M. Salichs, "Cooperation: Concepts and General Typology," *IEEE Intl. Conf. on Systems, Man and Cybernetics 2006 (SMC'06)*, 2006, pp. 1499–1505.

[39] Xuan, P., V. Lesser, and S. Zilberstein, "Communication Decisions in Multi-Agent Cooperation: Model and Experiments," *Proc. of 5th Intl. Conf. on Autonomous Agents*, 2001, pp. 616–623.

[40] Yu, H., et al., "A Survey of Multi-Agent Trust Management Systems," *IEEE Access,* Vol. 1, 2013, pp. 35–50.

[41] Hart, S., and A. Mas-Colell, *Cooperation: Game-Theoretic Approaches*, New York: Springer Science & Business Media, 2012.

[42] Yokoo, M., *Distributed Constraint Satisfaction: Foundations of Cooperation in Multi-Agent Systems*, New York: Springer Science & Business Media, 2012.

[43] Rasmussen, J., A. M. Pejtersen, and L. P. Goodstein, *Cognitive Systems Engineering*, New York: Wiley, 1994.

[44] Rasmussen, J., "Skills, Rules, and Knowledge; Signals, Signs, and Symbols, and Other Distinctions in Human Performance Models," *IEEE Trans. on Systems, Man and Cybernetics*, 1983, pp. 257–266.

[45] Chaudron, L., and N. Maille, "Generalized Formal Concept Analysis," in *Conceptual Structures: Logical, Linguistic, and Computational Issues*, New York: Springer, 2000, pp. 357–370.

[46] Jaoua, A., and S. Elloumi, "Galois Connection, Formal Concepts and Galois Lattice in Real Relations: Application in a Real Classifier," *Journal of Systems and Software*, Vol. 60, 2002, pp. 149–163.

[47] Schroeck, M., et al., "Analytics: The Real-World Use of Big Data," 2012. http://www-03.ibm.com/systems/hu/resources/the_real_world_use_of_big_data.pdf.

[48] Zikopoulos, P., and C. Eaton, *Understanding Big Data: Analytics for Enterprise Class Hadoop and Streaming Data*, New York: McGraw-Hill Osborne Media, 2011.

[49] Olson, M., "Hadoop: Scalable, Flexible Data Storage and Analysis," *IQT Quarterly*, Vol. 1, 2010, pp. 14–18.

[50] Laura, W., *A Practical Guide to Big Data: Opportunities, Challenges & Tools*, Paris, France: Dassault Systèmes, 2012.

[51] Lambert, D. A., "A Blueprint for Higher-Level Fusion Systems," *Inf. Fusion*, Vol. 10, 2009, pp. 6–24.

[52] Maupin, P., and A. -L. Jousselme, "Interpreted Systems for Situation Analysis," *Proc. of 10th Intl. Conf. on Information Fusion (FUSION2007)*, Quebec City, Canada, 2007.

[53] Farahbod, R., et al., "Integrating Abstract State Machines and Interpreted Systems for Situation Analysis Decision Support Design," *11th Intl. Conf. on Information Fusion*, 2008, pp. 1–8.

[54] Börger, E., and R. F. Stärk, *Abstract State Machines: A Method for High-Level System Design and Analysis; with 19 Tables*, New York: Springer, 2003.

[55] Valin, P., et al., "Testbed for Distributed High–Level Information Fusion and Dynamic Resource Management," *Intl. Conf. on Information Fusion*, 2010.

[56] Solano, M. A., S. Ekwaro-Osire, and M. M. Tanik, "High-Level Fusion for Intelligence Applications Using Recombinant Cognition Synthesis," *Information Fusion*, Vol. 13, 2012, pp. 79–98.

[57] Chandana, S., and H. Leung, "Distributed Situation Assessment for Intelligent Sensor Networks Based on Hierarchical Fuzzy Cognitive Maps," *IEEE Intl. Conf. on Systems, Man and Cybernetics*, 2008.

[58] Chandana, S., et al., "Fuzzy Cognitive Map Based Situation Assessment for Coastal Surveillance," *11th Intl. Conf. on Information Fusion*, 2008, pp. 1–6.

[59] Chandana, S., H. Leung, and J. Levy, "Disaster Management Model Based on Modified Fuzzy Cognitive Maps," *IEEE Intl. Conf. on Systems, Man and Cybernetics 2007 (ISIC)*, 2007, pp. 392–397.

[60] Solaiman, B., et al., "A Conceptual Definition of a Holonic Processing Framework to Support the Design of Information Fusion Systems," *Information Fusion*, 2013.

[61] Paggi, H., et al., "On the Use of Holonic Agents in the Design of Information Fusion Systems," *17th Intl. Conf. on Information Fusion (FUSION)*, 2014, pp. 1–8.

[62] Sulis, W. H., "Archetypal Dynamics: An Approach to the Study of Emergence," *Formal Descriptions of Developing Systems, NATO Science Series*, Vol. 121, 2003, pp. 185–228.

[63] *Cognitive maps*, https://en.wikipedia.org/wiki/Cognitive_map.

[64] Tolman, E. C., "Cognitive Maps in Rats and Men," *Psychological Review*, Vol. 55, 1948, p. 189.

[65] Kosko, B., "Fuzzy Cognitive Maps," *International Journal of Man-Machine Studies*, Vol. 24, 1986, pp. 65–75.

[66] Kosko, B., *Fuzzy Cognitive Maps: Advances in Theory, Methodologies, Tools and Applications (Studies in Fuzziness and Soft Computing)*, New York: Springer, 2010.

[67] Leung, H., S. Chandana, and S. Wei, "Distributed Sensing Based on Intelligent Sensor Networks," *IEEE Circuits and Systems Magazine,* Vol. 8, 2008, pp. 38–52.

[68] Maupin, P., and A. -L. Jousselme, "A General Algebraic Framework for Situation Analysis," *Proc. of 8th Intl. Conf. on Information Fusion (FUSION2005)*, Philadelphia, PA, 2005.

[69] Sulis, W., "Causal Tapestries," *Bulletin of the American Physical Society,* Vol. 56, 2011.

[70] Sulis, W., "Archetypal Dynamics, Emergent Situations, and the Reality Game," *Nonlinear Dynamics, Psychology, and Life Sciences,* Vol. 14, 2010, pp. 209–238.

[71] Farahbod, R., U. Glässer, and M. Vajihollahi, "Specification and Validation of the Business Process Execution Language for Web Services," in *Abstract State Machines: Advances in Theory and Practice*, New York: Springer, 2004, pp. 78–94.

[72] Farahbod, R., V. Gervasi, and U. Glässer, "CoreASM: An Extensible ASM Execution Engine," *Fundamenta Informaticae,* Vol. 77, 2007, pp. 71–104.

[73] Farahbod, R., U. Glasser, and M. Vajihollahi, "An Abstract Machine Architecture for Web Service Based Business Process Management," *International Journal of Business Process Integration and Management,* Vol. 1, 2006, pp. 279–291.

[74] Börger, E., *Specification and Validation Methods*, New York: Clarendon, 1995.

[75] Glässer, U., R. Gotzhein, and A. Prinz, "The Formal Semantics of SDL-2000: Status and Perspectives," *Computer Networks,* Vol. 42, 2003, pp. 343–358.

[76] Börger, E., "The ASM Refinement Method," *Formal Aspects of Computing,* Vol. 15, 2003, pp. 237–257.

[77] Stärk, R. F., J. Schmid, and E. Börger, *Java and the Java Virtual Machine: Definition, Verification, Validation*, New York: Springer Science & Business Media, 2012.

[78] Glaesser, U., Y. Gurevich, and M. Veanes, "Abstract Communication Model for Distributed Systems," *IEEE Trans. on Software Engineering*, Vol. 30, 2004, pp. 458–472.

[79] R. Farahbod, V. Gervasi, and U. Glässer, "Executable Formal Specifications of Complex Distributed Systems with CoreASM," *Science of Computer Programming,* Vol. 79, 2014, pp. 23–38.

CHAPTER 8

FIAT Prospective Research and Future Work

This chapter proposes prospective research activities on fusion of information and analytics technologies (FIAT) to improve quality of information (QoI) and dependability in cyber-physical and social systems (CPSS) facing information overload (e.g., big data, IoT) and complexity. The suggested list of research activities is the result of a continuum of thoughts, started in previous chapters, about where we stand on FIAT and its potential usage to address the problem of system complexity and information overload. The list is nonexhaustive, but, in the our opinion, if that research is conducted in part or in whole, significant progress could be achieved to apply FIAT to critical CPSS such as in defense and security, health, energy, and transport.

8.1 Introduction

This chapter presents some prospective research activities to progress the integration of FIAT for operational (real-time and near-real-time) decision support systems. The overall goals are to improve quality of information (QoI) (see Section 4.3) and system dependability (see Section 1.5) within CPSS where decision-makers face information overload and system complexity. The discussed approach is organized around three main themes that are common for every domain of application as illustrated in Figure 8.1: (1) domain knowledge representation and modeling, (2) processing of information and its imperfections, and (3) FIAT-based prognosis, diagnosis, and prescriptive decision support tools. In this chapter, some examples are provided in the context of future power networks Smart Grid, but the research activities are quite valid for the other three critical CPSS environments: defense and security, health, and transport.

8.2 Domain Knowledge Representation and Modeling

In the artificial intelligence (AI) domain, the notions of data, information, and knowledge are central notions. AI aims at providing computers with a degree of intelligence mimicking the human mind. This intelligence is mainly due to exploitation

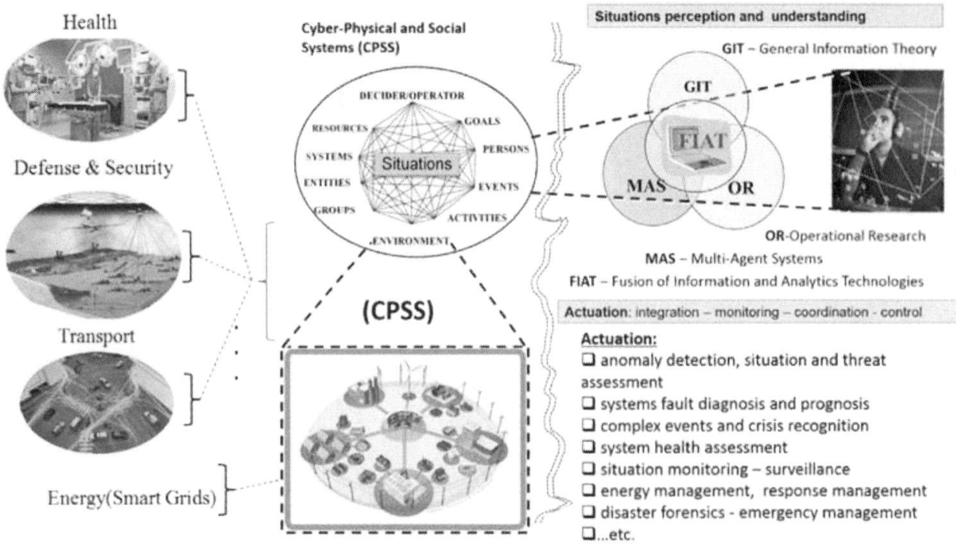

Figure 8.1 FIAT-based decision support tools for CPSS environments.

of contexts where intelligent reasoning relies on the understanding of the perceived world. Data is observed (or sensed) and provided to the intelligent reasoning engines to provide them with a representation model of the world. Once the data items are perceived by the systems, they become information which is interpreted according to the context of observation of the data items and the domain knowledge. The data or information coming from the different sources must be converted into a certain language or with other means (e.g., visualization, graphic, symbols). Knowledge representation is a crucial step needed to transform data into actionable knowledge.

Ontologies are proving to be effective tools for capturing and specifying categorical information about objects at various levels of granularity, as well as the various sorts of relations that hold between them. Ontologies can be used to model physical objects (including their material composition, attributes, and properties), nonphysical or psychological objects (e.g., concepts, plans, intentions), temporal events (sequential or scattered processes), and relations between such items (e.g., logical, causal, dependence, internal, external, and intentional kinds of relations). Ontologies have been used to exchange information and knowledge representation in a variety of domains: in the Semantic Web [1], in power system automation [2–5], for critical infrastructure dependencies [6], and in defense and security problems [7–9]. Ontologies are part of FIAT to represent domain contextual knowledge. They are a very important technology to support any making-sense process [7].

8.2.1 Ontology Management Methods in Power Network Domain Modeling

The power network is developing towards the Smart Grid with multiple devices and actors continuously exchanging data to provide user-oriented flexible services and products while operating a self-healing, economic, ecologically friendly, and secure network. Information generated from power system resources contains numerous concepts, documentations, and relevant domain knowledge. In addition, large amounts of data will be generated by meters, sensors, controllers, and synchrophasors. Techniques for managing, analyzing, and acting on this data are required. Power networks will ultimately involve networking in transmission and distribution facilities a vast numbers of sensors and data sources. The data or information coming from the different sources must be converted into a certain language (e.g., mathematical) or with other means (e.g., textual, visualization, graphic, symbols) so that they can be processed to make sense out of it. To make cooperative work between applications, multiple data sources must be linked and integrated.

The integration of formally represented knowledge from different sources would be easy if all ontologies would be similar, using the same vocabulary for the same intended meaning, the same background assumptions, and the same logical formalism. In large CPSS like Smart Grid, different networks (generation, transmission, distribution, consumption) may develop their own ontologies. To maximize benefit, these heterogeneous ontologies need to interwork, but are likely to have different vocabularies and even different underlying formalisms. The capacity to access data or information from multiple information systems is known as information interoperability (see Section 7.2). In future power grids, the capacity of different information systems, applications, and services to communicate, share, and interchange data, information, and knowledge in an effective and precise way, as well as to integrate with other systems, applications, and services to deliver services is simply mandatory (see Section 7.2).

As previously mentioned, ontologies have been used to exchange information and to represent knowledge in a variety of domains. In power systems [10], an ontology can be used to describe electrical concepts and their use as well as be capable of describing the automation system used for protection, control and monitoring of a plant. Description logics (DLs) [11] form an important and powerful class of logic-based knowledge representation languages being used to build ontologies such as Ontology Web Language (OWL) [12]. In power CPSS, all types of systems (discrete, discrete event, hybrid, human) as well as all sources of data [phasor measurement management units (PMU), power system stabilizers (PSS), flexible alternating current transmission systems, smart meters, smart controllers (FACTS), and energy management systems (EMS)] are part of a complex network. This emphasizes the challenge of heterogeneity and consequently the problem of interoperability.

The distributed nature of most contexts of application requires to relate heterogeneous ontological specifications and to integrate information from multiple diverse sources. The development of ontology management methods (OMM) is then essential to cope with heterogeneity and to enable systems interoperability. OMM are required for alignment [13–15], for merging and translation [16, 17], and for revision and refinement [18–20].

8.2.2 List of Research Activities for Domain Modeling in Power Networks

In view of applying FIAT to future power Smart Grids, Table 8.1 proposes some prospective research activities. These activities, when adjusted to a specific CPSS domain, could be also valid for other CPSS of Figure 8.1. The list of Table 8.1 will contribute to address the big data dimension Variety.

Table 8.1 List of FIAT Research Activities on Domain Modeling

Activities (Group 1): Development of Assessment Tools for the Degree of Interoperability in Power Networks
Prospective research approaches:
Conduct an analysis of existing power grid standards starting from works of Santodomingo et al. [4] on matching ontologies for power networks and then a review of the state of the art on automation in power systems [21] and then finally on interoperability issues such as in [22, 23].
Conduct an analysis of EPRI (Electric Power Research Insitute) or CIGRÉ (Conseil International des Grandes Réseaux Électroniques) power communities [24–26] case studies, formalisms, taxonomies, processes, and ontologies [27, 28] in the perspective of their current and future use in Smart Grids [29].
Define criteria and measures to establish the degree of interoperability/heterogeneity of a complex network by extending the works on interoperability maturity models of [30–33] to system of systems such as power networks.
Expected outcomes:
A system health assessment method to appraise the degree of interoperability or heterogeneity of a complex network using existing power networks (generation, transmission, distribution, and consummation) as a case study.
Activities (Group 2): Definition of Methods to Represent Domain Contextual Information
Prospective research approaches:
Develop representation methods for contextual information [34–38] and for high-level information also referred to as soft data (e.g., human experts judgments, textual data, speech signal, XML files, or databases) using methods such as ontologies [10], conceptual graphs, and semiotics techniques starting by extending, adapting, and applying works of [36, 39–44] and [34–38] to power networks domain.
Explore methods and languages to represent the various power network processes by first extending the work of [27, 28, 45–51].
Expected outcomes:
Methods to structure soft and contextual information to be used in any computing process (e.g., a fusion process).
Representations of power network business processes.
Activities (Group 3): Development of Ontology Management Methods (OMM)
Prospective research approaches:
Develop ontology management methods to cope with heterogeneity and enable interoperability in power grid by analyzing and adapting the following works: alignment [13, 46, 52], merging and translation [14, 53–55], and revision and refinement [18, 19, 44].
Explore the category theory [46, 56], which looks like a promising avenue to provide formal backgrounds to ontology management methods [15, 16]. The category theory is a conceptual framework that represents how structures of different kinds are interrelated [14, 57].
Expected outcomes:
Methods to manage ontologies: integration, matching, revision, and refinement in a power network domain.

8.3 Processing of Information and Its Imperfections

The previous section exemplifies the dimension of Variety of sources (e.g., physical sensors, measurement systems, and experts' opinions) and having different formats (e.g., numerical, statistical, and linguistic). This section emphasizes the Veracity dimension. FIAT provide techniques and tools to analyze data, enable situation awareness, and support decision-making. Characterizing and interpreting information imperfections are probably the most important and difficult task of the FIAT-based process. From this characterization is derived the choice of the most adequate decision-supporting concepts to the user. Uncertainty management and relevance of information are keystones behind any computer-assisted reasoning scheme.

A very crucial aspect of QoI is relevance of information; relevance should be behind any intelligent filtering and context-aware processing concepts and impacts heavily on decision quality and the problem of Volume of information in a big data environment. The analysis of information relevance [58–62] is a very important prerequisite to build any FIAT-based support system to cybernetics functions of integration, control, coordination, and monitoring. The objective is to evaluate the relevance of all available information sources to provide the decision-maker with the ones that are the most relevant for a given situation. This demands classification and prioritization of information required by a given task execution [63]: "As data become abundant, the main problem is no longer finding the information as such but laying one's hands on the relevant bits easily and quickly."

FIAT support a process to transform data into actionable knowledge for sense-making, to reduce uncertainty, and to improve dependability in systems. Meaning and making sense are intimately related to actions of any agent, human or machine. A goal-driven or an activity-based approach sounds appropriate for this process. The development of a framework in which knowledge, information, and uncertainty can be organized, structured, represented, combined, managed, reduced, and updated is a core requirement. Such integration FIAT framework (see Chapters 5–7) should:

1. Provide means to represent knowledge through well-defined notions of situation and awareness.
2. Support modeling uncertainty, belief, and belief change.
3. Provide the key integration model for FIAT and linkage to actions (users and machines) according to the four cybernetics functions but specifically coordination (Figure 8.2).
4. Provide practical support for system design through modularization, refinement, validation, and verification.
5. Offer a good compromise between operational and functional modeling in capturing systems behavior.
6. Enable rapid prototyping and experimental validation of fairly abstract models.
7. Support modeling of multi-agent organizational paradigms as pictured in Figure 8.2 (built from [64, 65]).

We propose prospective research activities (Table 8.2) to pursue (1) the work undertaken in the previous chapters on the definition and validation of a FIAT integrating framework and (2) on the QoI that is an integral part of FIAT. The list of activities in Table 8.2 is applicable to all CPSS of Figure 8.1.

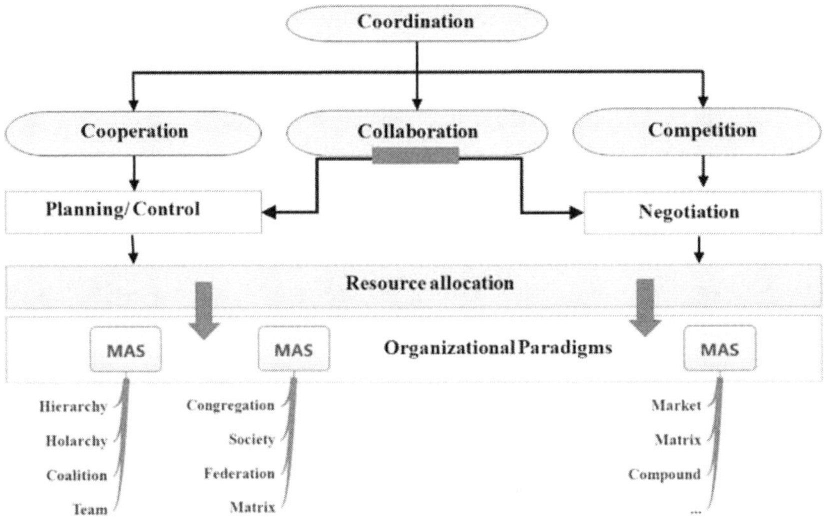

Figure 8.2 Organizational paradigms of multi-agent systems (MAS).

8.4 Prognosis/Diagnosis/Prescriptive FIAT-Based DSS

Dependability has been addressed extensively in physical systems and computer engineering [121–125]. However, the design of dependable cyber physical systems (CPS) [126] is very immature due to our poor understanding of the interface between the physical world and the cyber world. This is ever more complicated with the addition of a social computing dimension as in cyber physical and social systems. The dependability assessments of distributed physical systems have been particularly focused on the probabilistic modeling of random behaviors, given sufficient informative data. In the case of networked CPS, we face not only aleatoric, or randomness, uncertainty but epistemic or systematic uncertainty as discussed in the previous sections. Traditional ways to provide decision support to the four general functions of the cyber physical interface (integration, coordination, monitoring, and control) to make the overall system dependable must be revisited, adapted or even more often reinvented. That reinvention is possible with what we call the passage through FIAT-based decision support and very few papers have addressed the notion of dependability in the FIAT community [127, 128].

Decision support can be provided in whole or in parts to the four general functions of the cyber physical interface. Figure 8.1 lists some examples of concrete actuation like anomaly detection, fault diagnosis, and crisis recognition. The nature of that decision support has mainly three parts: prognosis, diagnosis, and adaptive prescriptive tools. Prognosis and diagnosis of complex situations (e.g., events, crises, major failures) [129–131] are pretty hard in large CPSS that present both aleatory and epistemic uncertainties. Prescriptive DSS is added here in the sense of dynamic resource management to cope with constraints (e.g., real-time, online) of the cyber physical interface and adapt to the language of analytics in the previous chapters.

A vast literature exists on application of AI for the diagnosis of faults of mechanisms and systems in general [132–136]. For instance, reports of AI-based systems

Table 8.2 FIAT Prospective Research Activities for Veracity Dimension (Big Data)

Activities (Group 4): Definition and Validation of a FIAT Framework

Prospective research approaches:

Define and validate an integrated information fusion framework that includes analytics, models, systems, and technologies.

Define the mathematical structures that can represent the information element discussed in previous chapters as well as the information fusion cells that can be common to analytics and information fusion fields. As with Group 3 in Table 8.1, category theory and the notions of monoids [66, 67], monads [68], and semigroups [69–71] can be of prime importance for that definition.

Expected outcomes:

FIAT integration framework to be exploited for the design of FIAT-based real-time and nonreal-time decision support for diagnosis and prognosis applications in power networks.

Activities (Group 5): Design Methodologies for FIAT-Based DSS

Prospective research approaches:

Conduct a multicriteria decision analysis (MCDA) [72–74] for power network domain to establish a set of criteria based upon system dependability and QoI [75–77] that can integrate situation analysis with the decision support layers (dynamic resource management) implied by Figure 7.1.

Based upon the MCDA above and in conjunction with multi-objective, multicriteria optimization, develop a design methodology for FIAT-based operational decision support systems [75–79].

Expected outcomes:

A set criteria (system dependability and QoI) on which one can define the integrating FIAT concepts to support risk analysis for instance.

A methodological foundation for the design of operational FIAT-based decision support systems.

Activities (Group 6): Simulation of Cyber Physical Systems (CPS)

Prospective research approaches:

Develop a generic system modeling method or a generic simulation tool representative of attributes and behaviors of cyber and physical components to understand, design, and analyze CPS. Multi-agent systems (MAS) have been proposed for that simulation due to their ability to model and simulate complex systems [80–83].

Study the integration of social computing with respect to simulations of CPS benefiting from works of [84, 85].

Expected outcomes:

Better understanding of complex issues related to the cyber physical interface as well as the difficulty to integrate the social computing dimension.

A generic simulation tool representative of attributes and behaviors of cyber and physical components

Activities (Group 7): Performance Evaluation Methods in CPSS

Prospective research approaches:

Study the integration of classical performance models such as queueing networks, stochastic Petri nets, or stochastic process algebras with model-based approaches such as model checking [86, 87] and the relationship with a FIAT framework.

Evaluate potential FIAT contribution to specific problems of a CPSS context such as Smart Grids: develop a FIAT testbed based upon extension of available testbeds such as in [88] with the performance evaluation models as discussed in [86, 87].

Expected outcomes:

Assessment of FIAT support to understand, design and analyze CPS and social computing in complex environments.

Table 8.2 (continued)

Activities (Group 8): Definition of Information Measures and Distances
Prospective research approaches:
Compare different concepts of uncertainty and develop uncertainty management methods (UMM) that take into account all aspects of uncertainty (aleatoric and epistemic) [89] and develop the associated distances (metrics) to UMM that can be used to evaluate FIAT [89–91].
Analyze the operational meaning of the different UMM and the concept of total uncertainty measure such as in [92]; integrate (make the link) with the theory of measurement [93] by extending previous works [94–97]; and consider the ontological perspectives of [7, 8] and the integration with UMMs to apply UMM in power networks domain.
Develop measures of similarity to support the soft fusion graphical approaches [13, 39, 98].
Develop methods to deal with noncommensurate measures [77, 99] as well as nonadditive measures [77, 100, 101]: fuzzy and Choquet integrals.
Expected outcomes:
Mathematical tools to characterize and represent imperfect information for the domain of power networks with potential application to other domains.
Activities (Group 9): Hard/Soft Information Fusion Methods and Algorithms
Prospective research approaches:
Develop computationally tractable hard/soft information fusion algorithms by extending the works on hard/soft data sets of [40, 41, 102], on conceptual graphs in [39], and on robust combination of unreliable sources [103, 104].
Exploit work of [105] on implementing soft fusion and extend to design hard/soft fusion methods.
Develop fusion rules for heterogeneous sources of imperfect information represented by different UMM [106] and enrich the fusion calculus by an ontological and contextual processing [7] for the domain of power networks.
Expected outcomes:
FIAT methods and algorithms to perform hard and soft fusion in power networks with potential application to other domains.
Activities (Group 10): Relevance of information
Prospective research approaches:
Develop criteria and a formal mathematical definition of information relevance to be used in the design of intelligent filtering algorithms.
Conceptualize an intelligent informational bandwidth based upon QoI criteria and demonstrate the concept for the design of operational decision support; and extend the FAIR framework [58] to incorporate QoI criteria.
Develop smart filtering and smart sensing approaches based on relevance of information to be integrated in fusion processes [107–110].
Expected outcomes:
Relevance information criteria and their exploitation in the domain of power networks with potential application to other domains.
Activities (Group 11): Fusion of Heterogeneous Classifiers
Prospective research approaches:
Develop approaches for the fusion of classifiers defined under various UMMs [111–114] and decision rules [115–118] to accompany possibilistic and evidential UMMs to support the fusion of classifiers.
Expected outcomes:
Know-how to combine results of heterogeneous classifiers.

Table 8.2 (continued)

Activities (Group 12): Autonomic Distributed Fusion Principles
Prospective research approaches:
Develop Level 4 of the JDL data fusion model as an autonomic process refinement level (self-healing, self-configure, self-protect, and self-optimize); and exploit works of [119, 120] on trustworthiness networks management and design.
Expected outcomes:
A network-centric data fusion model

and discrete event systems (e.g., finite automata, Petri nets) have appeared in many papers [137–139] to address fault diagnosis in power systems. These methods need refinement to be used for cyber physical systems since they suffer either from state explosion or to get heuristics in real time (knowledge acquisition problem); therefore, applications are limited to small systems.

Petri nets have been extensively used for deterministic recognition [140, 141]. However, they have been used to much less extent when uncertainty has to be taken into account [142–144]. Bayesian networks, temporal Bayesian networks, and hidden Markov models [145, 146] have been used extensively to successfully incorporate uncertainties of various types. Markov-type models are typically of sequential nature and therefore do not provide suitable tools for representing and visualizing concurrency at which Petri nets are good. Naturally, a combination of Petri nets and Markov models could provide the benefits of both for prognosis and diagnosis in CPSS [142, 147–150].

The methods mentioned above strongly depend on a good model of the system dynamics. If important aspects of the dynamics are genuinely unknown, then these methods are rarely effective. To learn the unknown (e.g., structures and characteristics), machine learning methods can be developed. Reinforcement learning [151–153] is a body of theory and techniques for optimal control that have been developed in the past twenty years primarily within the machine learning and operations research communities for approximating dynamic programming. Online pattern recognition approaches [154, 155] can also be adapted to complex environments such as power networks. Novel machine learning methods [156–159] must be developed to learn the unknown structures or behavior patterns that are hidden in the networked aspect of CPS. Table 8.3 presents a nonexhaustive list of some prospective research activities that will contribute to provide decision support in dynamic complex environments. The list of activities in Table 8.3 is applicable to all CPSS of Figure 8.1.

8.5 Conclusion

The reader certainly realizes that both problem space and solution space are quite vast and complex. There is no pretention from the authors to claim a full description of all aspects of the problem space neither to provide an exhaustive list of

Table 8.3 FIAT Prospective Research Activities for FIAT-Based DSS

Activities (Group 13): Machine Learning Methods and Algorithms

Prospective research approaches:

Construct higher-level dimension objects (e.g., high-dimensional graphs, structural properties, clustering) through feature extraction, data mining, and machine learning inspired from [136, 155, 160, 161, 162].

Model the relationship between the local and global structural characteristics of a CPS (e.g., power networks) [163–165]. Find ways to extract the network structural properties (static and dynamic cases) [166].

Develop the required optimization schemes to support the feature extraction process based on the most recent advances in multi-objective optimization and decision-making for pattern recognition [167, 168].

Expected outcomes:

Knowledge about the structural properties of a complex network.

Activities (Group 14): FIAT-Based Diagnosis/Prognosis/Prescriptive Tools

Prospective research approaches:

Combine current modeling approaches of discrete events systems (e.g., Petri nets) [144] with ontological-approaches inspired from [47, 143].

Develop techniques to cope with the state explosion problem with finite state machines: extend the work of [169, 170].

Study predictive models for anomalies [171–173] in power systems by extending Markov approaches [146, 174, 175] with uncertainty representation methods [175–178].

Develop Markov models and further extensions (e.g., Markov-like processes in the possibility framework) to model sequences of events or complex events arising in a crisis [147–149, 179].

Revisit some state-space methods (e.g., Petri nets, Markov chains) used for dependability in physical distributed systems and enrich them with UMM that are more suited to represent epistemic uncertainty.

Extend the predictive state representations [171, 173, 180–184] and temporal-difference networks [185–188] to develop FIAT for prediction of events combining temporally and spatially abstract knowledge that is grounded in experience (expert knowledge).

Expected outcomes:

Set of prognosis and diagnosis tools for power networks with potential applications to other domains.

Activities (Group 15): FIAT-Based Monitoring Tools

Prospective research approaches:

Study the use of distributed sensor networks (DSN) and the most advances in multicriteria decision-making and optimization, AI and smart sensing, and hybrid optimization [79, 189–192] to evaluate their potential contribution to a monitoring solution.

Develop strategies for sensors placement and asset monitoring inspired from [108, 156, 166, 193–195].

Adapt distributed sensor/information fusion [195–198] to input data from sensors directly into automated reasoning systems like FIAT to monitor CPS.

Develop cyber-sensing concepts and techniques based on high-dimensional objects, online learning, and smart sensing inspired, for instance, by [157] to monitor CPS.

Expected outcomes:

Monitoring tools for complex networks with potential application to all CPSS domains.

research activities that would define a solution space but to offer suggestions of research activities that will certainly contribute to a better understanding of that cyber-physical interface ubiquitous to CPSS of our today and future networked worlds. In a future book, the current authors plan to edit several contributions from authors of international diverse communities on FIAT applications that will be gathered along the CPSS domains of Figure 8.1: energy, health, transport, and defense and security.

References

[1] Berners-Lee, T., J. Hendler, and O. Lassila, "The Semantic Web," *Scientific american,* Vol. 284, 2001, pp. 28–37.

[2] Uslar, M., "Semantic Interoperability Within the Power Systems Domain," *Proc. of 1st Intl. Workshop on Interoperability of Heterogeneous Information Systems*, 2005, pp. 39–46.

[3] Wang, L., A. Hosokawa, and H. Murayama, "An Evolutive and Multilingual CIM Ontology Management System," *Energy Procedia,* Vol. 12, 2011, pp. 18–26.

[4] Santodomingo, R., J. Rodríguez-Mondéjar, and M. Sanz-Bobi, "Ontology Matching Approach to the Harmonization of CIM and IEC 61850 Standards," *1st IEEE Intl. Conf. on Smart Grid Communications (SmartGridComm)*, 2010, pp. 55–60.

[5] Buse, D., et al., "Agent-Based Substation Automation," *Power and Energy Magazine, IEEE,* Vol. 1, 2003, pp. 50–55.

[6] Pederson, P., et al., "Critical Infrastructure Interdependency Modeling: A Survey of US and International Research," *Idaho National Laboratory,* 2006, pp. 1–20.

[7] Kokar, M. M., C. J. Matheus, and K. Baclawski, "Ontology-Based Situation Awareness," *Information Fusion,* Vol. 10, 2009, pp. 83–98.

[8] Kokar, M. M., J. A. Tomasik, and J. Weyman, "Formalizing Classes of Information Fusion Systems," *Information Fusion,* Vol. 5, 2004, pp. 189–202.

[9] Little, E., and G. Rogova, "Designing Ontologies for Higher Level Fusion," *Information Fusion,* Vol. 10, 2009, pp. 70–82.

[10] Feng, J., Q. Wu, and J. Fitch, "An Ontology for Knowledge Representation in Power Systems," *Proc. of UKACC Control, University of Bath, UK,* Vol. 35, 2004, pp. 1–5.

[11] Horrocks, I., "Applications of Description Logics: State of the Art and Research Challenges," in *Conceptual Structures: Common Semantics for Sharing Knowledge*, New York: Springer, 2005, pp. 78–90.

[12] McGuinness, D. L., and F. Van Harmelen, "OWL Web Ontology Language Overview," *W3C Recommendation,* Vol. 10, 2004, p. 10.

[13] Blasch, E. P., et al., "Ontology Alignment in Geographical Hard-Soft Information Fusion Systems," *2010 13th Conference on Information Fusion (FUSION)*, 2010, pp. 1–8.

[14] Kalfoglou, Y., and M. Schorlemmer, "Ontology Mapping: The State of the Art," *The Knowledge Engineering Review,* Vol. 18, 2003, pp. 1–31.

[15] Shvaiko, P., and J. Euzenat, "Ontology Matching: State of the Art and Future Challenges," *IEEE Trans. on Knowledge and Data Engineering*, Vol. 25, 2013, pp. 158–176.

[16] Zimmermann, A., et al., "Formalizing Ontology Alignment and Its Operations with Category Theory," *Proc. 4th Intl. Conf. on Formal Ontology in Information Systems (FOIS)*, 2006, pp. 277–288.

[17] Noy, N. F., "Tools for Mapping and Merging Ontologies," in *Handbook on Ontologies*, New York: Springer, 2004, pp. 365–384.

[18] McNeill, F. J., A. Bundy, and M. Schorlemmer, "Dynamic Ontology Refinement," University of Edinburgh, 2006.

[19] Klein, M., et al., "Ontology Versioning and Change Detection on the Web," in *Knowledge Engineering and Knowledge Management: Ontologies and the Semantic Web*, New York: Springer, 2002, pp. 197–212.

[20] Heflin, J., and J. Hendler, "Dynamic Ontologies on the Web," *AAAI/IAAI*, 2000, pp. 443–449.

[21] Strasser, T., et al., "Review of Trends and Challenges in Smart Grids: An Automation Point of View," in *Industrial Applications of Holonic and Multi-Agent Systems*, New York: Springer, 2013, pp. 1–12.

[22] Von Dollen, D., "Report to NIST on the Smart Grid Interoperability Standards Roadmap," *Electric Power Research Institute (EPRI) and National Institute of Standards and Technology*, 2009.

[23] Britton, J. P.. and A. Devos, "CIM-Based Standards and CIM Evolution," *IEEE Trans. on Power Systems*, Vol. 20, 2005, pp. 758–764.

[24] EPRI (Electric Power Research Institute), http://www.smartgrid.epri.com/.

[25] CIGRE (International Council on Large Electric Networks), http://www.cigre.org/.

[26] "IEC Smart Grid Standardization Roadmap," June 2010. http://eee.iec.ch/smartgrid/downloads/sg3_roadmap.pdf.

[27] Allweyer, T., *BPMN 2.0: Introduction to the Standard for Business Process Modeling*, Second Edition, Norderstedt: BoD–Books on Demand, 2010.

[28] Berners-Lee, T., and D. Connolly, "Delta: An Ontology for the Distribution of Differences Between RDF Graphs," 2004. http://dipl-arb-tmdiff.googlecode.com/hg/Quellen/[TBLO4]Delta-anOntologyForTheDistributionOfDifferencesBetweenRDFGraphs.pdf.

[29] Mo, Y., et al., "Cyber–Physical Security of a Smart Grid Infrastructure," *Proc. of the IEEE*, Vol. 100, 2012, pp. 195–209.

[30] Alberts, D. S., R. K. Huber, and J. Moffat, "NATO NEC C2 Maturity Model," NATO SAS-065, 2010, www.dodccrp.org2010.

[31] Guédria, W., Y. Naudet, and D. Chen, "Interoperability Maturity Models–Survey and Comparison," *On the Move to Meaningful Internet Systems: OTM 2008 Workshops*, 2008, pp. 273–282.

[32] Widergren, S., et al., "Smart Grid Interoperability Maturity Model," *2010 IEEE Power and Energy Society General Meeting*, 2010, pp. 1–6.

[33] Ford, T. C., et al., "Survey on Interoperability Measurement," DTIC Document2007. *Twelfth International Command and Control Research and Technology Symposium (12th ICCRTS)*, 2007.

[34] Beigl, M., et al., "Modeling and Using Context," *Proc. 7th Intl. and Interdisciplinary Conf. (CONTEXT 2011)*, Karlsruhe, Germany, September 26–30, 2011, 2011.

[35] Bettini, C., et al., "A Survey of Context Modelling and Reasoning Techniques," *Pervasive and Mobile Computing*, Vol. 6, 2010, pp. 161–180.

[36] Dey, A., "Understanding and Using Context," *Personal and Ubiquitous Computing*, Vol. 5, 2001, pp. 4–7.

[37] Hong, J. -Y., E. -H. Suh, and S. -J. Kim, "Context-Aware Systems: A Literature Review and Classification," *Expert Systems with Applications*, Vol. 36, 2009, pp. 8509–8522.

[38] Zimmermann, A., A. Lorenz, and R. Oppermann, "An Operational Definition of Context," in *Modeling and Using Context*, New York: Springer, 2007, pp. 558–571.

[39] Laudy, C., "Introducing semantic knowledge in high level information fusion," Ph.D. Thesis, Université Pierre et Marie Curie, Paris, France, 2010.

[40] Pravia, M. A., et al., "Generation of a Fundamental Data Set for Hard/Soft Information Fusion," *11th Intl. Conf. on Information Fusion*, 2008, pp. 1–8.

[41] Hall, D. L., et al., "A Framework for Dynamic Hard/Soft Fusion," *11th Intl. Conf. on Information Fusion*, 2008, pp. 1–8.

[42] Llinas, J., et al., "A Multi-Disciplinary University Research Initiative in Hard and Soft Information Fusion: Overview, Research Strategies and Initial Results," *13th Conference on Information Fusion (FUSION)*, 2010, pp. 1–7.

[43] Bouquet, P., et al., "C-OWL: Contextualizing Ontologies," *2nd Intl. Semantic Web Conf. (The Semantic Web ISWC 2003)*, Sanibel Island, FL, 2003.

[44] Zurawski, M., "Distributed Multi-Contextual Ontology Evolution - A Step Towards Semantic Autonomy," *15th Intl. Conf. on Knowledge Engineering and Knowledge Management (EKAW 2006)*, Podebrady, Czech Republic, 2006.

[45] Koschmider, A., and A. Oberweis, "Ontology Based Business Process Description," *EMOI-INTEROP*, 2005.

[46] Kutz, O., D. Lücke, and T. Mossakowski, "Heterogeneously Structured Ontologies," *Proc. of the Advances in Ontologies, Knowledge Representation Ontology Workshop*, Sydney, Australia, 2008, pp. 41–50.

[47] Recker, J. C., and M. Indulska, "An Ontology-Based Evaluation of Process Modeling with Petri Nets," *IBIS: International Journal of Interoperability in Business Information Systems,* Vol. 2, 2007, pp. 45–64.

[48] Brockmans, S., et al., "Semantic Alignment of Business Processes," *ICEIS,* Vol. 3, 2006, pp. 191–196.

[49] Kopp, O., et al., "The Difference Between Graph-Based and Block-Structured Business Process Modelling Languages," *Enterprise Modelling and Information Systems Architecture*, 2009.

[50] McHugh, P., G. W. Merli, and W. A. Wheeler, *Beyond Business Process Reengineering: Towards the Holonic Enterprise*, New York: Wiley, 1995.

[51] Mendling, J., K. B. Lassen, and U. Zdun, "Transformation Strategies Between Block-Oriented and Graph-Oriented Process Modelling Languages," *Multikonferenz Wirtschaftsinformatik*, 2006, pp. 297–312.

[52] Wang, J., and L. Gasser, "Mutual Online Ontology Alignment," *Proc. of the Workshop on Ontologies in Agent Systems*, 2002.

[53] Chungoora, N., and R. Young, "Ontology Mapping to Support Semantic Interoperability in Product Design and Manufacture," *Proc. of the 1st International Workshop on Model Driven Interoperability for Sustainable Information Systems (MDISIS 2008)*, 2008.

[54] Gruber, T., "A Translation Approach to Portable Ontology Specifications," *Knowledge Acquisition,* Vol. 5, 1993, pp. 199–220.

[55] Heeps, S., et al., "Dynamic Ontology Mapping for Interacting Autonomous Systems." http://sdstrowes.co.uk/publications/heeps-ontology.pdf.

[56] Johnson, M., and C. N. Dampney, "On Category Theory as a (Meta) Ontology for Information Systems Research," *Proc. of Intl. Conf. on Formal Ontology in Information Systems*, Vol. 2001, 2001, pp. 59–69.

[57] Mijic, R., "Theoretical Approaches to Combining Heterogeneous Ontologies for Modular Knowledge Engineering," Technical Report, University of Edinburgh, School of Informatics, 2008.

[58] Breton, R., et al., "Framework for the Analysis of Information Relevance (FAIR)," *2012 IEEE Intl. Multi-Disciplinary Conf. on Cognitive Methods in Situation Awareness and Decision Support (CogSIMA)*, 2012, pp. 210–213.

[59] Hadzagic, M., et al., "Reliability and Relevance in the Thresholded Dempster-Shafer Algorithm for ESM Data Fusion," *15th Intl. Conf. on Information Fusion (FUSION)*, 2012, pp. 615–620.

[60] Pichon, F., D. Dubois, and T. Denœux, "Relevance and Truthfulness in Information Correction and Fusion," *International Journal of Approximate Reasoning,* Vol. 53, 2012, pp. 159–175.

[61] Saracevic, T., "Relevance: A Review of the Literature and a Framework for Thinking on the Notion in Information Science. Part III: Behavior and Effects of Relevance," *Journal of the American Society for Information Science and Technology,* Vol. 58, 2007, pp. 2126–2144.

[62] White, H. D., "Relevance Theory and Citations," *Journal of Pragmatics,* Vol. 43, 2011, pp. 3345–3361.

[63] "Data, Data Everywhere," *The Economist: A Special Report on Managing Information,* February 2010, http://www.economist.com/node/15557443.

[64] Horling, B., and V. Lesser, "A Survey of Multi-Agent Organizational Paradigms," *The Knowledge Engineering Review,* Vol. 19, 2004, pp. 281–316.

[65] Khamis, A. M., M. S. Kamel, and M. Salichs, "Cooperation: Concepts and General Typology," *IEEE Intl. Conf. on Systems, Man and Cybernetics 2006 (SMC'06)*, 2006, pp. 1499–1505.

[66] Krob, D., J. Mairesse, and I. Michos, "Computing the Average Parallelism in Trace Monoids," *Discrete Mathematics,* Vol. 273, 2003, pp. 131–162.

[67] Lin, J., "Monoidify! Monoids as a Design Principle for Efficient Mapreduce Algorithms," http://arxiv.org/pdf/1304.7544v1.pdf, 2013.

[68] Dzik, J., et al., "MBrace: Cloud Computing with Monads," *Proc. of 7th Workshop on Programming Languages and Operating Systems*, 2013, p. 7.

[69] Howie, J. M., *Fundamentals of Semigroup Theory*, Clarendon, U.K.: Oxford, 1995.

[70] Kuroki, N., "On Fuzzy Semigroups," *Information Sciences,* Vol. 53, 1991, pp. 203–236.

[71] White, D. R., and K. P. Reitz, "Graph and Semigroup Homomorphisms on Networks of Relations," *Social Networks,* Vol. 5, 1983, pp. 193–234.

[72] Tacnet, J. -M., et al., "Applying New Uncertainty Related Theories and Multicriteria Decision Analysis Methods to Snow Avalanche Risk Management," *Intl. Snow Science Workshop 2010 (ISSW)*, 2010.

[73] Greco, S., M. Ehrgott, and J. R. Figueira, *Trends in Multiple Criteria Decision Analysis,* New York: Springer, 2010.

[74] Figueira, J., S. Greco, and M. Ehrgott, *Multiple Criteria Decision Analysis: State of the Art Surveys*, New York: Springer, 2005.

[75] Wang, J. -J., et al., "Review on Multi-Criteria Decision Analysis Aid in Sustainable Energy Decision-Making," *Renewable and Sustainable Energy Reviews,* Vol. 13, 2009, pp. 2263–2278.

[76] Clivillé, V., L. Berrah, and G. Mauris, "Quantitative Expression and Aggregation of Performance Measurements Based on the MACBETH Multi-Criteria Method," *International Journal of Production Economics,* Vol. 105, 2007, pp. 171–189.

[77] Grabisch, M., and C. Labreuche, "A Decade of Application of the Choquet and Sugeno Integrals in Multi-Criteria Decision Aid," *Annals of Operations Research,* Vol. 175, 2010, pp. 247–286.

[78] Peng, Y., et al., "An Incident Information Management Framework Based on Data Integration, Data Mining, and Multi-Criteria Decision Making," *Decision Support Systems,* Vol. 51, 2011, pp. 316–327.

[79] Bansal, R., "Optimization Methods for Electric Power Systems: An Overview," *International Journal of Emerging Electric Power Systems,* Vol. 2, 2005.

[80] Ahmed, E., A. Elgazzar, and A. Hegazi, "An Overview of Complex Adaptive Systems," arXiv preprint nlin/0506059, http://arxiv.org/pdf/nlin/0506059.pdf, 2005.

[81] Lin, J., S. Sedigh, and A. R. Hurson, "An Agent-Based Approach to Reconciling Data Heterogeneity in Cyber-Physical Systems," *2011 IEEE Intl. Symp. on Parallel and Distributed Processing Workshops and PhD Forum (IPDPSW)*, 2011, pp. 93–103.

[82] Lin, J., S. Sedigh, and A. Miller, "Modeling Cyber-Physical Systems with Semantic Agents," *IEEE 34th Annual Computer Software and Applications Conf. Workshops (COMPSACW)*, 2010, pp. 13–18.

[83] Huang, J., et al., "Toward a Smart Cyber-Physical Space: A Context-Sensitive Resource-Explicit Service Model," *33rd Annual IEEE Intl. Computer Software and Applications Conf. 2009 (COMPSAC'09)*, 2009, pp. 122–127.

[84] Cook, D. J., and S. K. Das, "Pervasive Computing at Scale: Transforming the State of the Art," *Pervasive and Mobile Computing,* Vol. 8, 2012, pp. 22–35.

[85] Conti, M., et al., "Looking Ahead in Pervasive Computing: Challenges and Opportunities in the Era of Cyber–Physical Convergence," *Pervasive and Mobile Computing,* Vol. 8, 2012, pp. 2–21.

[86] Baier, C., and J. -P. Katoen, *Principles of Model Checking*: Cambridge, MA: MIT Press, 2008.

[87] Baier, C., et al., "Performance Evaluation and Model Checking Join Forces," *Communications of the ACM,* Vol. 53, pp. 76–85, 2010.

[88] Valin, P., et al., "Testbed for Distributed High–Level Information Fusion and Dynamic Resource Management," *Int. Conf. on Information Fusion*, 2010.

[89] Jousselme, A. -L., P. Maupin, and É. Bossé, "Uncertainty in a Situation Analysis Perspective," *Proc. of 6th Intl. Conf. of Information Fusion*, 2003.

[90] Jousselme, A. -L., and P. Maupin, "Distances in Evidence Theory: Comprehensive Survey and Generalizations," *International Journal of Approximate Reasoning,* Vol. 53, 2012, pp. 118–145.

[91] Costa, P. C., et al., "Towards Unbiased Evaluation of Uncertainty Reasoning: The URREF Ontology," *15th Intl. Conf. on Information Fusion (FUSION)*, 2012, pp. 2301–2308.

[92] Abellán, J., and A. Masegosa, "Requirements for Total Uncertainty Measures in Dempster–Shafer Theory of Evidence," *International Journal of General Systems*, Vol. 37, 2008, pp. 733–747.

[93] Salicone, S., *Measurement Uncertainty: An Approach Via the Mathematical Theory of Evidence*, New York: Springer, 2007.

[94] Liu, C., et al., "Reducing Algorithm Complexity for Computing an Aggregate Uncertainty Measure," *IEEE Trans. on Systems, Man and Cybernetics, Part A: Systems and Humans*, Vol. 37, 2007, pp. 669–679.

[95] Burkov, A., et al., "An Empirical Study of Uncertainty Measures in the Fuzzy Evidence Theory," *Proc. of 14th Intl. Conf. on Information Fusion (FUSION)*, 2011, pp. 1–8.

[96] Ferrero, A., M. Prioli, and S. Salicone, "Processing Dependent Systematic Contributions to Measurement Uncertainty," *Instrumentation and Measurement, IEEE Transactions on*, Vol. 62, Issue 4, 2013.

[97] Ferrero, A., and S. Salicone, "Uncertainty: Only One Mathematical Approach to Its Evaluation and Expression?" *IEEE Trans. on Instrumentation and Measurement*, Vol. 61, 2012, pp. 2167–2178.

[98] Gahegan, M., et al., "Measures of Similarity for Integrating Conceptual Geographical Knowledge: Some Ideas and Some Questions," *COSIT: Workshop on Semantic Similarity Measurements*, 2007.

[99] Labreuche, C., "Construction of a Choquet Integral and the Value Functions Without Any Commensurateness Assumption in Multi-Criteria Decision Making," *EUSFLAT Conf.*, 2011, pp. 90–97.

[100] Millet, I., and W. C. Wedley, "Modelling Risk and Uncertainty with the Analytic Hierarchy Process," *Journal of Multi-Criteria Decision Analysis,* Vol. 11, 2002, pp. 97–107.

[101] Chateauneuf, A., J. Eichberger, and S. Grant, "Choice Under Uncertainty with the Best and Worst in Mind: Neo-Additive Capacities," *Journal of Economic Theory,* vol. 137, 2007, pp. 538–567.

[102] Pravia, M. A., et al., "Lessons Learned in the Creation of a Data Set for Hard/Soft Information Fusion," *12th Intl. Conf. on Information Fusion 2009 (FUSION'09)*, 2009, pp. 2114–2121.

[103] Florea, M. C., et al., "Robust Combination Rules for Evidence Theory," *Information Fusion,* Vol. 10, 2009, pp. 183–197.

[104] Rogova, G., et al., "Context-Based Information Quality for Sequential Decision Making," *2013 IEEE Intl. Multi-Disciplinary Conf. on Cognitive Methods in Situation Awareness and Decision Support (CogSIMA)*, 2013, pp. 16–21.

[105] Kohler, H., et al., "Implementing Soft Fusion," *16th Intl. Conf. on Information Fusion (FUSION)*, 2013, pp. 389–396.

[106] Bowman, C. L., "Possibilistic Versus Probabilistic Trade-Off for Data Association," *Optical Engineering and Photonics in Aerospace Sensing*, 1993, pp. 341–351.

[107] Lemire, D., et al., "Collaborative Filtering and Inference Rules for Context-Aware Learning Object Recommendation," *Interactive Technology and Smart Education,* Vol. 2, 2005, pp. 179–188.

[108] Spencer, B. F., M. E. Ruiz-Sandoval, and N. Kurata, "Smart Sensing Technology: Opportunities and Challenges," *Structural Control and Health Monitoring,* Vol. 11, 2004, pp. 349–368.

[109] Tian, Y. -L., et al., "IBM Smart Surveillance System (S3): Event Based Video Surveillance System with an Open and Extensible Framework," *Machine Vision and Applications,* Vol. 19, 2008, pp. 315–327.

[110] Jain, P., "Intelligent Information Retrieval," *SETIT 2005 3rd International Conference: Sciences of Electronic Technologies of Information and Telecomunications*, Vol. 3, 2005.

[111] Altınçay, H., "On Naive Bayesian Fusion of Dependent Classifiers," *Pattern Recognition Letters,* Vol. 26, 2005, pp. 2463–2473.

[112] El-Bakry, H. M., "An Efficient Algorithm for Pattern Detection Using Combined Classifiers and Data Fusion," *Information Fusion,* Vol. 11, 2010, pp. 133–148.

[113] Kumar, S., J. Ghosh, and M. M. Crawford, "Hierarchical Fusion of Multiple Classifiers for Hyperspectral Data Analysis," *Pattern Analysis & Applications,* Vol. 5, 2002, pp. 210–220.

[114] Singh, S., et al., "Dynamic Fusion of Classifiers for Fault Diagnosis," *SMC*, 2007, pp. 2467–2472.

[115] Goebel, K., and W. Yan, "Choosing Classifiers for Decision Fusion," *Proc. of 7th Intl. Conf. on Information Fusion*, 2004, pp. 563–568.

[116] Gunes, V., et al., "Combination, Cooperation and Selection of Classifiers: A State of the Art," *International Journal of Pattern Recognition and Artificial Intelligence,* Vol. 17, 2003, pp. 1303–1324.

[117] Kuncheva, L. I., et al., "Is Independence Good For Combining Classifiers?" *15th Intl. Conf. on Pattern Recognition*, 2000, pp. 168–171.

[118] Aliev, R., et al., "Fuzzy Logic-Based Generalized Decision Theory with Imperfect Information," *Information Sciences,* Vol. 189, 2012, pp. 18–42.

[119] IBM, "Automatic Computing," white paper, June 2005, http://www-03.ibm.com/autonomic/pdfs/AC%Blueprint%20White%20Paper%20V7.pdf.

[120] Agoulmine, N., *Autonomic Network Management Principles: From Concepts to Applications*, New York: Elsevier Science, 2010.

[121] Avizienis, A., et al., "Basic Concepts and Taxonomy of Dependable and Secure Computing," *IEEE Trans. on Dependable and Secure Computing*, Vol. 1, 2004, pp. 11–33.

[122] Wan, K., and V. Alagar, "Dependable Context-Sensitive Services in Cyber Physical Systems," *IEEE 10th Intl. Conf. on Trust, Security and Privacy in Computing and Communications (TrustCom)*, 2011, pp. 687–694.

[123] Bobbio, A., et al., "Improving the Analysis of Dependable Systems by Mapping Fault Trees into Bayesian Networks," *Reliability Engineering & System Safety,* Vol. 71, 2001, pp. 249–260.

[124] Nicol, D. M., W. H. Sanders, and K. S. Trivedi, "Model-Based Evaluation: From Dependability to Security," *IEEE Trans. on Dependable and Secure Computing*, Vol. 1, 2004, pp. 48–65.

[125] Zio, E., "Reliability Engineering: Old Problems and New Challenges," *Reliability Engineering & System Safety,* Vol. 94, 2009, pp. 125–141.

[126] Lee, E. A., "Cyber Physical Systems: Design Challenges," *11th IEEE Intl. Symp. on Object Oriented Real-Time Distributed Computing (ISORC)*, 2008, pp. 363–369.

[127] Karlsson, A., "Dependable and Generic High-Level Information Fusion: Methods and Algorithms for Uncertainty Management," University of Skovde, Technical report HS-IKI-TR-07-0032010, http://www.diva-portal.org/smash/get/diva2:345982/FULLTEXT03, 2007.

[128] Karlsson, A., R. Johansson, and S. F. Andler, "Imprecise Probability as an Approach to Improved Dependability in High-Level Information Fusion," in *Interval/Probabilistic Uncertainty and Non-Classical Logics*, New York: Springer, 2008, pp. 70–84.

[129] Wang, J. -W., and L. -L. Rong, "Cascade-Based Attack Vulnerability on the US Power Grid," *Safety Science,* Vol. 47, 2009, pp. 1332–1336.

[130] Motter, A. E., and Y. -C. Lai, "Cascade-Based Attacks on Complex Networks," *Physical Review E,* Vol. 66, 2002, p. 065102.

[131] Chen, G., et al., "Attack Structural Vulnerability of Power Grids: A Hybrid Approach Based on Complex Networks," *Physica A: Statistical Mechanics and Its Applications,* Vol. 389, 2010, pp. 595–603.

[132] Chiang, L. H., R. D. Braatz, and E. L. Russell, *Fault Detection and Diagnosis in Industrial Systems*, New York: Springer, 2001.

[133] Simon, L. M., *Fault Detection: Theory, Methods and Systems,* Hauppauge, NY: Nova Science Publishers, 2011.

[134] Venkatasubramanian, V., et al., "A Review of Process Fault Detection and Diagnosis: Part III: Process History Based Methods," *Computers & Chemical Engineering,* Vol. 27, 2003, pp. 327–346.

[135] Venkatasubramanian, V., et al., "A Review of Process Fault Detection And Diagnosis: Part I: Quantitative Model-Based Methods," *Computers & Chemical Engineering,* Vol. 27, 2003, pp. 293–311.

[136] McArthur, S. D., et al., "Automating Power System Fault Diagnosis Through Multi-Agent System Technology," *Proc. of the 37th Annual Hawaii Intl. Conf. on System Sciences*, 2004.

[137] Davidson, E. M., et al., "Applying Multi-Agent System Technology in Practice: Automated Management and Analysis of SCADA and Digital Fault Recorder Data," *IEEE Trans. on Power Systems*, Vol. 21, 2006, pp. 559–567.

[138] Lin, X., et al., "A Fault Diagnosis Method of Power Systems Based on Improved Objective Function and Genetic Algorithm-Tabu Search," *IEEE Trans. on Power Delivery,* Vol. 25, 2010, pp. 1268–1274.

[139] Fritzen, P. C., et al., "Hybrid System Based on Constructive Heuristic and Integer Programming for the Solution of Problems of Fault Section Estimation and Alarm Processing in Power Systems," *Electric Power Systems Research,* Vol. 90, 2012, pp. 55–66.

[140] Ren, H., Z.-Q. Mi, and H.-S. Zhao, "Power System Fault Diagnosis Modeling Techniques Based on Encoded Petri Nets," *Zhongguo Dianji Gongcheng Xuebao (Proc. of the Chinese Society of Electrical Engineering)*, 2005, pp. 44–49.

[141] Wang, J.-Y., and Y.-C. Ji, "Application of Fuzzy Petri Nets Knowledge Representation in Electric Power Transformer Fault Diagnosis," *Proc. of the CSEE,* Vol. 23, 2003, pp. 121–125.

[142] Cardoso, J., R. Valette, and D. Dubois, "Possibilistic Petri Nets," *IEEE Trans. on Systems, Man, and Cybernetics, Part B: Cybernetics*, Vol. 29, 1999, pp. 573–582.

[143] Lavee, G., et al., "Building Petri Nets from Video Event Ontologies," in *Advances in Visual Computing*, New York: Springer, 2007, pp. 442–451.

[144] Dahlbom, A., L. Niklasson, and G. Falkman, "Situation Recognition and Hypothesis Management Using Petri Nets," in *Modeling Decisions for Artificial Intelligence*, New York: Springer, 2009, pp. 303–314.

[145] Aberdeen, D., "A (Revised) Survey of Approximate Methods for Solving Partially Observable Markov Decision Processes," *National ICT Australia, Canberra, Australia, Tech. Rep.*, 2003.

[146] Begleiter, R., R. El-Yaniv, and G. Yona, "On Prediction Using Variable Order Markov Models," *J. Artif. Intell. Res.(JAIR)*, Vol. 22, 2004, pp. 385–421.

[147] Soubaras, H., "On Evidential Markov Chains," in *Foundations of Reasoning Under Uncertainty*, New York: Springer, 2010, pp. 247–264.

[148] Soubaras, H., "An Evidential Measure of Risk in Evidential Markov Chains," in *Symbolic and Quantitative Approaches to Reasoning with Uncertainty*, New York: Springer, 2009, pp. 863–874.

[149] Sabbadin, R., "Possibilistic Markov Decision Processes," *Engineering Applications of Artificial Intelligence,* Vol. 14, 2001, pp. 287–300.

[150] Dubois, D., F. D. de Saintcyr, and H. Prade, "Updating, Transition Constraints and Possibilistic Markov Chains," *Advances in Intelligent Computing (IPMU'94)*, 1995, pp. 261–272.

[151] Kaelbling, L. P., M. L. Littman, and A. W. Moore, "Reinforcement Learning: A Survey," *Journal of Artificial Intelligence Research*, Vol. 4, 1996, pp. 237–285.

[152] Yu, T., and B. Zhou, "Reinforcement Learning Based CPS Self-Tuning Control Methodology for Interconnected Power Systems," *Power System Protection and Control,* Vol. 10, 2009.

[153] Ernst, D., M. Glavic, and L. Wehenkel, "Power Systems Stability Control: Reinforcement Learning Framework," *IEEE Trans. on Power Systems*, Vol. 19, 2004, pp. 427–435.

[154] Nguyen, T. T., and G. Armitage, "A Survey of Techniques for Internet Traffic Classification Using Machine Learning," *IEEE Communications Surveys & Tutorials*, Vol. 10, 2008, pp. 56–76.

[155] Shi, W.-W., H.-S. Yan, and K.-P. Ma, "A New Method of Early Fault Diagnosis Based on Machine Learning," *Proc. of 2005 Intl. Conf. on Machine Learning and Cybernetics*, 2005, pp. 3271–3276.

[156] Zhang, Y., M. D. Ilic, and O. K. Tonguz, "Mitigating Blackouts Via Smart Relays: A Machine Learning Approach," *Proc. of the IEEE,* Vol. 99, 2011, pp. 94–118.

[157] Harley, R. G., and J. Liang, "Computational Intelligence in Smart Grids," *IEEE SSCI 2011 Symp. Series on Computational Intelligence CIASG*, 2011.

[158] Jiang, Z., "Computational Intelligence Techniques for a Smart Electric Grid of the Future," *Advances in Neural Networks*, 2009, pp. 1191–1201.

[159] Guarracino, M. R., et al., "Supervised Classification of Distributed Data Streams for Smart Grids," *Energy Systems,* Vol. 3, 2012, pp. 95–108.

[160] Hady, M. F. A., and F. Schwenker, "Decision Templates Based RBF Network for Tree-Structured Multiple Classifier Fusion," in *Multiple Classifier Systems*, New York: Springer, 2009, pp. 92–101.

[161] Micalizio, R., "On-Line Monitoring and Diagnosis of a Multi-Agent System: A Model-Based Approach," *Citeseer*, 2007.

[162] Thorsley, D., and D. Teneketzis, "Diagnosability of Stochastic Discrete-Event Systems," *IEEE Trans. on Automatic Control*, Vol. 50, 2005, pp. 476–492.

[163] Doreian, P., V. Batagelj, and A. Ferligoj, *Generalized Blockmodeling*, Cambridge, U.K.: Cambridge University Press, 2005.

[164] Batagelj, V., et al., "Generalized blockmodeling with Pajek," *Metodoloski zvezki,* Vol. 1, 2004, pp. 455–467.

[165] Žiberna, A., "Generalized Blockmodeling of Valued Networks," *Social Networks,* Vol. 29, 2007, pp. 105–126.

[166] Flynn, D., J. Ritchie, and M. Cregan, "Data Mining Techniques Applied to Power Plant Performance Monitoring," *16th IFAC World Congress*, Prague, 2005.

[167] Khreich, W., et al., "Adaptive ROC-Based Ensembles of HMMs Applied to Anomaly Detection," *Pattern Recognition,* Vol. 45, 2012, pp. 208–230.

[168] Bose, A., "Smart Transmission Grid Applications and Their Supporting Infrastructure," *IEEE Trans. on Smart Grid*, Vol. 1, 2010, pp. 11–19.

[169] Ramírez-Treviño, A., et al., "Online Fault Diagnosis of Discrete Event Systems. A Petri Net-Based Approach," *IEEE Trans. on Automation Science and Engineering*, Vol. 4, 2007, pp. 31–39.

[170] Santoyo-Sanchez, A., et al., "Fault Diagnosis of Electrical Systems Using Interpreted Petri Nets," *IEEE Intl. Conf. on Emerging Technologies and Factory Automation 2008 (ETFA 2008)*, 2008, pp. 538–546.

[171] Littman, M. L., R. S. Sutton, and S. Singh, "Predictive Representations of State," *Advances in Neural Information Processing Systems,* Vol. 2, 2002, pp. 1555–1562.

[172] McCracken, P., and M. Bowling, "Online Discovery and Learning of Predictive State Representations," *Advances in Neural Information Processing Systems,* Vol. 18, 2006, p. 875.

[173] Singh, S., et al., "Learning Predictive State Representations," *ICML*, 2003, pp. 712–719.

[174] Dong, M., and D. He, "A Segmental Hidden Semi-Markov Model (HSMM)-Based Diagnostics and Prognostics Framework and Methodology," *Mechanical Systems and Signal Processing,* Vol. 21, 2007, pp. 2248–2266.

[175] Gupta, V., and S. Dharmaraja, "Semi-Markov Modeling of Dependability of VoIP Network in the Presence of Resource Degradation and Security Attacks," *Reliability Engineering & System Safety,* Vol. 96, 2011, pp. 1627–1636.

[176] Du, Y., H. Wang, and Y. Pang, "A Hidden Markov Models-Based Anomaly Intrusion Detection Method," *Fifth World Congress on Intelligent Control and Automation 2004 (WCICA 2004)*, 2004, pp. 4348–4351.

[177] Camci, F., and R. B. Chinnam, "Dynamic Bayesian Networks for Machine Diagnostics: Hierarchical Hidden Markov Models Vs. Competitive Learning," *Proc. of 2005 IEEE International Joint Conference on Neural Networks (IJCNN'05)*, 2005, pp. 1752–1757.

[178] Calderaro, V., et al., "Failure Identification in Smart Grids Based on Petri Net Modeling," *IEEE Trans. on Industrial Electronics,* Vol. 58, 2011, pp. 4613–4623.

[179] Pieczynski, W., "Multisensor Triplet Markov Chains and Theory of Evidence," *International Journal of Approximate Reasoning,* Vol. 45, 2007, pp. 1–16.

[180] Bowling, M., et al., "Learning Predictive State Representations Using Non-Blind Policies," *Proc. of the 23rd Intl. Conf. on Machine Learning*, 2006, pp. 129–136.

[181] Tanner, B., et al., "Grounding Abstractions in Predictive State Representations," *IJCAI*, 2007, pp. 1077–1082.

[182] Wingate, D., et al., "Relational Knowledge with Predictive State Representations," *IJCAI*, 2007, pp. 2035–2040.

[183] Wolfe, B., and S. Singh, "Predictive State Representations with Options," *Proc. of 23rd Intl. Conf. on Machine Learning*, 2006, pp. 1025–1032.

[184] Gammerman, A., and V. Vovk, "Hedging Predictions in Machine Learning The Second Computer Journal Lecture," *The Computer Journal,* Vol. 50, 2007, pp. 151–163.

[185] Berzuini, C., "Modeling Temporal Processes Via Belief Networks and Petri Nets, with Application to Expert Systems," *Annals of Mathematics and Artificial Intelligence,* Vol. 2, 1990, pp. 39–64.

[186] Serir, L., E. Ramasso, and N. Zerhouni, "Time-Sliced Temporal Evidential Networks: The Case of Evidential HMM with Application to Dynamical System Analysis," *2011 IEEE Conference on Prognostics and Health Management (PHM)*, 2011, pp. 1–10.

[187] Sutton, R. S., E. J. Rafols, and A. Koop, "Temporal Abstraction in Temporal-Difference Networks," *Advances in Neural Information Processing Systems,* Vol. 18, 2006, p. 1313.

[188] Moreira, M. V., T. C. Jesus, and J. C. Basilio, "Polynomial Time Verification of Decentralized Diagnosability of Discrete Event Systems," *IEEE Trans. on Automatic Control*, Vol. 56, 2011, pp. 1679–1684.

[189] AlRashidi, M. R., and M. E. El-Hawary, "A Survey of Particle Swarm Optimization Applications in Electric Power Systems," *IEEE Trans. on Evolutionary Computation*, Vol. 13, 2009, pp. 913–918.

[190] Kezunovic, M., L. Xie, and S. Grijalva, "The Role of Big Data in Improving Power System Operation and Protection," *2013 IREP Symp. Bulk Power System Dynamics and*

Control-IX Optimization, Security and Control of the Emerging Power Grid (IREP), 2013, pp. 1–9.

[191] Kimotho, J. K., et al., "Machinery Prognostic Method Based on Multi-Class Support Vector Machines and Hybrid Differential Evolution–Particle Swarm Optimization," *Chemical Engineering Transactions*, Vol. 33, http://www.aidic.it/cet/13/33/104.pdf, 2013.

[192] Okhravi, H., F. Sheldon, and J. Haines, "Data Diodes in Support of Trustworthy Cyber Infrastructure and Net-Centric Cyber Decision Support," in *Optimization and Security Challenges in Smart Power Grids*, New York: Springer, 2013, pp. 203–216.

[193] McGuire, P., et al., "The Application of Holonic Control to Tactical Sensor Management," *IEEE Workshop on Distributed Intelligent Systems: Collective Intelligence and Its Applications (DIS 2006)*, 2006, pp. 225–230.

[194] Peng, C., H. Sun, and J. Guo, "Multi-Objective Optimal PMU Placement Using A Non-Dominated Sorting Differential Evolution Algorithm," *International Journal of Electrical Power & Energy Systems*, Vol. 32, 2010, pp. 886–892.

[195] Ilic, M. D., et al., "Modeling of Future Cyber–Physical Energy Systems for Distributed Sensing and Control," *IEEE Trans. on Systems, Man and Cybernetics, Part A: Systems and Humans*, Vol. 40, 2010, pp. 825–838.

[196] Nakamura, E. F., and A. A. Loureiro, "Information Fusion in Wireless Sensor Networks," *Proc. of the 2008 ACM SIGMOD Intl. Conf. on Management of Data*, 2008, pp. 1365–1372.

[197] Wu, F.-J., Y.-F. Kao, and Y.-C. Tseng, "From Wireless Sensor Networks Towards Cyber Physical Systems," *Pervasive and Mobile Computing*, Vol. 7, 2011, pp. 397–413.

[198] Linda, O., et al., "Towards Resilient Critical Infrastructures: Application of Type-2 Fuzzy Logic in Embedded Network Security Cyber Sensor," *2011 4th International Symposium on Resilient Control Systems (ISRCS)*, 2011, pp. 26–32.

CHAPTER 9

Summary

Information overload and complexity are core problems to both military and civilian modern networked environments referred as cyber physical and social systems (CPSS). This book is devoted to fusion of information and analytics technologies (FIAT) to contribute solutions for these core problems. FIAT represents hundreds of methods and technologies (see Figures 2.22 and 3.11). For each of these methods and technologies, there are one or several existing and wholly dedicated books (see references of previous chapters). For fusion of information and analytics alone, there are more than 50 books available (e.g., [1–6]). The motivation for writing this book was not to add another one, but to insist upon the necessity to develop principles for integration at the algorithmic level (e.g., specific methods such as Dempster-Shafer theory or Bayes) and at the system level (e.g. cooperation of systems). We are pointing out integration because it is believed that the fundamental problem of networked military operations is that of information and system integration as NATO studies on systems interoperability have revealed [7–9]. Same can be commented for civilian CPSS such as Smart Grids, intelligent transport, and health systems. The design of FIAT-based systems, to support either situation analysis (SASS) or decisions (DSS), depends entirely on contexts and the domain of application. The challenge is to integrate appropriate technologies that will support measuring, organizing, understanding, and reasoning about situations in a given CPSS complex environment and support actuation (cybernetics functions).

Big data and data deluge are contextual to CPSS complex dynamic environments. Addressing the big data Vs (Value, Volume, Veracity, Variety, and Velocity) requires that FIAT be integrated. A kind of integration was proposed in Chapters 4 to 6 of this book via a FIAT holonic computational model in accordance with Sulis' archetypal dynamics framework. Other integrating frameworks and integration notions were discussed in Chapter 7 for making system interoperable: applications of openness and cooperability notions.

The objectives of FIAT are to transform data to information into actionable knowledge to support decisions and actions in complex dynamic environments. These complex environments have been described in Chapter 1, as well as in several dedicated books (e.g., [10–12]), using a terminology that comprises CPSS, cyber physical systems (CPS), Internet of things (IoT), and big data. The problem is tremendously complex to be the solution. Obviously, the current book has addressed only a portion of the problem space. For instance, FIAT-based issues have been considered mainly under the Veracity and Variety dimensions of big data.

The problem is multidisciplinary by nature and several books (actual and future) are required to explore the vast solution space. In this book, the FIAT discussion focuses more on situation analysis perspective than on decision support aspects. As an example, the reader is referred to the abundant literature on important topics, such as multicriteria decision-making and multi-objective optimization [13–18].

To position the contribution of this book, with respect to CPSS, CPS, and IoT, let us use the most elaborated map regarding CPS. That map, reproduced in Figure 9.1 [19], has been slightly modified to accommodate the concepts addressed in the previous chapters. Figure 9.1 defines CPS as feedback systems that require system dependability and design tools and methodologies and have applications in CPSS: energy, transport, health, defense, and security. Chapter contributions of this book have been linked to that CPS conceptual map. It shows how vast the problem and solution spaces can be. Only a small portion or few aspects have been covered with our book on FIAT integration. Portions not covered such as real-time issues, social computing, human-systems interactions, and networks are huge fields. Other books [10, 11] are and will be required to cover those aspects.

With respect to the big data dimensions, this book on FIAT applications, limited its contributions mainly to Variety and Veracity dimensions of big data as illustrated by the dotted lines in Figure 9.2. In this book, IoT has been considered only as a contributor to big data (Figure 9.2) and not in the sense of a CPS as pictured in Figure 9.1 where feedback and control are present. Recently, some authors [20, 21] suggested adding Visualization as another V dimension to big data problem (Figure 9.2). We agree to add it, since Visualization has proven effective for presenting essential information in vast amounts of data and for driving complex analyses. A large scientific community on visualization can contribute significantly to provide solutions to big data.

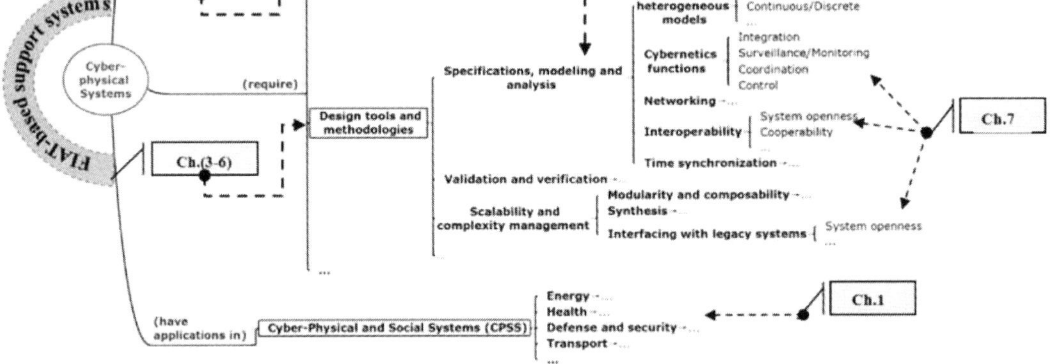

Figure 9.1 Structure and contribution of the book versus CPS: a concept map [19].

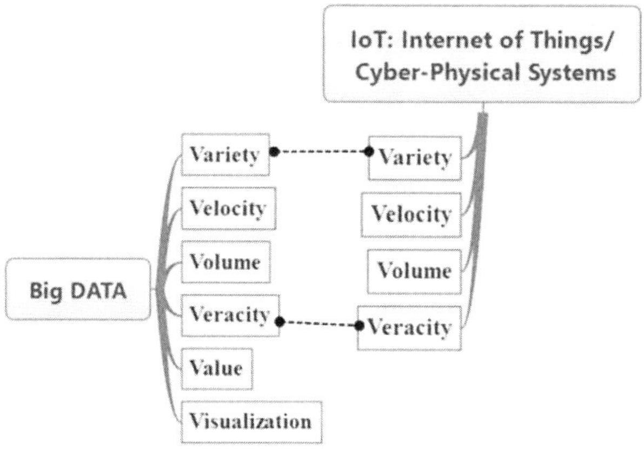

Figure 9.2 IoT versus big data dimensions.

References

[1] Blasch, E., E. Bossé, and D. A. Lambert, *High-Level Information Fusion Management and Systems Design*, Norwood, MA: Artech House, 2012.

[2] Das, S., *Computational Business Analytics*, New York: Taylor & Francis, 2013.

[3] Liggins, M., D. Hall, and J. Llinas, *Handbook of Multisensor Data Fusion: Theory and Practice*, 2nd ed., New York: Taylor & Francis, 2008.

[4] Hall, D. L., and J. M. Jordan, *Human-Centered Information Fusion*, Norwood, MA: Artech House, 2010.

[5] Waltz, E., *Knowledge Management in the Intelligence Enterprise*, Norwood, MA: Artech House, 2003.

[6] Bossé, É., J. Roy, and S. Wark, *Concepts, Models, and Tools for Information Fusion*, Norwood, MA: Artech House, 2007.

[7] Alberts, D. S., *The Agility Advantage: A Survival Guide for Complex Enterprises and Endeavors*, CCRP Publication Series, Washington DC, 2010.

[8] D. S. Alberts, "Agility, Focus, and Convergence: The Future of Command and Control," DTIC Document, *The International C2 Journal*, Vol. 2, No. 1, 2007.

[9] Alberts, D. S., R. K. Huber, and J. Moffat, "NATO NEC C2 Maturity Model," NATO SAS-065, www.dodccrp.org, 2010.

[10] Lee, E. A., and S. A. Seshia, *Introduction to Embedded Systems: A Cyber-Physical Systems Approach*, Berkeley, CA: Lee & Seshia, 2011.

[11] Marwedel, P., *Embedded System Design: Embedded Systems Foundations of Cyber-Physical Systems*, New York: Springer Science & Business Media, 2010.

[12] Minelli, M., M. Chambers, and A. Dhiraj, *Big Data, Big Analytics: Emerging Business Intelligence and Analytic Trends for Today's Businesses*, New York: John Wiley & Sons, 2012.

[13] Ho, W., X. Xu, and P. K. Dey, "Multi-Criteria Decision Making Approaches for Supplier Evaluation and Selection: A Literature Review," *European Journal of Operational Research,* Vol. 202, 2010, pp. 16–24.

[14] Wang, J.-J., et al., "Review on Multi-Criteria Decision Analysis Aid in Sustainable Energy Decision-Making," *Renewable and Sustainable Energy Reviews,* Vol. 13, 2009, pp. 2263–2278.

[15] Greco, S., M. Ehrgott, and J. R. Figueira, *Trends in Multiple Criteria Decision Analysis*, New York: Springer, 2010.

[16] Tzeng, G.-H., and J.-J. Huang, *Multiple Attribute Decision Making: Methods and Applications*, Boca Raton, FL, CRC Press, 2001.

[17] Deb, K., "Multi-Objective Optimization," in *Search Methodologies*, New York: Springer, 2014, pp. 403–449.

[18] Marler, R. T., and J. S. Arora, "Survey of Multi-Objective Optimization Methods for Engineering," *Structural and Multidisciplinary Optimization,* Vol. 26, 2004, pp. 369–395.

[19] *Cyber-Physical Systems: A Concept Map*, http://cyberphysicalsystems.org/.

[20] Keim, D., H. Qu, and K.-L. Ma, "Big-Data Visualization," *IEEE Computer Graphics and Applications*, Vol. 33, 2013, pp. 20–21.

[21] Wenxue, H., and W. Jinjia, "Survey on Visualization and Visual Analytics," *Journal of Yanshan University,* Vol. 34, 2010, pp. 95–99.

List of Acronyms

AASS Agonistic Antagonistic System Sciences

ACWA Applied cognitive work analysis

AD Archetypal dynamics

AI Artificial intelligence

ASM Abstract state machines

ATR Automatic target recognition

BM Belief models

BPA Basic probability assignment

CBSS Computer-based support system

CEP Complex event processing

C2 Command Control

C4ISR Command Control, Communications, Computer, Intelligence, Surveillance, and Reconnaissance

CCO Command and Control Organizations

CCRP Command and Control Research Program

CIGRÉ Conseil International des Grands Réseaux Électriques

COA Course of action

CPS Cyber physical systems

CPSS Cyber physical and social systems

CSE Cognitive system engineering

CTA Cognitive task analysis

CWA Cognitive work analysis

CWR Cognitive work requirements

DASM Distributed abstract state machines

DF Data fusion

DFD Data feature decision

DIFS Distributed information fusion system

DL Description logics

DM Decision-making

DNN Dual-node network

DoD Department of Defense

DoDAF Department of Defense Architecture Framework

DSN Distributed sensor networks

DSS Decision support systems

DS Dempster-Shafer

DST Dempster-Shafer's theory

DTM Digital terrain map

EA Enterprise architecture

EMS Energy Management Systems

EPRI Electric Power Research Institute

ETURWG Evaluation of Technologies for Uncertainty Representation Working Group

FACTS Flexible alternating current transmission systems

FAIR Framework Analysis for Information Relevance

FAN Functional abstraction network

FCM Fuzzy cognitive map

FE Fuzzy element

FIAT Fusion of Information and Analytic Technologies

FPMFWG Fusion Process Model and Frameworks Working Group

GIT Generalized information theory

HLIF High-level information fusion

IBM International Business Machines

ICT Information and communication technology

ID Identification

IF Information fusion

IFC Information fusion cell

IG Interoperable groups

INFORMLab Information Fusion and Resource Management Laboratory

Intel Military intelligence cycle

IoT Internet of things

IRR Information/relationship requirements

ISIF International Society of Information Fusion

ISSA Interpreted systems situation analysis

JDL Joint Directors of Laboratories

JDL DIFG Joint Directors of Laboratories' Data and Information Fusion Group

KID Knowledge, information, and data

MAPE Monitor, analyze, plan, and execute

MAS Multi-agent systems

MBES Multibeam echo sounder

MCDA Multicriteria decision analysis

ME Multiple elements

MRU Motion reference unit

NATO North Atlantic Treaty Organization

NATO SAS RG NATO Systems Analysis and Studies Research Group

NDM Naturalistic decision-making

NEC Network-enabled capability

NIT Networked information technology

NSF National Sciences Foundation

OODA Observe-orient-decide-act

OMM Ontology management methods

OPP Operational planning process

Ops Military operations cycle

OR Operational research

OWL Ontology Web language

PDC Presentation design concepts

PMU Phasor measurement management units

PSS Power system stabilizers

QoF Quality of fusion

QoI Quality of information

RDR Representation design requirements

RGB Red-green-blue

RPD Recognition-primed decision

RM Resource management

SA Situation analysis

SAW Situation awareness

SE Single element

SES Socioecological systems

SLS Side-looking sonar

SM Sense-making

SO Self-organization

STDF State transition data fusion

STO Sociotechnical organizations

TU Total uncertainty

UMM Uncertainty management methods

URREF Uncertainty Representation and Reasoning Evaluation Framework

About the Authors

Éloi Bossé received a B.A.Sc. in 1979, an M.Sc. in 1981, and a Ph.D. in 1990 from Université Laval, Québec City, in electrical engineering. In 1981, he joined the Communications Research Centre, Ottawa, Canada, where he worked on signal processing and high-resolution spectral analysis. In 1988, he was transferred to the Defence Research Establishment Ottawa to work on radar target tracking in multipath. In 1992, he moved to Defence Research and Development Canada Valcartier (DRDC Valcartier) to lead a group of four to five defense scientists on information fusion and resource management. He has published more than 200 papers in journals, book chapters, conference proceedings, and technical reports. Dr. Bossé has held adjunct professor positions at several universities from 1993 to 2013 (Université Laval, University of Calgary and McMaster University). He headed the C2 Decision Support Systems Section at DRDC Valcartier from 1998 until 2011. Dr. Bossé was the executive chair of the 10th International Conference on Information Fusion (FUSION'07), held in July 2007 in Québec City. He represented Canada (as a DRDC member) in numerous international research forums under the various cooperation research programs (NATO, TTCP, bilaterals, and trilaterals) in his area of expertise. He is the coauthor of *Concepts, Models, and Tools for Information Fusion* (Artech House, 2007), which won the 2006 NAMRAD Principals Achievement Award, and contributed to several other books and chapters. Namely, he coedited *High-Level Information Fusion Management and Systems Design* (Artech House, 2012). He retired from the DRDC in September 2011. Since then, he conducted some research activities under the NATO Peace and Security Programme, as a researcher in the Mathematics and Industrial Engineering Department at Polytechnic of Montreal, as an associate researcher at Telecom Bretagne, and, since 2015, as president of Expertise Parafuse Inc., a consultant firm on the fusion of information, decision support, and analytics technologies.

Basel Solaiman received a telecommunication engineering degree in 1983 from the Ecole Nationale Supérieure des Télécommunications de Bretagne and a D.E.A. in 1983, a Ph.D. in 1988, and an H.D.R. in 1997 from the Université de Rennes-I, France. In 1983, he joined the Laboratoire d'Electronique de Philips (L.E.P.), Paris, France, where he worked on image compression standards. In 1988, he joined the Institute of Industrial Informatics, Brest, France, as the R&D engineer-in-chief to work on new technologies to transfer into the industrial regional tissue. In 1992, he joined the Ecole National Supérieure des Télécommunications de Bretagne (Telecom Bretagne), Brest, France, where he is a full professor and the head of the

Image and Information Processing Department. He has published several academic books, more than 180 journal papers, several book chapters, and more than 230 conference proceedings and technical reports. Professor Solaiman has held visiting professor positions at several universities: University of Bristol (United Kingdom), University of Carabobo (Venezuela), University of Michigan (Ann Arbor, Michigan), Technical University of Budapest (Hungary), E.T.S. (Montréal, Canada), M.U.T. (Warsaw, Poland), and Technical University of Melaka (UTEM, Malaysia). Professor Solaiman chaired several international conferences focusing on his research activities and related to signal and image processing, data and information fusion, and pattern recognition. The application domains of his research activities range from civilian into military ones: medical, remote sensing, underwater imaging, and knowledge mining.

Index

The Artech House Intelligence and Information Operations Series

Ed Waltz, Series Editor

The Art and Science of Military Deception, Hy Rothstein and Barton Whaley

Aviation Security Engineering: A Holistic Approach, Rainer Kölle, Garik Markarian, and Alex Tarter

Concepts, Models, and Tools for Information Fusion, Éloi Bossé, Jean Roy, and Steve Wark

Data Fusion Support to Activity-Based Intelligence, Richard T. Antony

Electronic Intelligence: The Analysis of Radar Signals, Second Edition, Richard G. Wiley

Electronic Warfare for the Digitized Battlefield, Michael R. Frater and Michael Ryan

Electronic Warfare in the Information Age, D. Curtis Schleher

Electronic Warfare Target Location Methods, Second Edition, Richard A. Poisel

EW 101: A First Course in Electronic Warfare, David Adamy

High-Level Information Fusion Management and Systems Design, Erik Blasch, Éloi Bossé, and Dale A. Lambert, editors

Homeland Security Technology Challenges: From Sensing and Encrypting to Mining and Modeling, Giorgio Franceschetti and Marina Grossi, editors

Information Fusion and Analytics for Big Data and IoT, Éloi Bossé and Basel Solaiman

Information Warfare and Organizational Decision-Making, Alexander Kott, editor

Information Warfare Principles and Operations, Edward Waltz

Introduction to Communication Electronic Warfare Systems, *Second Edition,* Richard A. Poisel

Knowledge Management in the Intelligence Enterprise, Edward Waltz

Mathematical Techniques in Multisensor Data Fusion, Second Edition, David L. Hall and Sonya A. H. McMullen

Modern Communications Jamming Principles and Techniques, Second Edition, Richard A. Poisel

Modern Communications Receiver Design and Technology, Cornell Drentea

Principles of Data Fusion Automation, Richard T. Antony

Strategem: Deception and Surprise in War, Barton Whaley

Statistical Multisource-Multitarget Information Fusion, Ronald P. S. Mahler

Tactical Communications for the Digitized Battlefield, Michael Ryan and Michael R. Frater

Target Acquisition in Communication Electronic Warfare Systems, Richard A. Poisel

For further information on these and other Artech House titles, including previously considered out-of-print books now available through our In-Print-Forever® (IPF®) program, contact:

Artech House
685 Canton Street
Norwood, MA 02062
Phone: 781-769-9750
Fax: 781-769-6334
e-mail: artech@artechhouse.com

Artech House
16 Sussex Street
London SW1V 14RW UK
Phone: +44 (0)20-7596-8750
Fax: +44 (0)20-7630-0166
e-mail: artech-uk@artechhouse.com

Find us on the World Wide Web at: www.artechhouse.com